Synthesis Lectures on Intelligent Technologies

Series Editor

Janusz Kacprzyk, Systems Research Institute, Polish Academy of Science, Warsaw, Poland

Synthesis Lectures on Intelligent Technologies provides highly interdisciplinary research with the potential to change the fundamental principles of our society. It covers applications such as Intelligent Transportation, Humanoids, Self-Driving Cars, IoT, Ambient Intelligence, Smart Cities, Human-computer Interaction, Computational Intelligence, Industry 4.0, Medical Robotics, or Data Science. Synthesis Lectures on Intelligent Technologies brings together up-to-date resources from trusted authors working around the world in all aspects of Intelligent Systems.

Mohamed Abdelkader · Anis Koubaa
Editors

Unmanned Aerial Vehicles Applications: Challenges and Trends

Springer

Editors
Mohamed Abdelkader
Robotics and Internet of Things Lab
Prince Sultan University
Riyadh, Saudi Arabia

Anis Koubaa
Director of the Research and Initiatives Center
Prince Sultan University
Riyadh, Saudi Arabia

ISSN 2731-6912 ISSN 2731-6920 (electronic)
Synthesis Lectures on Intelligent Technologies
ISBN 978-3-031-32036-1 ISBN 978-3-031-32037-8 (eBook)
https://doi.org/10.1007/978-3-031-32037-8

This Springer imprint is published by the registered company Springer Nature Switzerland AG
The registered company address is: Gewerbestrasse 11, 6330 Cham, Switzerland

Contents

Design and Development

Tetrahedral and Dodecahedral UASs, Structured Designs

Bilal Maassarani, Jeremy Epps, Kévin Garanger, Mohamad T. Shahab, Obadah Wali, and Eric Feron

Abstract

Modular robotic systems offer an attractive edge in terms of adaptability, configurability, scalability and reliability; they are easy to maintain, manufacture, store, and transport. We introduce a variety of modular designs for unmanned aerial robotic systems based on tetrahedral and dodecahedral shapes. These modules, once connected with each other, allow the operator to build customizable vehicles with desirable structural and aerodynamic characteristics. Some of these designs are directly influenced by fractal geometry, following and building upon the early works of Alexander Graham Bell and his kites shaped after Sierpiński tetrahedra. Combining the possibility of building modules that contain both propulsive elements and lifting surfaces, a vertical take-off and landing

B. Maassarani (✉) · M. T. Shahab · O. Wali · E. Feron
King Abdullah University of Science and Technology, Thuwal 23955, Saudi Arabia
e-mail: bilal.maassarani@kaust.edu.sa

M. T. Shahab
e-mail: mohamad.shahab@kaust.edu.sa

O. Wali
e-mail: obadah.wali@kaust.edu.sa

E. Feron
e-mail: eric.feron@kaust.edu.sa

J. Epps
Georgia Institute of Technology, Atlanta, GA 30332-0250, USA
e-mail: jeremy.epps@gatech.edu

K. Garanger
University of California, Irvine, CA 92697, USA
e-mail: kevin.garanger@uci.edu

M. Abdelkader and A. Koubaa (eds.), *Unmanned Aerial Vehicles Applications: Challenges and Trends*, Synthesis Lectures on Intelligent Technologies,
https://doi.org/10.1007/978-3-031-32037-8_1

modular UAV structure is introduced. In addition to their aerodynamic and structural characteristics, control strategies for large assemblies of powered modules are discussed, including ways to leverage the presence of multiple, redundant avionics provided by the multiple modules.

1 Introduction

Human fascination with things that fly is undeniable. It can be observed in various daily social flight-related activities from museums, air shows, blockbuster movies, fashion trends, and people simply looking and pointing at sky dwellers. The current state of unmanned aerial vehicles (UAVs) has a strong influence from, and resemblance to, the historic development of UAVs and flight. As a result, we begin by briefly introducing a history of UAVs followed by that of modular UAVs. Some mention of manned flight is necessary due to the intimate relationship between the two.

1.1 A Brief History of Unmanned Aerial Vehicles

Flight, an early aspiration of the human race as we gazed in admiration at winged beings sailing freely through the skies with various techniques, speeds and altitudes. The speculated dates of the earliest successful attempts of flight date back to the invention of the **kite**, Fig. 1, in China at around 400 BC, or possibly before. Regardless, kites inaugurated a vivid future of flying machines. One may argue that this first successful attempt at flight was also the first success of unmanned aerial vehicles, considering that the kite is controlled from the ground. In China, the initial uses included colorful versions displayed in religious ceremonies and more sophisticated ones to test weather conditions [1].

For centuries, the desperate dream of flight lead to foolish trials to mimic bird flight. Individuals in different parts of the world built human-sized wings, wore them, and lept from heights to prove their concept. This lead to grave injuries and deaths. Sometimes great minds were lost such as lexicographer Ismail ibn Hammad al-Jawhari [3]. Abbas ibn Firnas may be the first human to fly, in the year 875, Fig. 2. His wings consisted of a wooden frame covered with silk and feathers. He was able to glide for some distance, allegedly for around ten minutes, before aggressively landing and injuring his back. He attributed the unsuccessful landing to the lack of a tail [4]. Fortunately, he survived his daring feat and lived for another twelve years in which he proposed several flying machines that were not realized.

In 1485, another polymath, Leonardo da Vinci, conceptualized the Ornithopter flying machine. Although it was never built, he supported the concept with studies and over a

Fig. 1 Artistic depiction of kite flying in ancient China [2]

Fig. 2 Widely circulated depiction of Ibn Firnas's flight from an unknown artist

hundred drawn illustrations of his flight theories, some of which are shown in Fig. 3. Almost a millennium after the first flying human, brothers Joseph-Michel and Jacques-Étienne Montgolfier invented the **hot air balloon** in 1783. Fire heated the air that is trapped in a silk balloon. The hot air is less dense than atmospheric air, and creates lift as a result. The first test flight was led by a sheep, a rooster and a duck. These passengers demonstrated balloon flight publicly with an audience that included King Louis XVI of France and Queen Marie Antoinette at the royal palace of Versailles, as depicted in Fig. 4. The brave clueless animals flew for a distance of three kilometers at an altitude of 460 meters for eight minutes. One may argue that this is the second success of unmanned aerial vehicles! The brothers proceeded to send passengers Jean-François Pilâtre de Rozier and François Laurent d'Arlandes in the first untethered manned balloon flight, Fig. 5, on November 21, 1783. The pioneers survived the flight although de Rozier perished two years later as he crashed with a balloon of his design.

The field of aeronautics began to solidify with Sir George Cayley in 1799 who designed a functional piloted **glider**, Fig. 6, and discovered the four aerodynamic forces of flight: weight, lift, drag and thrust. In 1849, balloon bombs were used to attack Venice by Austrian forces, marking the beginning of a trend of utilizing UAVs in the military. Another unprecedented

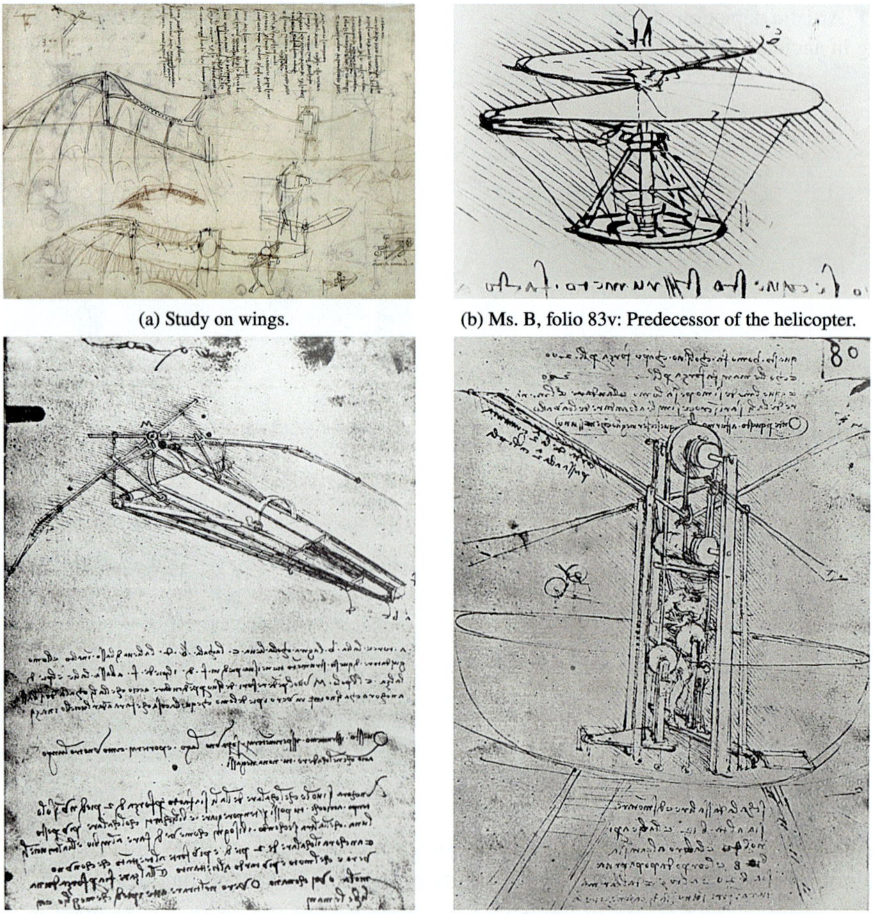

(a) Study on wings.

(b) Ms. B, folio 83v: Predecessor of the helicopter.

(c) Ms. B 2173, folio 74v: Study for a flying machine.

(d) Flying machine with a man operating it.

Fig. 3 Some of Leonardo da Vinci's drawings for his flying machines (c. 1487–1490) [5]

event was an aerial photograph, Fig. 7, taken from a hot air balloon of the French capital, Paris, in 1858. Aviation pioneer Otto Lilienthal gave his life after losing control of a glider of his invention in 1896. He left a legacy of several well-documented, functional and self-flown gliders that the world had not seen before. One of his photographed flights is shown in Fig. 8. After this list of various contributions, and many others, another pair of brothers, Wilbur and Orville Wright, flew their Wright Flyer in 1903. The maiden flight is known as the first powered, piloted, heavier-than-air and controlled **airplane** flight. This historic breakthrough propelled the development of modern aviation and was captured in the photo included in Fig. 9.

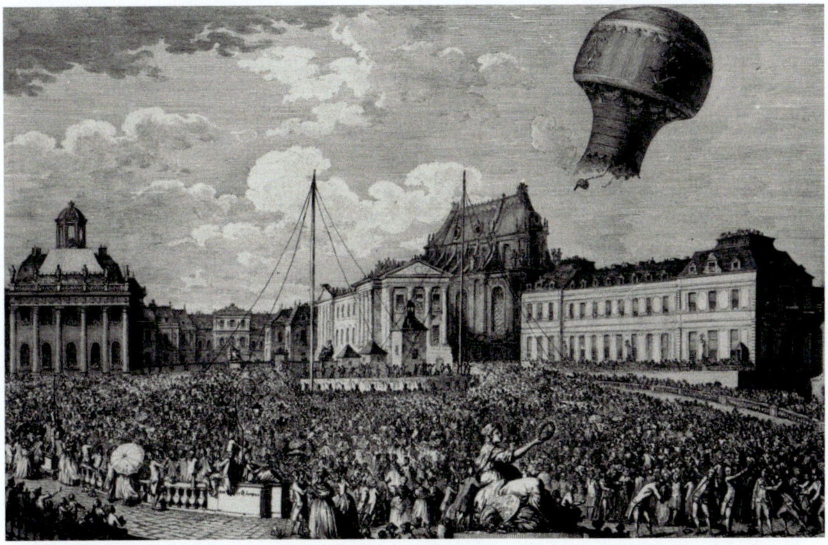

Fig. 4 The first balloon flight with passengers: a cock, a duck, and a sheep [Image credits: 2001 National Air and Space Museum, Smithsonian Institution (SI Neg. No. A-4691)]

Fig. 5 The first free ascent of a hot-air balloon with human passengers [Image credits: 2001 National Air and Space Museum, Smithsonian Institution (SI Neg. No. 93-2342)]

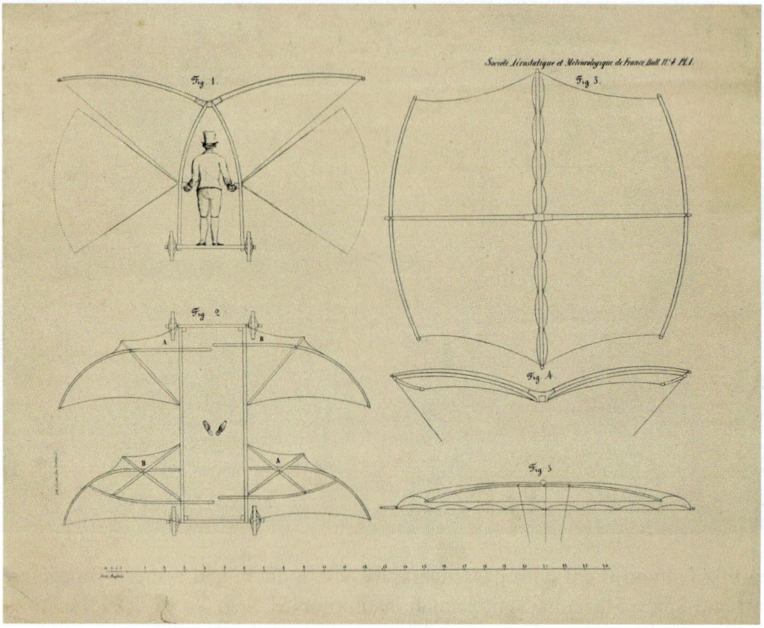

Fig. 6 George Cayley's glider, 1853 [Image credits: Library of Congress, Washington, D.C. (neg. no. LC-DIG-ppmsca-02521)]

The interest and application of UAVs in the military continued throughout the nineteenth century and till modern times. Drones were not publicly accessible otherwise. The United States allowed civilian deployment of UAVs to provide disaster relief during the aftermath of Hurricane Katrina. Thermal cameras were mounted on the military Predator drones to detect survivors to be rescued. Upon this 2006 event, the consumer drone industry began to emerge. Another significant factor in the surge of drones is the maturation of relevant technologies such as electronics, batteries, electric motors and control systems. Following the footsteps of its compatriots, French UAV manufacturer Parrot unveiled the AR Drone in 2010, a small smartphone-controlled consumer quadcopter. The Phantom drone further impacted the market by introducing consumer camera-equipped drones in 2013. The producing company, DJI, quickly took charge of the consumer drone industry. At that point, the spur attracted delivery titans to study the deployment of UAVs for payload transport, and the growth of the industry continues till today.

The COVID-19 pandemic had a significant impact on the world and our vision of the future. One of the consequences of the pandemic's safety measures, enforced to contain the spread of the coronavirus, was a shift towards the popularization and normalization of autonomy. As a result, the UAV industry is seeing further progress in terms of technology and application.

Fig. 7 First aerial photo, of Paris, taken from a hot air balloon [Photograph credits: by Gaspard–Félix Tournachon, known as Nadar]

There are numerous potential use cases of UAVs. Following the categorization of [6], the applications can be divided into four major fields:

1. **Logistics**: There is no doubt that drones can revolutionize the world of logistics. With speed and versatility, items ranging from life-saving first aid kits and donor organs to life-threatening bombs can be delivered. In addition to the health and military domains, there are agricultural opportunities to deliver pesticides, fertilizers, nutrients and to collect crops in a targeted and efficient manner. Passenger drones have the potential to offer the long-anticipated everyday flying vehicle. Autonomous air taxis offer safe aerial transport reducing travel time, road congestion, and possibly mitigating negative environmental impact. Another area of high interest is the delivery of food and goods due to the enormity of the electronic commerce industry. UAVs are appearing in different stages of the supply chain not limited to collecting raw materials, distribution centers, and last mile deliveries. The broad range of UAV transport use cases lead to research areas that examine the feasibility of these ideas in terms of managing the generated UAV air traffic. Regulators

Fig. 8 Otto Lilienthal with a large biplane in flight [Photograph credits: Fliegeberg 19.10.1895 by Dr. Richard Neuhauss, photo-no.: OLM F0123]

Fig. 9 Iconic photo of the Wright Flyer's maiden flight, restored [Photograph credits: 17.12.1903 by John T. Daniels]

and lawmakers have some challenges to address as well that will strongly influence technology adoption.

2. **Monitoring, inspection and data acquisition**: As we continue to reveal the portfolio of the UAV, a pattern of advantages begins to emerge that will be evident throughout this chapter and perhaps this book! Simply put, drones are, with a lower cost, more capable at reaching locations. The immediate consequences are that human risk is minimized, hazardous and/or remote areas become more accessible, and the environmental degradation is mitigated directly and indirectly. Directly, drones replace existing methods that have a larger environmental footprint and, in some cases, surpass these methods to collect unprecedented information. Indirectly, in the age of the fourth industrial revolution, data has become a significant asset. A drone-captured asset that provides farmers with crop health status and consequently limits the use of harmful pesticides and fertilizers. An asset that powers mine rehabilitation, and gathering samples from nature to be processed by scientists for environmental assessment purposes. Strictly collecting physical samples may send this application to the logistics field. However, since drones can carry a variety of sensing and analysis systems, the analysis can be performed onboard and communicated to the user. Timely reports are particularly beneficial in inspection tasks, such as the drone inspection of pipes, energy lines, roads, railways, bridges, construction sites, and other structures. In a more advanced form of communications, there are several projects examining the deployment of networks of drones to bring telecommunication networks to remote areas. Other time-critical monitoring uses include search and rescue missions, surveillance, humanitarian aid, fires, and natural disasters. With drone surveillance various illegal activities can be detected, with the power of Artificial Intelligence, and reported such as illegal construction, deforestation, hunting, immigration, trespassing, weapon deployment, vandalism, gatherings, and assault.

   Moreover, the availability of data facilitates the optimization and management of processes. In mining, for example, UAVs monitor ore stockpiles, leeching pads for maintenance issues, and ore blast preprocessing analysis. Unlike in the major UAV application field of logistics, many of the target areas in the preceding scenarios tend to be remote. As a result, the significance of concern about air traffic management is less.

3. **Photography**: Since the 1858 aerial photo of Paris, the world is being seen from a new perspective. With the development of the aeronautics and aerospace industry, including that of UAVs, access to aerial photography is becoming increasingly accessible. While the value of drone footage may overlap with the preceding field of monitoring/inspection/data acquisition and the upcoming field of recreation, there are some distinct opportunities of benefit. Indeed, photos are valuable forms of data as well as sources of entertainment. However, as the adage goes, "a picture is worth a thousand words". Aerial photos are documenting personal events, sporting events, nature, landscapes, geological sites, archeological sites, building interiors and building exteriors in ways that would not be otherwise possible given the confines of space and cost. They also serve as effective communication tools that efficiently convey information, elicit emotion and market ideas.

4. **Recreation**: the aesthetic of bird's-eye view photography is often pleasing. There are more distinct drone-provided sources of recreation, however. Remote-controlled (RC) drones are a welcome addition to the recreational RC world. Communities have formed for hobbyist and competitive drone racing. RC UAVs are sold as tools for family, friends and touristic fun activities. Three-dimensional flying art installations are increasingly seen around the world in visually stunning displays of drone technology.

1.2 A Brief History of Modular Flying Machines

The scope of this chapter when it comes to modular robotics is limited to the *sectional* subtype of modular architecture as described in [7]. In a sectional architecture, components are self-contained and similar, yet composable into an assembly via identical and compatible interfaces. An early instance of modular aerial robots, following the scope just defined, are Alexander Graham Bell's tetrahedral kites, patented in 1904 [8]. Widely known as a telecommunications pioneer, the polymath had a broad set of interests. Towards the end of his life, his contributions to the field of aeronautics began with silk tetrahedral box kites. Smaller child kites arranged in tetrahedral cells formed a larger flying parent structure, as shown in Fig. 10. These experiments ultimately lead him to contributions to gliders and the *Silver Dart*, the first powered airplane to fly in Canada. The designs presented in this chapter by its authors are strongly linked to the tetrahedral kites. Over a century later, modular flight appeared again with ETH Zürich's Distributed Flight Array (*DFA*) introduced in 2009 [9] and detailed later in [10, 11]. The array consists of wheeled single-rotor units that assemble on land in numerous possible formations. The resulting multi-rotor vehicle, Fig. 11, is a flying hovercraft. Similar designs using single-rotor ducted units are found in [12–14], published between 2011 and 2015. Another project that took place in those years is the *LIFT!* project

(a) Four cell kite. (b) Kite in flight. (c) Sixty-four cell kite.

Fig. 10 Photos of Alexander Graham Bell's tetrahedral kites [Photo credits: Internet Archive/Library of Congress]

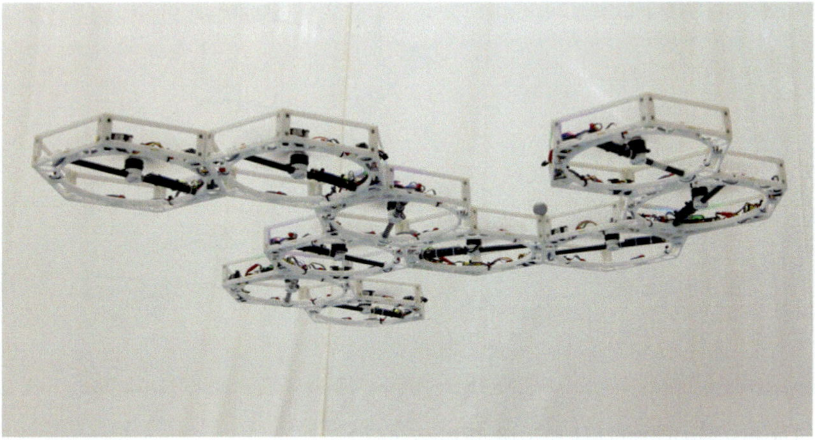

Fig. 11 The distributed flight array [11]. ©2014, IEEE

Fig. 12 The Lift project [15]. ©2015, AIAA

by The Boeing Company documented in [15, 16]. As its name indicates, the aim was to lift heavy payloads, Fig. 12. This goal was achieved by designing a system of quadcopters that are attachable to each other with rigid interconnections. The final machine, which is tethered for power supply, is scalable to match the payload weight with the vehicle's lifting capacity. This idea of scalability is similar to what will be seen in upcoming sections of this chapter.

After 2015, a number of projects and publications covering modular aerial robotics appeared. The aerial robots mentioned, thus far, have rigid inter-module connections. An additional member of that family is the *ModQuad* from the University of Pennsylvania's GRASP lab. The concept was introduced in a 2018 publication [17] as a cuboid-framed

Fig. 13 The first ModQuad [17]. ©2018, IEEE

quadrotor capable of self-assembling during hover into larger structures via magnetic links, Fig. 13. The lab group further explored the concept in a number of directions. In [18], it was shown that one version of the *ModQuad* could utilize vision techniques to self-assemble in a more decentralized manner. Reference [19] demonstrated *ModQuad*s that could connect and disconnect. Both capabilities are essential for self-reconfiguration in the air. These *ModQuad*s could efficiently self-reconfigure to attenuate any rotor failure disruptions [20]. In the spirit of configurability, authors from Brazilian universities proposed a modular UAV architecture entitled Drone Reconfigurable Architecture (*DRA*) [21]. Mechanically, there are two main elements of the *DRA*: a ring-shaped enclosure for a single rotor and a rigid ring-to-ring connector, Fig. 14. Since each ring has six possible connector docks, a large variety of vehicles can be constructed with co-planar rotors.

For the remainder of the subsection, we will continue to cover works within the same post-2015 timeframe. In that era, publications appeared highlighting another family of modular UAVs. One with flexible and combined (both flexible and rigid) inter-module connections. Montalvo and Costello studied the aerodynamics of fixed-wing modular aircraft [22]. An aircraft formed from several modules was described as a meta aircraft in the paper, and the modules analyzed were those of typical dual-winged airplanes. Meta airplanes considered had modules with flexible joints between modules placed either exclusively at the wings tips or exclusively at the noses and tails. Carlson and authors [23] had a resembling machine consisting of dual-winged tilt-rotor vertical take-off and landing (VTOL) modules. The identical components were inflexibly joined at the wingtips, Fig. 15. Yet another resemblance to this multi-VTOL machine is found in a NASA patent [24].

Another set of publications came from the University of Tokyo. The authors fabricated a system where a single rotor was in a ring-shaped cage and mounted on a rigid rod. Several of these rotor assemblies were chained with servo-actuated pin joints, Fig. 16. They explored this system in a number of ways in [25–28]. The researchers developed models, controllers,

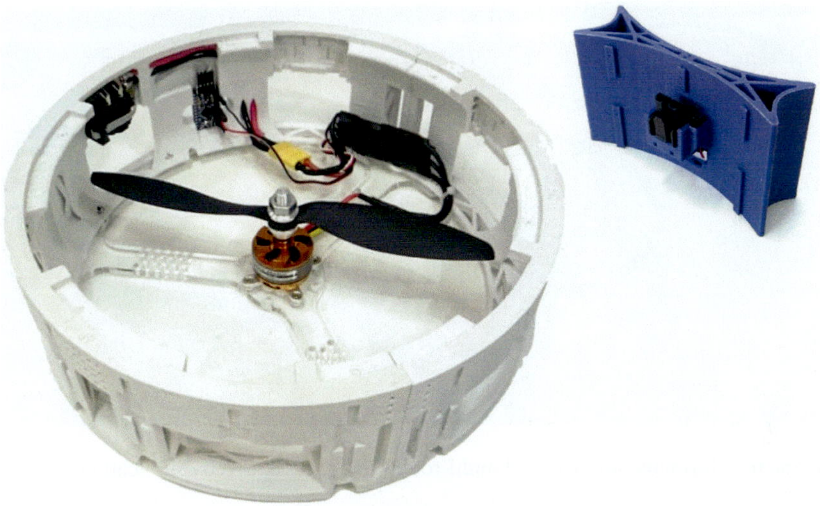

Fig. 14 A rotor unit (left) and a rotor-to-rotor connector (right) of the DRA [21]. ©2020, Springer

Fig. 15 The aspect ratio-modular VTOL reconfigurable aerial robot [23]. ©2022, Springer

internal communications, and planners to demonstrate a number of in-flight abilities: flight maneuvers in narrow areas, transforming the robot into a gripper to grasp and carry objects, and simultaneous transport of multiple objects that were lifted with electromagnetic grippers. The challenge of the last goal was to sustain stable flight given the load-induced disturbances

Fig. 16 The transformable multi-linked multi-rotor [28]. ©2018, SAGE Publications

Fig. 17 DRAGON [29]. ©2018, IEEE

in the center of gravity. The group advanced the maneuverability with a new vehicle called *DRAGON* [29, 30]. *DRAGON* was made by linking four modules with joints that have yaw and pitch actuation, Fig. 17. An individual module consists mainly of two ducted fans mounted on either sides of a rod. Another rod rigidly connects the fan mounting to the overall frame. This fan-mount subassembly is free to rotate along the axis of the mounted rod and along that of the supporting one. The resulting degree-of-freedom (DoF) additions equip the aerial robot with the ability to traverse tight spaces in a snake-like fashion. Revisiting

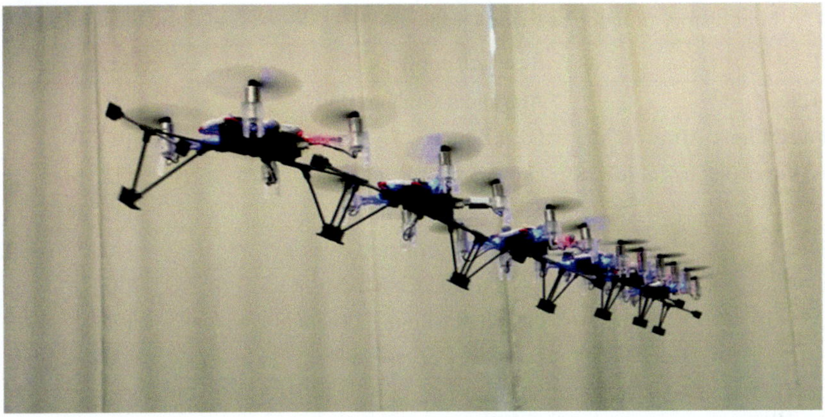

Fig. 18 ModQuad with additional DoFs [32]. ©2020, IEEE

the *ModQuad*, similar work was done. Replacing the previously rigid magnetic docks with hinge ones, the quadcopters were controlled to grasp external objects [31]. The *ModQuad* took a new form in [32] to improve the maneuverability of the aggregate as well, Fig. 18. An issue that arises from aerial assemblies is some loss of yaw authority. The cuboid frame was replaced by cross-shaped inter-quadcopter joints. Adhesion was again accomplished with magnets embedded in the crosses. Two of these joints were attached on opposing sides of the drone via a one DoF link. A novel controller was then formulated, and improved yaw actuation was proved. With one and two DoF linkages explored, three DoF ones are left! Nguyen and authors [33] proposed a rigid robotic platform with three DoF ball-joint quadcopter docking stations. However, this concept could take a form that falls outside the *sectional* modular framework. Coincidently, an author with the same last name discussed, with others, yet another reconfigurable aerial robotic chain design in [34, 35]. It is comparable to the works done at the University of Tokyo in the sense that quadcopters, instead of single rotors, were linked via rods. Another difference was that the joints were passive three DoF ball joints, which required separate control studies.

Table 1 summarizes the works covered in Sect. 1.2.

1.3 The Structured Modular UAV Solution

Concepts from the robotics field, such as modularity, naturally diffuse into the subfield of aerial robotics. Various benefits of aerial modular drone can be extracted from the history covered in Sects. 1.2 and 1.1. In certain applications, such as the optimization of a modular drone delivery system, modularity can bring further benefits [36]. When managing a modular fleet compared to a non-modular one, it is possible to reduce the size of the modular fleet and deliver items faster. Consequently, energy efficiency and fleet readiness are enhanced.

Table 1 A brief history of modular flying machines

Project	Research entity	References
Bell's Kites	Alexander Graham Bell	[8]
Distributed Flight Array	ETH Zürich	[9–11]
A class of modular aerial robots	University of Bologna	[12–14]
LIFT!	The Boeing Company	[15, 16]
ModQuad	University of Pennsylvania	[17–20, 31, 32]
DRA	UNESP and UNICAMP	[21]
Meta Aircraft	Georgia Institute of Technology	[22]
ARM-VTOL	University of Nevada	[23]
Modular UAS with multi-mode propulsion (patent)	NASA	[24]
Transformable multi-linked multi-rotor	University of Tokyo	[25–28]
DRAGON	University of Tokyo	[29, 30]
SmQ platform	Seoul National University	[33]
Reconfigurable Aerial Robotic Chain	University of Nevada	[34, 35]

Moreover, it is self-evident that a system with numerous identical components can be seen as more fault-tolerant, provided that the controller takes advantage of the redundancy to recover from failures. In addition, assemblies offer a few design opportunities. Modules designed to form increasingly larger systems can provide structure, scalability, configurability, and portability. The aggregated vehicle is reconfigurable in that different structural shapes may be formed, some of which are quite unusual as seen in the *Shapeshifter* [37] concept, presented in Fig. 19. Also, each module may carry components unique to the overall conglomerate.

All the modular aerial flying machines mentioned thus far, except for Bell's kites, are single layered in the sense that modules do not stack vertically. In systems that will be presented in the remainder of this chapter, the hovercraft modules are designed to be stacked into multiple layers of subassemblies in a structured fashion. In one set of systems, the resulting vehicles scale, as desired, in terms of size and payload without significant deterioration in aerodynamic efficiency due to wake interactions. Moreover, the structure remains rigid compared to its single-layered counterparts. In another set of systems, an additional goal of increasing the degrees of freedom is targeted, improving the maneuverability of the vehicles. The first set is expanded to include a vertical take-off and landing (VTOL) version with both fixed-wing and hover flight modes.

Fig. 19 Artist render of the Shapeshifter [37]. ©2020, IEEE

2 Prototypes

As foreshadowed in Sect. 1, the aim of the work highlighted in the rest of this book chapter is to offer structured designs of modular aerial robotic systems. A choice of structurally efficient three dimensional shapes of equal proportions is bound to yield uniform stiffness in every direction. Additionally, building vehicles from modules demands that these modules connect in a way that is feasible and distributes loads as evenly as possible. This leads to regular (of equal proportion) convex (connectable and load distributing) polyhedra (intrinsically stiff geometry). There is proof that the five Platonic solids, summarized in Table 2, are the only regular solids in existence [38].

Table 2 The five Platonic solids

Name	Tetrahedron	Cube	Octahedron	Dodecahedron	Icosahedron
	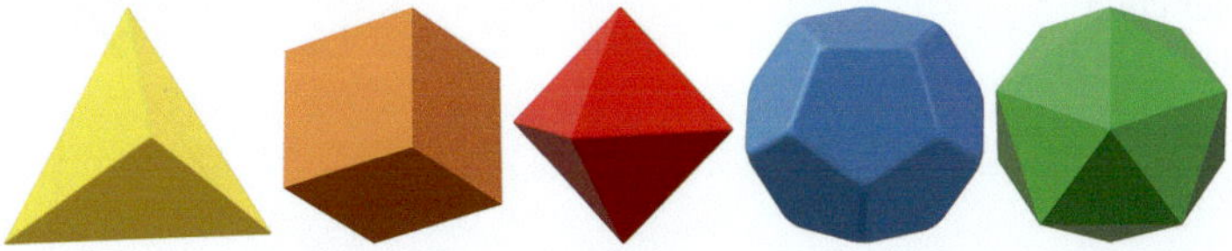				
Faces	4	6	8	12	20
Edges	6	12	12	20	30
Vertices	4	8	6	20	12

Triangles are staple shapes when it comes to strong stable, yet relatively simple, structures such as struts and trusses integrated into bridges, roofs, buildings, and various types of vehicles. The regular tetrahedron with its triangular faces subsequently inherits these properties. In addition, the tetrahedron is the simplest three-dimensional truss. As a result, it is the first selected shape in this work. The vehicle prototypes in Sects. 2.1 and 2.2 are based on this polyhedron. In a trend resembling that of the research in Sect. 1.2, the authors of this book chapter selected the dodecahedron to expand the degrees of freedom and configurability of modular vehicles. The solid has a high number of vertices and symmetries making it the best candidate to build a plethora of configurations of the vehicle in Sect. 2.3.

2.1 Tetracopter

The *Tetracopter* is a scalable multi-rotor aerial robot that flies in hovering mode. Its child cells are fitted into regular tetrahedral shapes. These polyhedra are shown in theory [39] and in application [40, 41] to be structurally efficient. To preserve the shape of the parent cell under heavy loads and to minimize aerodynamic interactions between the propellers, the *Tetracopter* is designed based on the Sierpiński tetrahedral fractal [42]. Each child cell, illustrated in Fig. 20, is a self-contained module consisting of a regular tetrahedral frame and a propeller placed close to one of the faces of the frame with its center overlapping with the center of gravity of the face. The propeller is fixed-pitch and may be built to rotate either clockwise or counter-clockwise. The purpose of propeller's position is to allow the largest propeller diameter given the geometric constraints, which has some aerodynamic advantages. The building blocks are connected rigidly at their vertices to form a vehicle of custom size. Drones with four, sixteen and sixty-four modules are drawn in Fig. 21.

By simply examining the top view of these drone drawings, Fig. 22, it is evident that there is no overlap, geometrically, between the rotors, represented with colored circles, as the fractal grows.

The minimum number of modules to have controllable flight of the assembly is four. This minimal prototype was fabricated by mounting components on a frame of 5 mm-diameter carbon fiber reinforced polymer, CFRP, and rods that are rigidly connected via 3D-printed

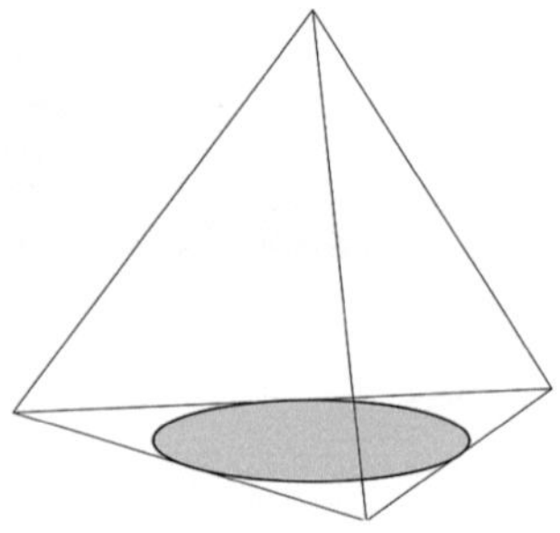

Fig. 20 Tetracopter child module representation [43]

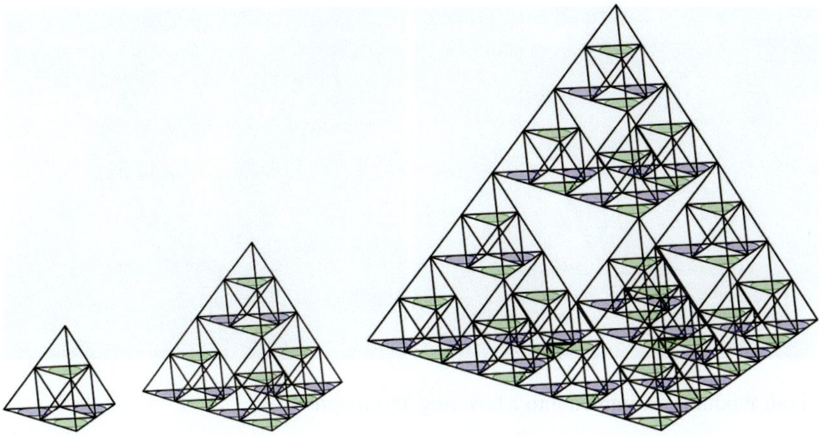

Fig. 21 Dimetric view of 4, 16 and 64 module tetrahedron fractals, respectively [43]

Fig. 22 Top view of 4, 16 and 64 module tetrahedron fractals, respectively from left to right [43]

polylactic acid, PLA, parts, as seen in Fig. 23. The components include a battery, a power distribution board, and a flight controller and are placed on a 3D-printed PLA plate close to the center of gravity, CG, of the hovercraft. The proof-of-concept prototype was successfully flown in hover mode with minimal wake interaction power loss, although the yaw authority was observed to be low.

In the second design iteration of the *Tetracopter*, it is desirable to validate that a *Tetracopter* design can scale into a functional 16-propeller rotorcraft. This larger prototype introduces new challenges not seen in the 4-propeller prototype. Some of these will be addressed in upcoming sections and future research. A new prototype is built with a simplistic and minimalistic mindset. An individual module, Fig. 24 (left), is created with a base of carbon fiber plate cutouts that are connected with mechanical fasteners. The top part of the frame consists of plastic rods and the entire hull takes shape with 3D-printed rigid joints. Four

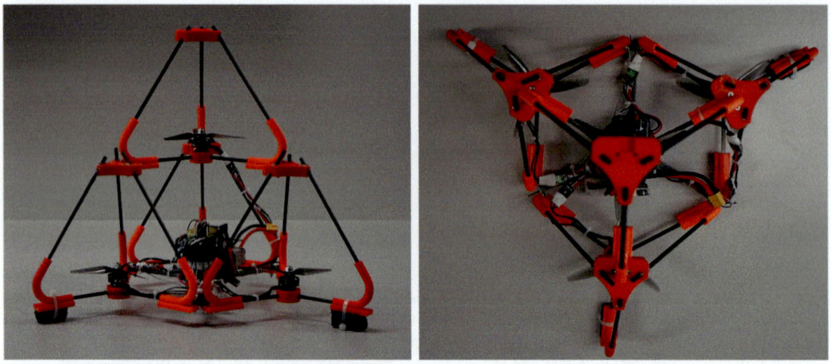

Fig. 23 Four modules assembled into a hovering *Tetracopter*

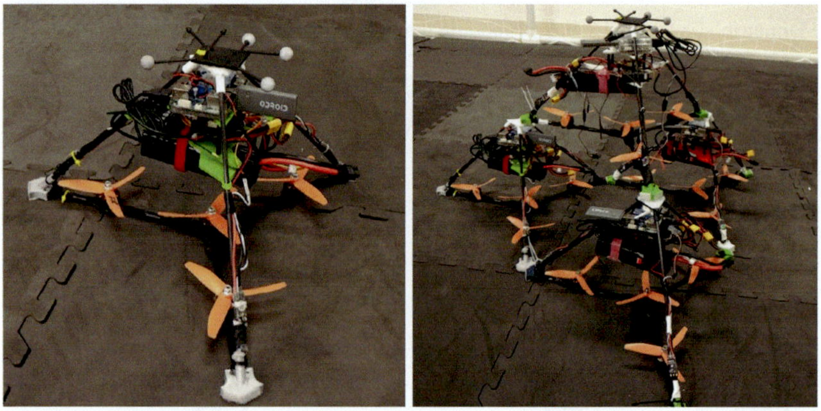

(a) Individual module. (b) Four-module assembly.

Fig. 24 Second design iteration of the *Tetracopter*

rotors per module, instead of one, are fastened mechanically into the cutouts such that one of the side rotors spins in the same direction as the center rotor, while the other two side rotors spin in the opposing direction. The electronics include, as before, a battery, a power distributor and a flight controller. Additionally, an onboard computer is added due to the increased control complexity. The components are attached to the top connections of the frame and within its tetrahedral shape. Four of the described modules are linked to form a large tetrahedral drone via a combination of adhesion, hook-and-loop fasteners and cable ties, as shown to the right in Fig. 24. There are two main differences between the first and second iterations. First, the former has one rotor per module while the latter has four. Second, in a four-rotor system, the former has two layers while the latter has co-planar rotors. As a result of this variation, some structural elements are traded for a lighter and simpler build. Both drones in Fig. 24 were also flown successfully.

2.2 TetraheDrone

The modules of the latest *Tetracopter* are each capable of independent operation. Thus, autonomous assembly and disassembly is possible, and with a sufficient number of modules, it becomes a robotic swarm! Heiko Hamann describes in his book [44] what he calls "swarm robotics 2.0", where he identifies a need for swarms to escape the confines of laboratories into practical applications. In line with that vision and with inspiration from Alexander Graham Bell's original tetrahedral kites [8], the *TetraheDrone* was born.

The *TetraheDrone* is essentially a hybrid fixed-wing vertical take-off and landing (HFW-VTOL) version of the *Tetracopter*. The advantage of VTOL aircraft compared to fixed-wing (FW) aircraft is maneuverability that renders runways redundant. On the other hand, FW planes are generally more efficient than their rotary counterparts. This can be justified by simply examining the lift mechanism of the two. The rotors of a rotary aircraft are responsible for both lift and forward propulsion, while in the FW case they only need to push forward as the aerodynamics around the wing result in lift. In fact, FW planes can continue to cruise by simply overcoming the drag forces. To further support the claim of the efficiency superiority of FW aircraft, the lift to drag ratio (L/D) is considered. This ratio is the amount of lift created with a wing compared to its drag. A higher ratio indicates better power efficiency. The helicopter, a popular hovercraft, has an L/D ratio that does not exceed 4.5 [45]. A typical quadrotor has a lower ratio, while FW jetliners such as the Boeing 777-200 have higher cruise L/D ratios of around 19.3 [46]. Some planes boast higher number, as high as 37 for the Virgin Atlantic GlobalFlyer according to some expert opinions. The HFW-VTOL category aims to combine the fuel efficiency of FW with the versatility of rotorcraft bringing electrical VTOL (eVTOL) closer to a popular science-fiction idea, the personal flying vehicle.

The *TetraheDrone* project is a unique study on the feasibility of HFW-VTOL vehicles that can form a larger self-similar structured UAS with the ambition to bring the idea of such UAS closer to commercial applications. The current design concept, at the time of writing, to implement the *TetraheDrone* is that of a tail-sitter VTOL without control surfaces. The concept was chosen in line with a minimally viable design strategy. A tail-sitter is a type of VTOL flyer where only the tail assembly makes contact with the ground during landing and take-off; the nose of the flyer points vertically upwards. With the spirit of modularity, an aerodynamic hull is 3D-printed from Markforged's Onyx™ material, a fusion of engineering nylon and chopped carbon fiber. The hull contains all the components except wings, rotors and landing feet. This permits convenient swapping of wings in case of inevitable damage or re-design. The decoupling of wings from other elements has other benefits for design, analysis and testing. The nose cone of the hull has a Von Kármán shape to minimize drag. The fuselage contains similar main components as the *Tetracopter* in addition to a GPS receiver for outdoor flight. Once again visualizing a regular tetrahedron, three individual identical triangular wings are evenly distributed around its longitudinal axis, such that dihedral angles between the wings are 120 degrees. This choice is to maximize wing size, and consequently lift, given that the wings are confined in the tetrahedral airframe. The airfoils are simply cut

(a) Isometric view.

(b) Top view.

(c) Hover flight.

(d) Fixed wing flight.

Fig. 25 An individual *TetraheDrone* module

from foamboard sheets for a few reasons. First, foamboard is readily available, lightweight as it consists of polystyrene foam sandwiched between paper, inexpensive and easy to cut. Also, if the aerodynamic performance of sheets is sufficient, then selecting a suitable airfoil shape can only result in improving the proof-of-concept. The rotors are positioned in a similar fashion to the *Tetracopter* except they act as pushers in an attempt to further improve aerodynamic performance. Finally, landing feet, constructed from 3D-printed parts, are adhered to the wings. The resulting *TetraheDrone* prototype, shown in Fig. 25, was successfully flown in both FW and hover flight modes, including the flight mode transitions.

Individual *TetraheDrone* modules are joined with 3D-printed connections to form a larger UAS as before. In Fig. 26, two versions of such a vehicle are shown. One has four simplified

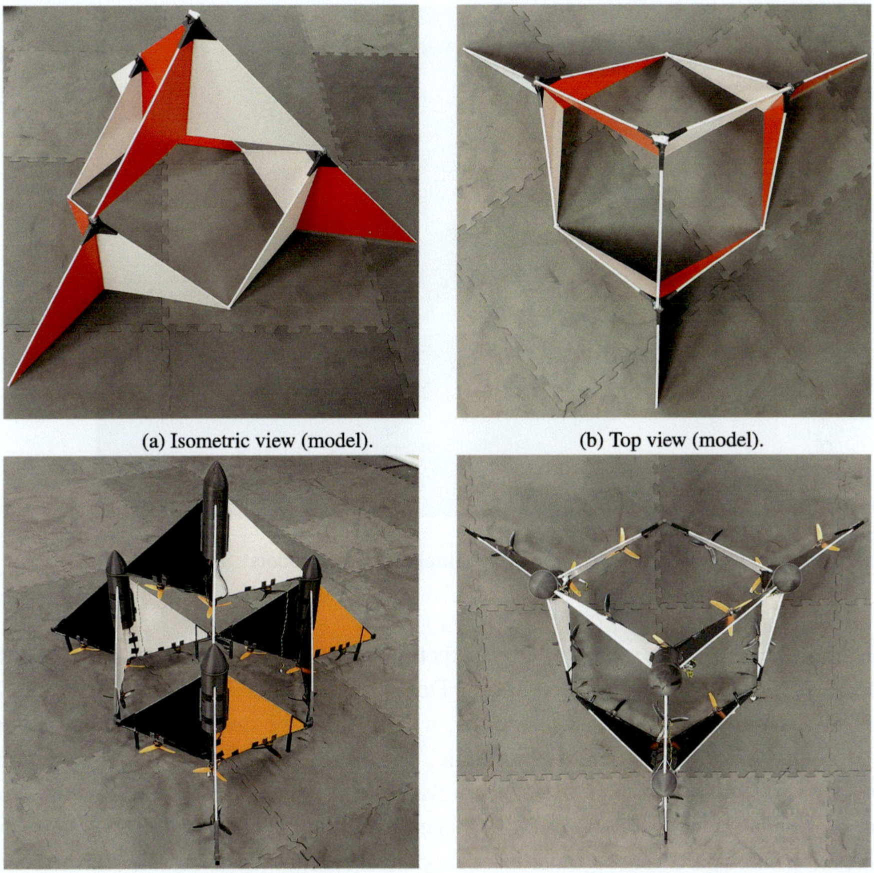

(a) Isometric view (model). (b) Top view (model).

(a) Isometric view. (b) Top view.

Fig. 26 An assembly of four *TetraheDrone* single-modules

modules consisting of foamboard wings only, and the other is the full untested prototype. Figure 27 has a foam model of an aggregate of sixteen.

2.3 Dodecacopter

The *Dodecacopter* is a highly configurable structured hovercraft UAS with 6-DoF potential. Note that in this book chapter, DoF corresponds to the number of independently controllable configurations of a system, as widely used in the robotics community. For example, a classical quadcopter in three-dimensional space has six representational DoF; three translations along its x, y and z axes, and three rotations along those axes (roll, pitch, yaw). However, there are only four controllable DoF: thrust (z-axis translation), roll (x-axis rotation with y-axis

Fig. 27 An assembly of sixteen *TetraheDrone* single-module models

translation), pitch (y-axis rotation with x-axis translation), and yaw (z-axis rotation). If the control DoF is equal to the representational DoF, a system is described as holonomic. This is not the case for the classical quadcopter, which would be described as nonholonomic. A 6-DoF *Dodecacopter* vehicle is holonomic. The prototype of the *Dodecacopter* module, shown in Fig. 28, has two main elements: a frame in the shape of the dodecahedron and a rotor, represented by a disc. The frame is, once again, shaped with CFRP tubes joined with plastic 3D-printed parts. These joints serve to connect the tubes of an individual frame and/or to connect modules to each other, Fig. 29. For internal connections, the ends of tubes are tightly fit and closed with soft thermoplastic polyurethane (TPU) material. Then, three end tubes are clamped in between a pair of compatible polyethylene terephthalate glycol (PETG) parts. The PETG parts are held in place with metal fasteners and are equipped with holes to fasten modules together. Similar to the other prototypes, the remaining components include brushless motors, three-blade 39.3 cm-diameter propellers, controllers, communications modules and power sources.

Increasing the degrees of freedom to six is enabled by adjusting the tilt angles, and consequently thrust, of the rotors, as drawn in Fig. 30. Due to the symmetries of the dodecahedral shape, the frame is unaltered in terms of how it can be assembled into different UAS structures.

The *Dodecacopter* took flight in six different configurations listed in Table 3. The table describes the number of layers, modules, DoFs and the corresponding figures for each UAS configuration. Controlled flight was achieved by manual tuning of the ArduPilot flight controller software [47] for each UAS. The classical quadrotor and hexarotor formations were the easiest to tune. However, equipping the hexarotor with 6 DoFs resulted in maximum

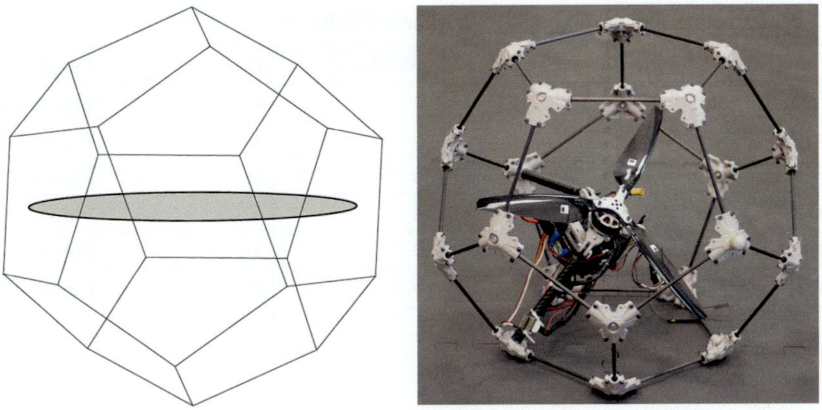

(a) Illustration. (b) Prototype.

Fig. 28 A single *Dodecacopter* module [43]

Fig. 29 Three different *Dodecacopter* modules connected at their joints [43]

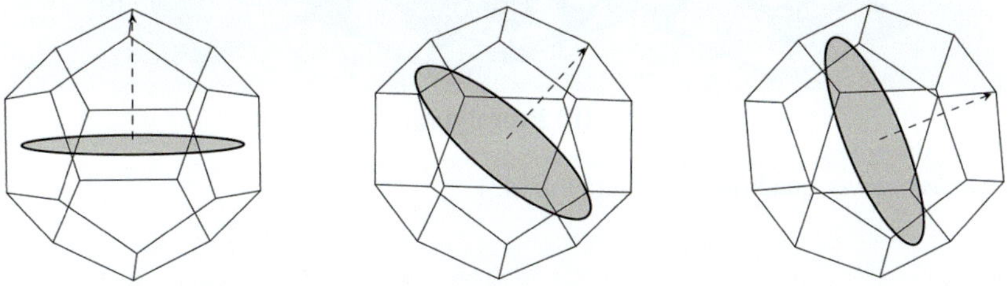

Fig. 30 Tilting rotor angles to allow different thrust directions [43]

Table 3 Flown configurations built with dodecahedral cells [43]

Name	Layers	Modules	DoF	Figures
Quadrotor	1	4	4	Figure 31
Hexarotor	1	6	4	Figure 32
6-DoF Hexarotor	1	6	6	Figure 33
Tetrahedron Quadrotor	2	4	4	Figure 34
Tetrahedron Decarotor	3	10	4	Figure 35
Tetrahedron Hexadecarotor	4	16	4	Figure 36

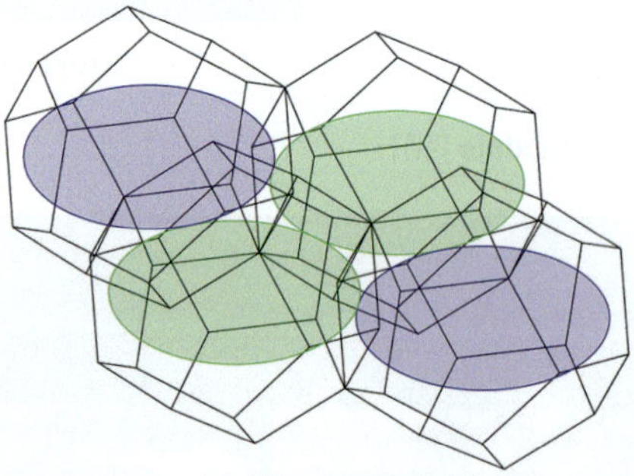

(a) Illustration.

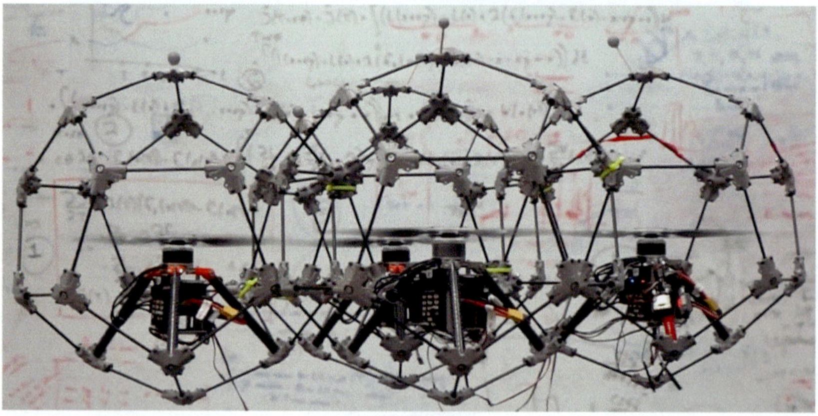

(b) Prototype.

Fig. 31 *Dodecacopter* quadrotor configuration [43]

thrust at hover due to high weight-to-thrust ratio. In that situation, maneuvering while sustaining hover is inconceivable. The tetrahedral quadrotor also suffered from saturation in its top rotor while attempting to control yaw. Thus, the yaw control authority was low. Due to

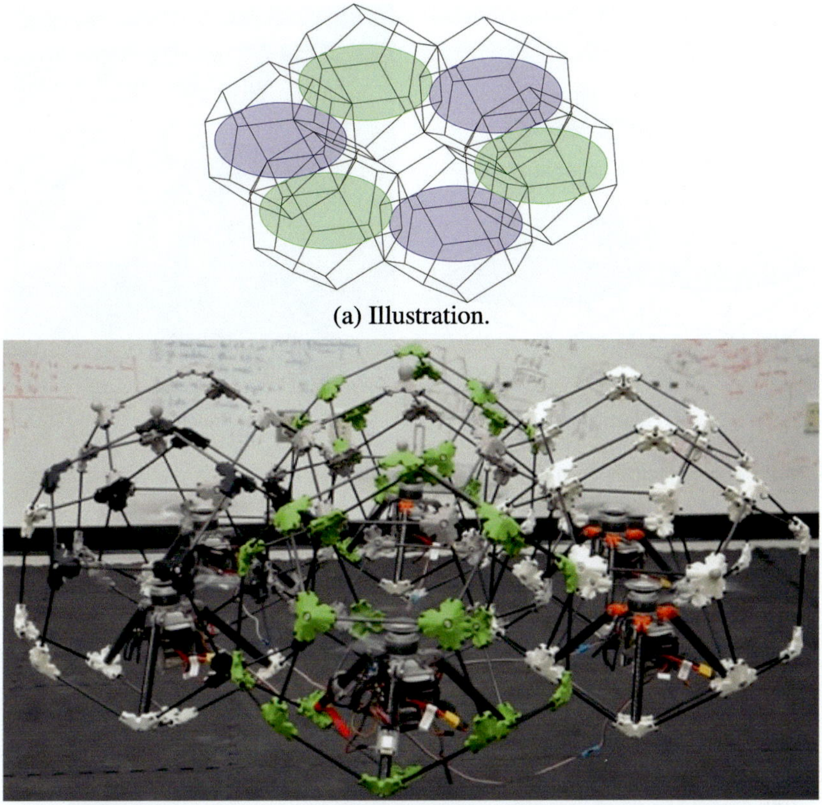

(a) Illustration.

(b) Prototype.

Fig. 32 *Dodecacopter* hexarotor configuration [43]

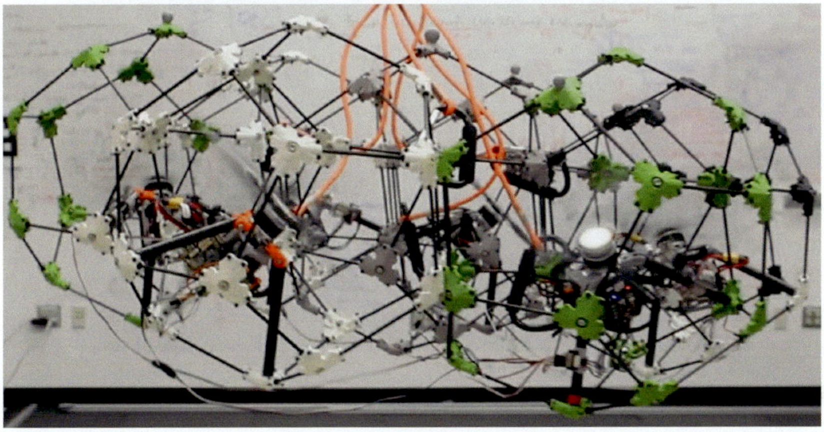

Fig. 33 6-DoF *Dodecacopter* hexarotor configuration [43]

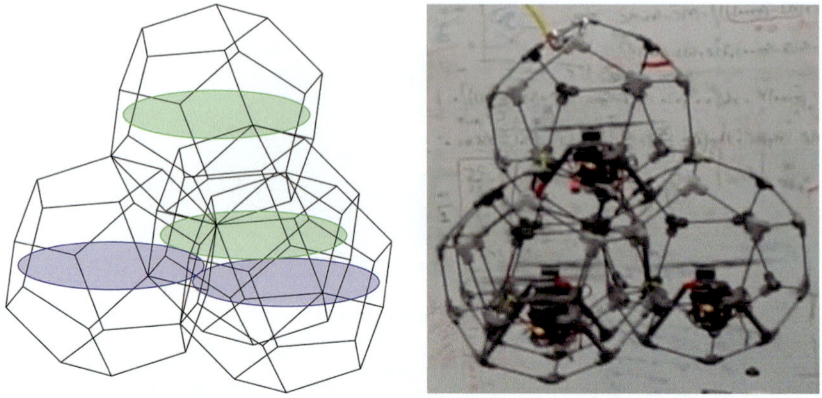

(a) Illustration. (b) Prototype.

Fig. 34 *Dodecacopter* tetrahedron quadrotor configuration [43]

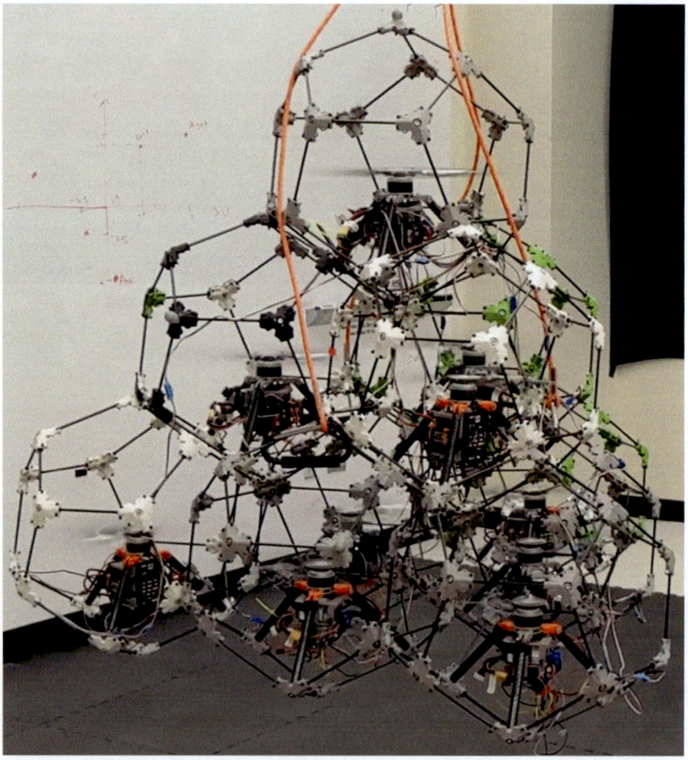

Fig. 35 *Dodecacopter* tetrahedron decarotor configuration [43]

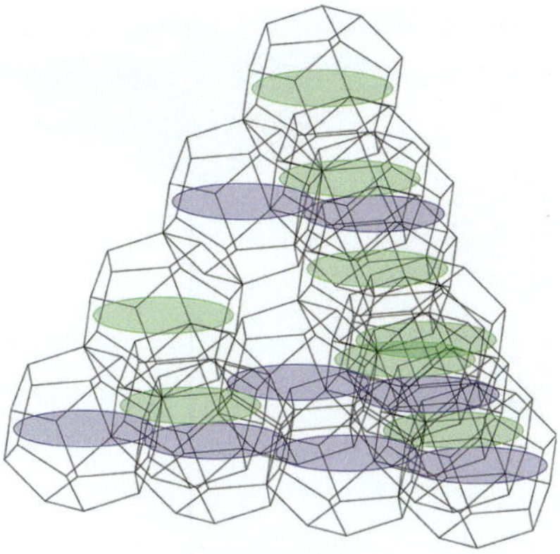

(a) Illustration.

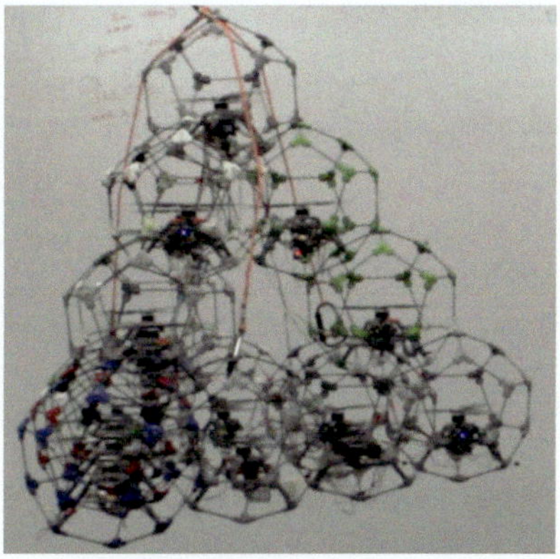

(b) Prototype.

Fig. 36 *Dodecacopter* tetrahedron hexadecarotor configuration [43]

Table 4 Comparing the *Tetracopter*, *TetraheDrone* and *Dodecacopter*

Feature	*Tetracopter*	*TetraheDrone*	*Dodecacopter*
Hover flight	Yes	Yes	Yes
Fixed-wing flight	No	Yes	No
Fractal Design	Yes	Yes	No
Configurability	Low	Low	High
Thrust-to-weight ratio	Medium	High	Low
Max DoF	4	4	6

wake interactions, the overall flight also lost efficiency. Fortunately, the 10-rotor and 16-rotor tetrahedral copters did not experience yaw and efficiency issues.

Table 4 summarizes the features of the prototypes of Sect. 2.

3 Structural Considerations

One of the main motivations for the tetrahedral and dodecahedral UASs is the mitigation of rigidity loss in scalable modular aerial systems, as introduced in Sect. 1.3. When UAV child units can only be assembled in a co-planar fashion, they tend to deform under vertical loads. The magnitude of this deformation is exacerbated with larger aerial parent vehicles.

It is possible to qualitatively illustrate this effect by examining the deformation of the horizontal square grids in Fig. 37. Three grids are shown with 16, 64 and 256 tiles respectively subject to a vertical downwards force at the center. If the tiles and grids all have uniform material properties, then the force represents gravity and is proportional to the weight, and

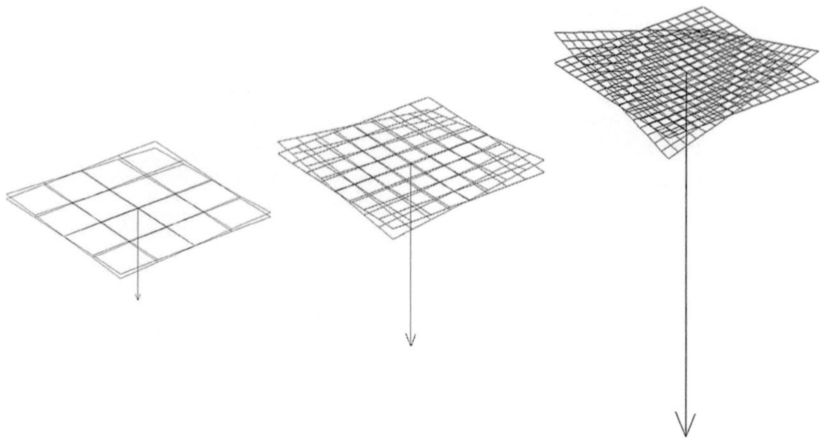

Fig. 37 Deformations of grids of 16, 64, 256 tiles under proportional load [43]

subsequently, the number of tiles in each grid. The grids in dark blue are in the unloaded state, while those in dark red show the deformations induced by gravity. Each tile is comparable to a child UAV unit, and the grid to a parent UAS. Clearly, the severity of compliance grows with the size of the UAS as observed in the figure. This qualitative structural analysis relies on the hypothesis of linear elasticity. The hypothesis can show that the maximum deflection of simply supported plates, subject to a perpendicular force concentrated at its center, is proportional to the multiple of force and plate area [48, Eq. 5.3.46]. While the analysis is not an accurate model of the modular UASs, it mimics the deformation effects of horizontal parent vehicle growth as the number of child modules in the system increases.

There are several undesirable consequences of structural deformations in a modular UAS. First, propulsion efficiency is lost as the directions of thrust forces deviate. Second, high amplitude vibrations are generated causing mechanical stresses, controller disturbances and possible flutter. Finally, the cyclic loading may result in nonlinear geometric softening further deteriorating structural strength. The last two effects are particularly critical since a complete structural failure in UAVs while flying is a significant safety concern.

As described in Sect. 2, the *Tetracopter* and the *TetraheDrone* overcome these issues by adopting a fractal tetrahedral structure. The framework that can model this construct is the truss in the three-dimensional frame, also known as the space frame. A truss is a rigid assembly of members and nodes. In this case, the members are the rods of the drone frame and the nodes are the connecting mechanisms. The space truss behaves as a single object with its members bearing exclusively axial forces, namely either tension or compression. It is stable and kinematically determinate; a junior engineering student taking a statics course may calculate the forces on all members just with the static equilibrium equations at the joints. The property of being kinematically determinate is preserved for fractal trusses. The stiffness equation relates linear displacement to forces and gives an initial estimate of stresses in a tetrahedral truss. When tetrahedra are aggregated in a non-fractal fashion, a different and more general representation is more suitable, the frame description. A frame is a structure whose rigid joints and members can withstand three-dimensional forces and/or moments.

Similarly, for the *Dodecacopter*, the structure can be modelled as a frame. However, it should be noted that the dodecahedral structure is not fractal and not ideal for a strong rigid structure. It is powerful to produce a variety of vehicle configurations and to unlock six degrees of freedom.

4 Aerodynamics

Another primary motivation of tetrahedral and dodecahedral UASs is to preserve aerodynamic efficiency of individual modular units when placed in the overall vehicle. By design, and as was demonstrated in Fig. 22, this was targeted by placing rotors such that there is minimal overlap in their footprints. This section discusses the results of an investigation on whether this strategy is effective.

The primary cause of hover efficiency reduction in 3D flying structures can be attributed to the interactions between the upstream rotors with the downstream ones. To avoid this in fractal tetrahedral UAVs, one may suggest increasing the distance between rotors to eliminate the operation of rotors in the wake or slipstream of others. However, separation entails a gain in size, resulting in hover efficiency drops related to rigidity losses, and weight, causing higher energy consumption. Thus, optimal vehicle hover efficiency is accomplished via a choice of an ideal combination of airframe rigidity, rotor position, and vehicle weight. A thrust stand was built to experimentally collect data essential for the optimization.

In the first set of experiments, the efficiency of overlapping multi-planar tandem rotor configurations is quantified as a function of the relative position of the rotors. This relationship is then compared to that estimated using the momentum theory. After comparison, it is found that the theory provides good estimates despite its assumptions that limit aerodynamic effects. The second conclusion is that the vertical distance between overlapping rotors has insignificant performance impact. Applying this data to design a quadrotor *Tetracopter* and a quadrotor tetrahedral *Dodecacopter*, vertical and horizontal rotor separations associated with ideal hover efficiency are found. In both drones, there are three downstream rotors on the same horizontal plane with one upstream rotor in another horizontal plane (placed at the top). Let the distance between the two planes be z and the diameter of the four identical propellers be D. The distances in the horizontal plane between the centers of the upstream rotor and each downstream rotor is equal and expressed as x. Based on the experimental data of the thrust stand, the ratios of $x/D = 0.93$ and $z/D = 1.0$ provide optimally efficient hover of the *Tetracopter*. This translates to minimal overlap of rotors when observed from the top, and the vertical separation in that case, once again, does not have a considerable performance impact. In the case of the aforementioned *Dodecacopter*, the values are $x/D = 0.80$ and $z/D = 1.13$.

Next, flight test data are collected from the quadrotor tetrahedral *Dodecacopter* and the hexadecarotor tetrahedral *Dodecacopter*. Data analysis demonstrates that performance degradation can be recorded by measuring rotor angular velocity while in flight. Furthermore, the analysis suggests that downstream rotors are influenced by only the closest upstream rotor when it comes to performance. This means that if four optimally efficient quadrotor tetrahedral *Dodecacopter*s are aggregated into a hexadecarotor tetrahedral *Dodecacopter*, the latter will also be optimally efficient.

4.1 TetraheDrone considerations

The *TetraheDrone* flies in two modes, one of which involves wing-generated lift. The success of this fixed-wing flight mode relies mainly on whether the wings can generate enough lift and on the position of the center of lift (CL) relative to the center of gravity (CG). As indicated in Sect. 2.2, if flat wings are capable of generating sufficient lift, then a well-chosen airfoil can only improve the design. Thus, the proof-of-concept prototype attempts to accomplish FW flight with simple flat wings, depicted in Fig. 38. Assuming inviscid and incompressible airflow, and remembering the choice of symmetric airfoils, the magnitude of the lift generated by a wing i at a small angle of attack α_i in *degrees*, as labelled in Fig. 39 and is less than 15°, is equal to

$$L_i = qSC_{l_\alpha}\alpha_i \tag{1}$$

where $q = \frac{1}{2}\rho V^2$ is dynamic pressure, ρ is air density, V is the velocity, S is wing i's surface, and C_{l_α} is the stability derivative coefficient for the wing under consideration. For the flat wing in Fig. 38, C_{l_α} can be taken as 0.1, and $S = sl/2$. With Eq. (1), it is possible to select the dimensions of the wings for target velocities and weights, such that the lift generated is sufficient for controllable flight. In other words, the lift force generated by the wings is higher than the overall gravitational force.

For a symmetric airfoil, the center of aerodynamic forces is known to coincide with the center of pressure and to be at the quarter-chord point, as shown on Fig. 38. It means that the resultant of aerodynamic forces for each wing cross-section applies at that point regardless of angle of attack. Given s and l, respectively the width and root chord length of the wing, the center of pressure c^i_p for the whole wing is computed by taking the average of center of pressure for each cross-section weighted by that cross-section's contribution to the total aerodynamic force:

$$\begin{aligned} c^i_{Px} &= \frac{\int_{t=0}^{s} \frac{3l(s-t)^2}{4s} dt}{\int_{t=0}^{s}(s-t)dt} = \frac{l}{2} \\ c^i_{Py} &= \frac{\int_{t=0}^{s} t(s-t)dt}{\int_{t=0}^{s}(s-t)dt} = \frac{s}{3}. \end{aligned} \tag{2}$$

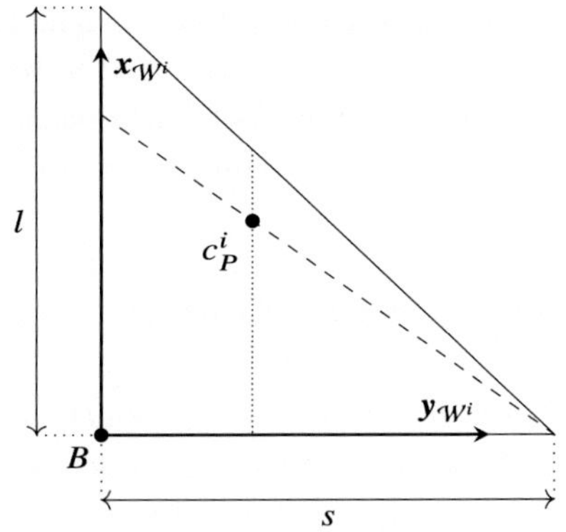

Fig. 38 Wing geometry

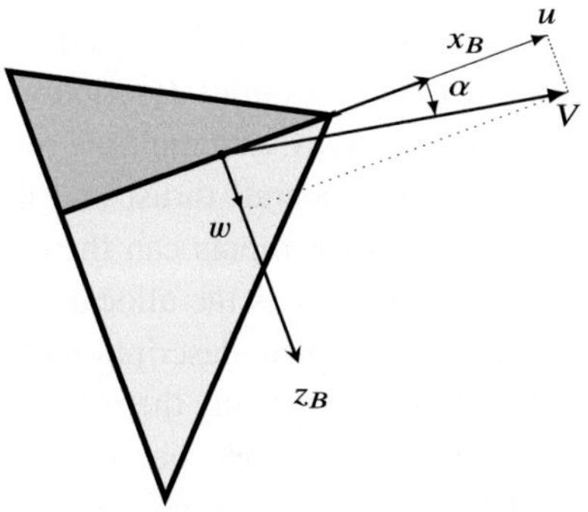

Fig. 39 Angle of attack

Considering that the drone lacks control surfaces, it is useful to align the center of gravity (CG) with that of aerodynamic lift (CL) along the longitudinal axis. In forward flight, a CL that is ahead of the CG results in unstable flight in the longitudinal, or pitching, plane. When the CG is ahead, flight in the longitudinal plane is stable, but the controls have to compensate for different deviating moments. The location of the center of gravity can be tuned by careful positioning of the vehicle's electronics, especially the relatively heavy battery.

In a fractal tetrahedral assembly (as sketched in Fig. 21) of n drones of equal mass m, where n is a multiple of 4, if the CG of module $i \in \{1, \ldots, n\}$ is located at $(x_{cg_i}, y_{cg_i}, z_{cg_i})$, the overall CG of the system is

$$x_{CG} = \frac{\sum m_i x_{cg_i}}{\sum m_i} = \frac{m \sum x_{cg_i}}{nm} = \frac{1}{n} \sum x_{cg_i}$$

The same follows for the y and z coordinates. The center of lift is defined as the point where the sum of the lift forces act. Thus, for the same assembly, the location of the overall CL is

$$x_{CL} = \frac{\sum L_i x_{cl\,i}}{\sum L_i} = \frac{L \sum x_{cl\,i}}{nL} = \frac{1}{n} \sum x_{cl\,i}$$

where L is the lift force magnitude acting on each module i at its CL, located at $(x_{cli}, y_{cli}, z_{cli})$. It is assumed to be equal for all modules in the assembly. The y and z coordinates can be found in the same fashion. It can be concluded that the distance between the CG and CL remain constant for successive generation of fractal assemblies.

5 System Control

Flown *Tetracopter* and *TetraheDrone* prototypes utilized flight controller hardware powered by PX4 flight controller software [49]. *Dodecacopter* vehicles are flown with ArduPilot [47] software instead. Both software allow customizations to tailor the firmware for the vehicle's properties. At the time of writing, no novel controllers were produced by the authors to enable these flights. PX4 and ArduPilot are well-documented and, as a result, this section will be limited to a brief description of the modifications done to fly the structured vehicles.

In aerial robots with multiple rotors, control commonly revolves around a control vector concatenating overall thrust and three torque values. A relationship between the vector and direct motor inputs can then be written with a matrix multiplication. This matrix is usually described as the allocation or mixing matrix. In the flight software, it is possible to input a geometric description of the aerial vehicle, as well as experimentally measured constants characterizing the rotors. With those files, the software is then compiled to solve the custom allocation problem, uploaded to the vehicle flight controllers, and the vehicles are flown to manually tune the underlying PID controllers. In the case of the *Dodecacopter* vehicles, a single controller is used for all rotors. While for the second design iteration of the *Tetracopter*, a single controller is responsible for each four-rotor module. With an assembly of four such *Tetracopter*s, the overall system still flew successfully despite the lack of direct communications between the modules. For the *TetraheDrone*, each quadrotor module has one controller. Strategies to control larger assemblies is still under investigation.

5.1 Control Under Uncertainty

There are potentially an arbitrary number of modules in an assembly with each module relying on their own sensors and actuators. In order to achieve the desired behavior at the assembly level, uncertainty has to be accounted for. Some examples where uncertainties may occur include: the number of modules after an in-air assembly or disassembly; the configuration of the assembly system; or the relative position of the other modules in cases where connections between modules have some degree of freedom. In a modular system, it is reasonable to assume that we do not have the full measurement of the assembly state, and we only have module sensor measurements. Moreover, we assume we control the assembly only indirectly via the modules' actuator inputs. So in order to control the assembly system under suitable assumptions, ideally we should be able to perform control allocation and state estimation. On the other hand, under uncertainty, we have to account for it as part of the control process to achieve the desired behavior. We discuss a couple of examples dealing with uncertainties in the configuration of tetrahedron assemblies. In the following, for each module j, its position in the tetrahedron assembly structure Θ_j is assumed to be unknown but belongs to a finite set:

$$\Theta_j \in \{\Theta_1, \Theta_2, \ldots, \Theta_N\}.$$

The following approaches are inspired by adaptive control approaches.

In [50], the problem of controlling a tetrahedron fractal flying assembly under uncertainty is considered. An adaptive switching controller is proposed; a decentralized control law, where each module only uses its own measurements for feedback, is developed, and a switching algorithm is applied to determine which parameters to plug into the corresponding control law. Given a stabilizing controller for the case when there is no uncertainty, it is shown that this adaptive approach achieves the control objective.

An automatic identification of the unknown configuration of the tetrahedron assembly before flight is considered in [51]. To this end, the approach consists of two parts: (1) a calibration motion is performed by the assembly system, and (2) an identification algorithm is applied. A user-defined calibration motion is implemented by the assembly system automatically with minimal human intervention; using the available actuators, the goal of the calibration motion is to make rotational movements about the three axes. Using the collected module measurements, the identification algorithm tries to identify the unknown configuration. The proposed approach is verified experimentally where it was able to provide the correct parameters with simple calibration motions.

6 Summary

This book chapter highlights a new class of innovative structured flying machines. The *Tetracopter* and *TetraheDrone* vehicles are based on tetrahedral frame shapes in order to prioritize structural strength and power efficiency. Child tetrahedral frames form a collective parent frame with a similar tetrahedral shape. This modular style design has additional pros for the adaptability, configurability, maintenance, manufacturing, reliability, scalability, storage, and transport of the vehicles and their elements. The *Dodecacopter* aims to enhance configurability, in terms of vehicle shapes and degrees-of-freedom, at the expense of power consumption.

For further details and readings, refer to published works by the authors [43, 50–56] and follow their work for expected future publications!

Acknowledgements This book chapter was written by Bilal Maassarani based on the input, works and publications of all the authors. Section 5.1 was written by Mohamad T. Shahab and edited by Bilal. All research done by the authors and presented in this work was supervised by Professor Eric Feron. The authors are grateful for the help and support received from numerous entities including sponsors, colleagues, peers, friends and family. Special thanks for Thanakorn Khamvilai and Kuat Telegenov.

References

1. R.J. Shaw, History of flight. NASA. Accessed 10 Aug 2022. https://www.grc.nasa.gov/www/k-12/UEET/StudentSite/historyofflight.html. (Visited on 05/13/2021)
2. A.E. Moule, *New China and Old: Personal Recollections and Observations of Thirty Years* (Seeley, 1892)
3. P. Boitani et al., *Winged Words: Flight in Poetry and History* (Read-HowYouWant.com, 2010)
4. D. Zaelke, Aviation partners the future is on the wing, in *Industry Genius* (Routledge, 2017), pp. 32–52
5. A. Marinoni, C. Pedretti et al., Il codice atlantico della Biblioteca Ambrosiana di Milano, vol. 2 (Giunti, 2000)
6. R. Merkert, J. Bushell, Managing the drone revolution: a systematic literature review into the current use of airborne drones and future strategic directions for their effective control. J. Air Trans. Manag. **89**, 101929 (2020). ISSN: 0969-6997. https://doi.org/10.1016/j.jairtraman.2020.101929. https://www.sciencedirect.com/science/article/pii/S0969699720305123
7. K. Ulrich, The role of product architecture in the manufacturing firm. Res. Policy **24**(3), 419–440 (1995) ISSN: 0048-7333. https://doi.org/10.1016/0048-7333(94)00775-3. https://www.sciencedirect.com/science/article/pii/0048733394007753
8. A.G. Bell, Aerial vehicle or other-structure, US Patent No. A770626 (1904)
9. R. Oung, A. Ramezani, R. D'Andrea, Feasibility of a distributed flight array, in *Proceedings of the 48h IEEE Conference on Decision and Control (CDC) Held Jointly with 2009 28th Chinese Control Conference* (IEEE, 2009), pp. 3038–3044
10. R. Oung, R. D'Andrea, The distributed flight array. Mechatronics **21**(6), 908–917 (2011)
11. R. Oung, R. D'Andrea, The distributed flight array: design, implementation, and analysis of a modular vertical take-off and landing vehicle. Int. J. Robot. Res. **33**(3), 375–400 (2014). https://doi.org/10.1177/0278364913501212
12. R. Naldi, F. Forte, L. Marconi, A class of modular aerial robots, in *2011 50th IEEE Conference on Decision and Control and European Control Conference* (IEEE, 2011), pp. 3584–3589
13. R. Naldi et al., A modular aerial vehicle with redundant actuation, in *2013 IEEE/RSJ International Conference on Intelligent Robots and Systems* (IEEE, 2013), pp. 1393–1398
14. R. Naldi et al., Modeling and control of a class of modular aerial robots combining under actuated and fully actuated behavior. IEEE Trans. Control Syst. Technol. **23**(5), 1869–1885 (2015). https://doi.org/10.1109/TCST.2015.2391173
15. M.J. Duffy, A. Samaritano, The lift! project-modular, electric vertical lift system with ground power tether, in *33rd AIAA Applied Aerodynamics Conference* (2015), p. 3013
16. M.J. Duffy, S.H. Bouwer, J.J. Mattero, Modular vehicle lift system, US Patent No. 9457899 (2016)
17. D. Saldana et al., Modquad: the flying modular structure that self-assembles in midair, in *2018 IEEE International Conference on Robotics and Automation (ICRA)* (IEEE, 2018), pp. 691–698
18. G. Li et al., ModQuad-Vi: a vision-based self-assembling modular quadrotor, in *2019 International Conference on Robotics and Automation (ICRA)* (IEEE, 2019), pp. 346–352
19. D. Saldana, P.M. Gupta, V. Kumar, Design and control of aerial modules for inflight self-disassembly. IEEE Robot. Autom. Lett. **4**(4), 3410–3417 (2019)

20. N. Gandhi et al., Self-reconfiguration in response to faults in modular aerial systems. IEEE Robot. Autom. Lett. **5**(2), 2522–2529 (2020)
21. M.A. da Silva Ferreira et al., Drone reconfigurable architecture (DRA): a multipurpose modular architecture for unmanned aerial vehicles (UAVs). J. Intell. & Robot. Syst. **99**(3), 517–534 (2020)
22. C. Montalvo, M. Costello, Meta aircraft flight dynamics. J. Aircraft **52**(1), 107–115 (2015)
23. S.J. Carlson, P. Arora, C. Papachristos, A multi-VTOL modular aspect ratio reconfigurable aerial robot, in *2022 International Conference on Robotics and Automation (ICRA)* (2022), pp. 8–15. https://doi.org/10.1109/ICRA46639.2022.9811542
24. M.D. Patterson, J.R. Quinlan, W.J. Fredericks, Modular unmanned aerial system with multi-mode propulsion (2019)
25. M. Zhao et al., Transformable multirotor with two-dimensional multilinks: modeling, control, and motion planning for aerial transformation. Adv. Robot. **30**(13), 825–845 (2016)
26. T. Anzai et al., Multilinked multirotor with internal communication system for multiple objects transportation based on form optimization method, in *2017 IEEE/RSJ International Conference on Intelligent Robots and Systems (IROS)* (IEEE, 2017), pp. 5977–5984
27. M. Zhao et al., Whole-body aerial manipulation by transformable multirotor with two-dimensional multilinks, in *2017 IEEE International Conference on Robotics and Automation (ICRA)* (IEEE, 2017), pp. 5175–5182
28. M. Zhao et al., Transformable multirotor with two-dimensional multilinks: modeling, control, and whole-body aerial manipulation. Int. J. Robot. Res. **37**(9), 1085–1112 (2018)
29. M. Zhao et al., Design, modeling, and control of an aerial robot dragon: a dual-rotor-embedded multilink robot with the ability of multi-degree-of freedom aerial transformation. IEEE Robot. Autom. Lett. **3**(2), 1176–1183 (2018)
30. M. Zhao et al., Flight motion of passing through small opening by dragon: transformable multi-linked aerial robot, in *2018 IEEE/RSJ International Conference on Intelligent Robots and Systems (IROS)* (IEEE, 2018), pp. 4735–4742
31. B. Gabrich et al., A flying gripper based on cuboid modular robots, in *2018 IEEE International Conference on Robotics and Automation (ICRA)* (IEEE, 2018), pp. 7024–7030
32. B. Gabrich, G. Li, M. Yim, Modquad-dof: a novel yaw actuation for modular quadrotors, in *2020 IEEE International Conference on Robotics and Automation (ICRA)* (IEEE, 2020), pp. 8267–8273
33. H.-N. Nguyen et al., A novel robotic platform for aerial manipulation using quadrotors as rotating thrust generators. IEEE Trans. Robot. **34**(2), 353–369 (2018)
34. M. Kulkarni, H. Nguyen, K. Alexis, The reconfigurable aerial robotic chain: shape and motion planning. IFAC-PapersOnLine **53**(2), 9295–9302 (2020)
35. H. Nguyen, T. Dang, K. Alexis, The reconfigurable aerial robotic chain: modeling and control, in *2020 IEEE International Conference on Robotics and Automation (ICRA)* (IEEE, 2020), pp. 5328–5334
36. J. Lee, Optimization of a modular drone delivery system, in *2017 Annual IEEE International Systems Conference (SysCon)* (2017), pp. 1–8. https://doi.org/10.1109/SYSCON.2017.7934790
37. A. Tagliabue et al., Shapeshifter: a multi-agent, multi-modal robotic platform for exploration of Titan, in *2020 IEEE Aerospace Conference* (2020), pp. 1–13. https://doi.org/10.1109/AERO47225.2020.9172739
38. H.S. Macdonald Coxeter, *Regular Polytopes* (Courier Corporation, 1973)
39. J. Chilton, *Space Grid Structures* (Routledge, 2007)
40. V.S. Deshpande, M.F. Ashby, N.A. Fleck, Foam topology: bending versus stretching dominated architectures. Acta Mater. **49**(6), 1035–1040 (2001)
41. G. Zhang et al., Mechanical behaviour of CFRP sandwich structures with tetrahedral lattice truss cores. Compos. Part B: Eng. **43**(2), 471–476 (2012)

42. H. Jones, A. Campa, Fractals based on regular polygons and polyhedra, in *Scientific Visualization of Physical Phenomena* (Springer, 1991), pp. 299–314
43. K. Garanger, Versatile and structurally efficient aerial systems assembled from polyhedral rotorcraft modules. Ph.D. Thesis, Georgia Institute of Technology (2022)
44. H. Hamann, *Swarm Robotics: A Formal Approach*, vol. 221 (Springer, 2018)
45. G.J. Leishman, *Principles of Helicopter Aerodynamics with CD Extra* (Cambridge University Press, 2006)
46. R. Martinez-Val, E. Perez, J. Palacin, Historical perspective of air transport productivity and efficiency (2005). ISBN: 978-1-62410-064-2. https://doi.org/10.2514/6.2005-121
47. ArduPilot Dev Team. ArduPilot documentation. Accessed 28 Aug 2022. https://ardupilot.org/ardupilot/. (Visited on 08/28/2022)
48. J.N. Reddy, *Theory and Analysis of Elastic Plates and Shells* (CRC Press, 2006)
49. PX4 Dev Team. PX4 User Guide. Accessed 6 Sept 2021. https://docs.px4.io/main/. (Visited on 09/28/2022)
50. M.T. Shahab, K. Garanger, E. Feron, Control of an assembly of aerial vehicles under uncertainty, in *2022 American Control Conference (ACC)* (2022), pp. 514–519
51. M.T. Shahab, O. Wali, E.M. Feron, Automatic identification of a modular unmanned aerial system (UAS) with experimental verification, in *AIAA SCITECH 2023 Forum* (2023), p. 2160
52. K. Garanger, J. Epps, E. Feron, Modeling and experimental validation of a fractal tetrahedron uas assembly, in *IEEE Aerospace Conference* (IEEE, 2020), pp. 1–11
53. J. Epps, K. Garanger, E. Feron, Wake interactions of a tetrahedron quadcopter, in *2020 International Conference on Unmanned Aircraft Systems (ICUAS)* (IEEE, 2020), pp. 1852–1859
54. J.T. Epps et al., An empirical analysis of the wake interactions of a novel multi-agent UAS, in *AIAA SCITECH 2022 Forum* (2022), p. 0280
55. J.T. Epps, Rotor-rotor interactions in the design of unmanned aerial systems. Ph.D. Thesis, Georgia Institute of Technology (2022)
56. O. Wali, M.T. Shahab, E. Feron, A non-planar assembly of modular tetrahedral-shaped aerial robots, in *2023 IEEE International Conference on Robotics and Automation (ICRA)* (2023), to appear

Bilal Maassarani is a Ph.D. candidate at the Robotics, Intelligent Systems and Control (RISC) lab located in King Abdullah University of Science and Technology (KAUST), Saudi Arabia. He worked full-time as an FPGA Developer for Advanced Antenna Systems at Ericsson. Bilal has a multidisciplinary background and interest in different elements of Mechatronic systems. His current research interests include mechatronics, control, aerial robotics and multi-robot systems. Bilal received a M.A.Sc. in Electrical Engineering and a B.A.Sc in Mechatronics Engineering both at the University of Waterloo, in 2017 and 2014. During his undergraduate studies, he completed six internships at Broadcom, Blackberry, General Motors, Ciena, Natural Resources Canada and IDT. Since his undergraduate studies, he managed to co-found three failed startups, so far!

Jeremy Epps is currently an aerospace engineer at optimAero, an engineering consulting firm that focuses on the design, development, and control of manned and unmanned aircraft. Jeremy completed a Ph.D. in Aerospace Engineering at the Georgia Institute of Technology in the Summer of 2022. During this time, he led and participated in designing and fabricating a wide range of unmanned aerial systems, which include group 3, heavy-lift, UAS, and smaller hobby grade UAS. The focus of Jeremy's thesis research was the aerodynamic interaction of closely spaced rotors on small unmanned aerial systems and the optimization of a vehicle's hover efficiency with respect to these interactions. Prior to entering Georgia Tech's Ph.D. program, Jeremy earned a Master of Science in Aerospace Engineering from the Georgia Institute of Technology and a Bachelor of Science in Mechanical Engineering at North Carolina A&T State University.

Kévin Garanger is a postdoctoral scholar in the department of Mechanical and Aerospace Engineering at the University of California, Irvine. His research interests include autonomous vehicles, control theory, structural mechanics, and mathematical optimization. In his research, Kevin combines these areas to support the design and analysis of novel aerospace vehicles. Specifically, a focus of his work has been to use methods of optimization for the design of modular aerospace vehicles based on their dynamical and structural properties. Before joining UC Irvine, Kevin attended Georgia Tech where he obtained a Ph.D. in robotics in 2022 and a master in aerospace engineering in 2016. He completed his undergraduate studies at École polytechnique in France. He is also an aircraft pilot and holds a commercial pilot certificate with an instrument rating.

Mohamad T. Shahab has held postdoctoral positions at King Abdullah University of Science and Technology. He previously worked as a research and development engineer with Baker Hughes. His current research interests include adaptive control, control under uncertainty, autonomous systems, and optimization & control. His general research interests fall into the areas including and in between robotics and control theory. Shahab received the Ph.D. degree in Electrical and Computer Engineering from the University of Waterloo in 2020. He received the M.Sc. degree in Systems Engineering and the B.Sc. degree in Control and Instrumentation Systems Engineering, both from King Fahd University of Petroleum and Minerals, in 2015 and 2007, respectively.

Obadah Wali is a Ph.D. candidate in Mechanical Engineering, specialized in robotics at King Abdullah University of Science and Technology (KAUST) in the Kingdom of Saudi Arabia. He has worked as a Mechanical Engineer for three years at SABIC petrochemical company, where he led the reliability and inspection teams. Currently, Wali's research is focused on designing and controlling novel autonomous modular aerial systems. Such devices aim to enrich the development of aerial system control. Wali obtained his master's degree from KAUST, where he worked on designing deformable aerial systems used for navigating through narrow paths. The motivation behind the system was to transport urgent aid to incapacitated individuals. He received his bachelor's degree in Mechanical Engineering from King Fahd University of Petroleum and Minerals.

Eric Feron is a professor of Electrical, Biological, Computer, and Mechanical Engineering at King Abdullah University of Science and Technology, Saudi Arabia. Feron has been on the faculties of École Nationale de l'Aviation Civile, France, the Georgia Institute of Technology, USA, the Institut Supérieur de l'Aéronautique et de l'Espace - SUPAERO, France, and the Massachusetts Institute of Technology, USA. Eric Feron was also a research engineer at Office national d'études et de recherches aérospatiales, France. Eric Feron's interests are using fundamental concepts of control systems, optimization theory and computer science to address problems in aerospace and transportation engineering, including air transportation, aerial robotics, software and system certification, and human-machine interaction. Feron was educated at Lycée Louis-le-Grand, École polytechnique, École navale, École normale supérieure, and Stanford University. He has operational experience with small airplanes, large submarines and sailboats of all sizes.

Top-Down Design Approach for the Customization and Development of Multi-rotors Using ROS

Fabeha Raheel, Hassan Mehmood, and Muhammad Bilal Kadri

Abstract

Drones, Unmanned Aerial Vehicles (UAVs), and particularly multi-rotors have become the talk-of-the-century, especially among robotics engineers and researchers. People are now interested in finding industrial and commercial solutions to problems arising in sectors such as agriculture, defense, transportation and entertainment with the help of aerial robots. However, developing a customized drone solution is not a trivial matter especially since researchers need to consider different methods of customization as well as developing complex algorithms to make drones capable of tackling various kinds of challenging tasks autonomously. The objective of this chapter is to formulate a simple top-down design process for developing and controlling customized multi-rotor UAVs for any desired application using the widely popular Robot Operating System (ROS) framework. The various stages of a simplistic and systematic design approach, into which any research objective pertaining to drones can be divided, are discussed in order to assist researchers towards a quick, cost-effective and fail-proof process for the development of customized UAVs.

F. Raheel (✉) · H. Mehmood
Department of Mechatronics Engineering, National University of Sciences and Technology, College of Electrical and Mechanical Engineering, Rawalpindi, Pakistan
e-mail: fraheel.mts22ceme@student.nust.edu.pk

H. Mehmood
e-mail: hmehmood.mts22ceme@student.nust.edu.pk

M. B. Kadri
College of Computer and Information Sciences, Prince Sultan University, Riyadh, Kingdom of Saudi Arabia
e-mail: mkadri@psu.edu.sa

M. Abdelkader and A. Koubaa (eds.), *Unmanned Aerial Vehicles Applications: Challenges and Trends*, Synthesis Lectures on Intelligent Technologies,
https://doi.org/10.1007/978-3-031-32037-8_2

Keywords

Drones • Multi-rotors • ROS • Aerial robots • Top-down design • Customization

1 Introduction

1.1 Unmanned Aerial Vehicles in Society

Unmanned Aerial Vehicles (UAVs), also sometimes referred to as 'drones', have recently become the new focus of attention among researchers, scientists, industrialists, military, tactical experts as well as students. UAVs are aerial robots capable of moving in a 3-dimensional space and performing a number of application-specific tasks without the presence of an onboard pilot. They come in various shapes and sizes with varying capabilities. One way of categorizing UAVs and drones is based on their structural design which includes Fixed-wing, Multi-rotors, Helicopters (single-rotor), Ornithopters (Flapping Wing) and Hybrid UAVs. These different types of drones are shown in Fig. 2.1.

Among the types in Fig. 2.1, Fixed-wing and Multi-rotor UAVs are considered to be the most popular drones for various research applications. Fixed Wing UAVs are built with rigid wings and an aerodynamic structure similar to that of an airplane. They are preferred for combat or long-distance surveillance missions because of their high payload-carrying

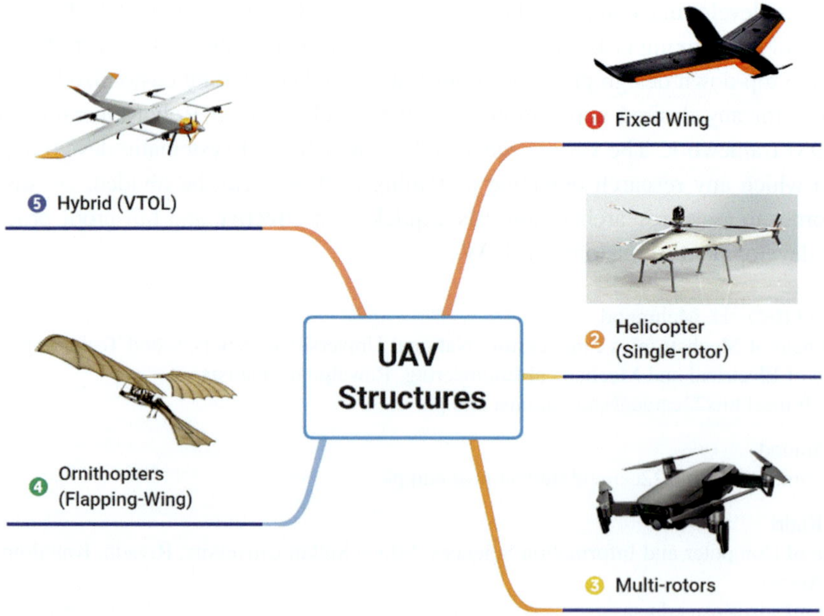

Fig. 2.1 Classification of UAVs based on Structural design [11, 15, 16, 61]

capacity, high endurance (long flight time), stability as well as the ability to reach higher airspeeds. However, Fixed-wing UAVs are incapable of hovering and can perform only limited maneuvers in small spaces due to their relatively large sizes (large wing spans) and limited agility.

On the other hand, multi-rotor UAVs are those that have more than one rotor placed symmetrically apart in different configurations. Typical configurations of multi-rotors are Quadcopters (with four rotors), Hexacopters (with six rotors) and Octocopters (with eight rotors), among several others. Multi-rotors are widely favored for a significantly large number of applications such as aerial photography, videography, inspection, and mapping due to their capabilities such as hovering, vertical takeoff and landing as well as omnidirectional flying. They also offer better control, agility and maneuverability compared to Fixed-wing UAVs. However, these advantages usually come at the expense of endurance, speed, and/or payload capacity. Multi-rotor UAVs are also simpler and more cost-effective in terms of manufacturing and deployment. Due to their overwhelming popularity, this chapter will focus solely on Multi-rotor UAVs, specifically Quadcopters.

Hybrid UAVs, sometimes known as VTOLs (Vertical Takeoff and Landing), amalgamate the advantages of both fixed-wing and multi-rotor UAVs but they are often quite expensive and require extensive training for both fabrication and control. Other types such as helicopters or ornithopters are not as widely favored for most applications because of the complexities and inherent instabilities involved in their system design.

UAVs can also be broadly divided into two main types: Remotely Piloted Aircrafts (RPAs) and Autonomous Aerial Vehicles (AAVs) as displayed in Fig. 2.2. RPAs are drones that are manually controlled by a pilot or user from a remote location while autonomous UAVs are capable of controlling their movement during missions and making decisions based on environmental perceptions up to a degree of independence from human intervention. According to [39, 51], UAVs vary in terms of different Levels of Autonomy (LOA) from automated drones to semi- and fully autonomous UAVs as shown in Fig. 2.3.

In the study of UAVs, one may also frequently come across the term UAS or Unmanned Aircraft System. A UAV, as a single independent unit, is incapable of performing any significant or noteworthy feats. It requires an entire framework supporting its functionality including not only the aircraft/vehicle itself but also the Ground Control Station (GCS), the pilot (user), sensors, data links, telemetry, supporting software, applications, algorithms, etc. Hence, the UAS includes everything that is essential for the UAV to successfully complete application-specific goals. Throughout this chapter, the focus will be on the development of semi- and fully autonomous UAVs as a complete system for any desired application.

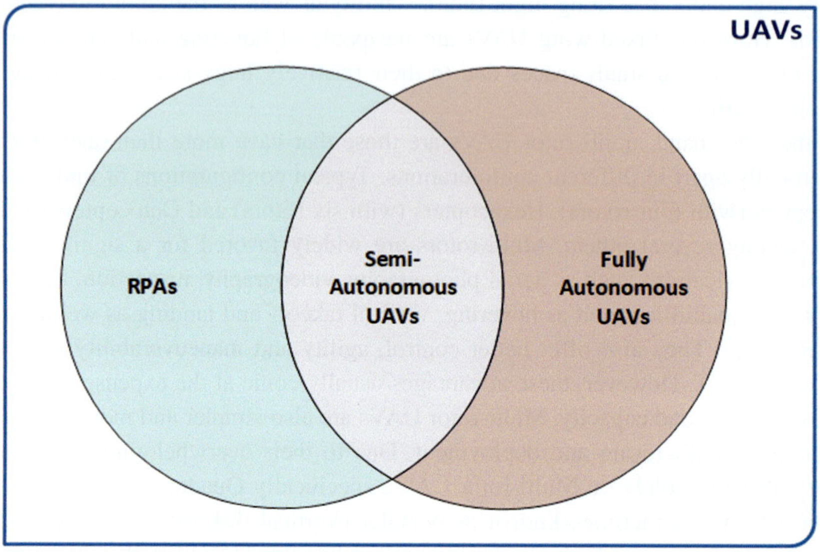

Fig. 2.2 Classification of UAVs based on control

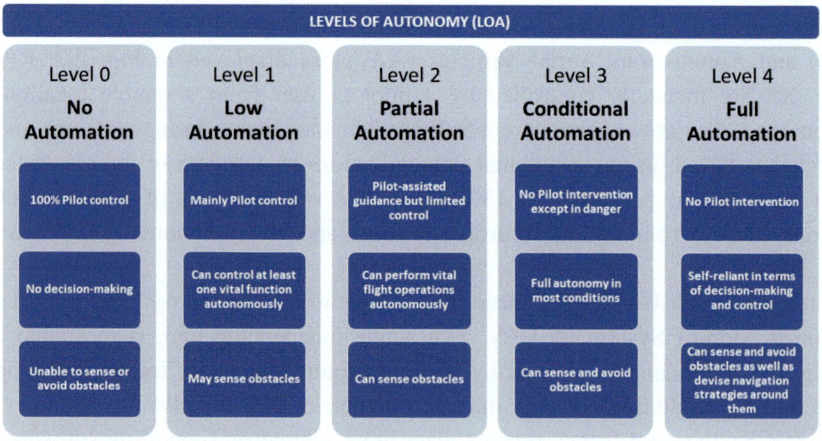

Fig. 2.3 Levels of drone autonomy

1.2 Robot Operating System (ROS)

The Robot Operating System (ROS) is an open-source set of software tools and libraries that help developers easily build robotic applications. It was first developed and introduced by Willow Garage [53]. Several modules, libraries, and drivers are available for users in the form of ROS packages. These are codes developed by other developers in the community that can be used generally by anyone anywhere working with similar hardware resources. A number of programs (written in C++or Python) can be developed independently of one another as modules and then launched together as ROS nodes. Communication between two or more ROS nodes can either be achieved by sending ROS messages over ROS Topics (that utilize the publish/subscribe model) or through ROS Services (that utilize the client/server model).

ROS provides a general, all-in-one framework for developers to program missions and algorithms for autonomous robots and drones with ease. It can be considered a meta-operating system as it offers some features of an OS such as:

- **Hardware Abstraction Layer (HAL)**—this refers to the fact that ROS deals with different hardware components at a general or abstract level rather than the detailed hardware level
- **Low-level device control**—it allows us to control the basic functionality of various robotic devices, sensors, and actuators and communicate with them
- **Inter-process Message-based Communication**—various different processes can communicate with each other by sending and receiving ROS messages at the software level
- **Reusability over Multiple Devices**—a code written on one computer is almost always likely to produce the same results when running on another computer provided that all the required packages are available
- **Supporting Applications**—ROS provides a variety of its own applications such as Gazebo, RQT, etc. which are highly beneficial for robotic developers.

These are only some of the reasons for the growing popularity of ROS in Robotics. This chapter will provide a framework and guidelines for both the hardware and software development of Customized Multi-rotor UAVs using ROS.

1.3 Applications of UAVs

The application of UAVs has drastically spread out due to the increase in performance and miniaturization of sensors and electronics as well as advancements in computational abilities and intelligence. From hobbyist drones to Unmanned Combat Aerial Vehicles (UCAVs), UAVs are being used all over the world in a diverse range of industries from entertainment to agriculture to defense through customization [37]. Customization of drones refers to the process of systematically selecting the different hardware components and peripherals after careful calculations and considerations of the design objectives. Services such as package delivery and transportation using UAVs are already being launched by Amazon Prime Air, DHL Parcel, and Google's parent company, Alphabet [41, 60]. Military UAVs such as Black Hornet [18] and InstantEye Mk-3 series are being used for

Intelligence, Search and Reconnaissance (ISR), stealth operations for strategic and tactical missions as well as for law enforcement.

Remote sensing, disaster management, and search-and-rescue are modern applications for UAV research. UAVs are suitable for these applications as they can safely reach and monitor affected areas that are hazardous or inaccessible to humans. Drone imaging can be used to locate disaster hot spots and survivors while also providing quick situational awareness (SA) to assist rescuers and concerned authorities [26, 56]. UAVs are also being developed to identify disaster-stricken areas with the help of deep learning techniques and predict disasters early in order to mitigate large-scale destruction and potential loss of human life [3, 32]).

Additionally, UAVs are also becoming increasingly popular in the agriculture and horticulture industries for a variety of applications such as crop health monitoring [24], environmental monitoring, precision farming [50], landscape mapping [14], irrigation management and livestock monitoring (also called aerial mustering) [4].

UAVs are also being used in the healthcare industry for the quick transportation of vaccines and other emergency medical supplies [60]. The applications for UAVs in vastly different domains are endless and impossible to cover within the scope of this chapter. Many organizations around the world are encouraging the development of Autonomous UAVs under different themes in order to redirect fresh talent to make new contributions in this field of emerging technology. Some of the popular competitions include the Student UAS Competition which is organized annually in St. Mary's County, Maryland [54], Teknofest International UAV Competition in Turkey [58], and the World Drone Competition [28].

1.4 Challenges in the Development of UAVs

While it is fascinating to think about developing drone applications, most researchers and students require a lot of prior hardware and software experience to get started with a customized, autonomous UAV project. UAVs are multi-peripheral systems that are built by the integration of a number of different sub-systems, sensors as well as controllers that are all functioning simultaneously in order to maintain stable flight and to ensure quick reaction of the drone to external stimuli. Since drones are inherently unstable systems as they possess six degrees of freedom (6-DOF), a number of advanced controllers are implemented to ensure their stability during flight operations [29]. These challenges make developing their customized hardware architecture and control algorithms for different applications extremely difficult. The performance and coordination of these various subsystems with minimum latency [30] are crucial to the overall success of UAV missions.

Furthermore, the design process of UAVs also requires consideration of various tradeoffs. Most UAVs designed for real-world applications require long-range operation which calls for extended flight time which in turn requires more power resources. For battery-powered multi-rotor drones, a bigger battery will increase the weight of the drone, thus, depreciating the UAV's thrust-to-weight ratio. Considering the high economic cost, time and valuable resources invested in the development of UAVs, the risk of failures and crashes is also quite substantial and serious [62].

Apart from this, most hobbyists and researchers prefer to use Commercial Off-the-Shelf (COTS) products to quickly develop aerial robots and get them running with minimum effort. However, customized UAV applications require sensors and devices of various types that usually have different manufacturers and often transmit data over different communication protocols and message types that other devices may not comprehend. The integration of these devices over a single platform and ensuring a fast exchange of information at the subsystem level is another major challenge.

These days an extensive amount of research is being done to improve the efficiency and robustness of UAVs in performing complex tasks using contemporary Artificial Intelligence (AI) approaches [13]. These applications are based on effective path planning, collision detection, and avoidance as well as navigation and localization. Researchers generally prefer to spend a minimum amount of time developing customized UAVs so that more resources can be utilized in the development of the overall system comprising novel solutions to real-world problems. The aim of this chapter is to aid researchers in the process of developing multi-rotor UAVs by circumventing most of these challenges discussed so far.

1.5 Organization of Content

The following sections of this chapter are organized as follows:

- Section 2 discusses related work and cites some literature sources that have contributed towards developing systematic approaches for UAV development along with their weaknesses.
- Section 3 introduces the Top-down Design Methodology as a solution to the challenges discussed in Sect. 1.4 as well as the weaknesses discussed in Sect. 2.
- Section 4 discusses an in-depth and thorough process for the hardware development of Multi-rotor UAVs from the ground up.
- Section 5 explains the process of developing a ROS-based software framework for customized Autonomous UAVs.
- Section 6 discusses an example case where the Top-down design methodology is used to develop an autonomous UAV for water transportation.
- Section 7 summarizes and concludes the chapter.

2 Background and Related Work

The importance and appeal of UAV technology in diverse applications have already been discussed in the previous sections. A myriad of students, academic scholars, researchers, and commercialists are interested in developing UAV solutions to solve real-world problems and a significant volume of contribution has been made in this regard.

Renault [51] discusses a model for generating a hardware architecture for UAVs based on known standards, requirements, and protocols in order to meet design specifications as well as both customer and mission requirements. A model for the assessment of the UAV architectures is explained that uses design heuristics along with some qualitative and quantitative measures to ascertain whether or not the UAV architecture will meet its design specifications. However, the model only proves valid for UAVs systems up to certain LOA. The model also does not explain the intricacies involved in the process of selecting individual UAV hardware components or the overall assembly of the drone. It also does not consider the assessment of the UAV software framework.

Kekec et al. [33] presents a generic software architecture for UAVs that enables effective communication between various low-level devices as well as analysis of the flight data mid-operation. The architecture is divided into three modules; namely, the core module that governs inter and intra-system communications, the application module that is responsible for application-specific operations, and finally, the Ground Station module that allows the pilot or users to monitor and analyze the data. The results prove the working and effectiveness of the proposed idea but the development of this complex software framework from scratch is a tedious and time-consuming task and also requires the expertise of a professional developer. Furthermore, the paper does not shed light on ways that the software framework can be validated using simulations prior to being deployed on the real hardware.

Gustave et al. [22] introduce a hierarchical architecture for UAVs using ROS. The architecture is divided into three functional layers; namely, the perception layer which uses different sensors to gain information about the environment, the control layer which is responsible for controlling the motion of the UAV in space, and finally the top decision layer which uses the data from the perception layer to plan strategies and compute actions to be taken by the UAV. The validation of this framework is implemented using the ROS-based Gazebo simulator. However, the paper is limited only to the development of the software aspect of autonomous UAVs and lacks to provide any insight regarding the implementation of the proposed framework over actual hardware.

Other literature such as [27] provide a detailed overview of a number of UAV control architectures proposed by researchers over time. Sivakumar and Naga Malleswari [56] share a description of how UAVs can be customized with a range of payloads for a variety of applications. Ahmed and Jenihhin [2] discuss the pros and cons of a range of UAV computing devices for various system performance requirements. Patrinopoulou et al. [45] use a ROS-based software framework for area surveillance and intruder detection using a swarm of UAVs. Meyer et al. [40] provide details regarding how research and development of UAVs can be sped up using ROS and tools such as the Gazebo simulator. These are a few examples among a large number of others that benefit from ROS tools, applications, and libraries for rapid software development, algorithm testing, and validation. However, very little research is available regarding the implementation of these software frameworks on real customized UAVs. It is also very difficult to find literature guiding toward a simplified and systematic approach for rapid UAV development, testing, and deployment for real-world applications.

3 Proposed Methodology

The main objective of this chapter is to propose a guideline to students, researchers, and commercialists with limited prior knowledge of UAV systems, components, and/or software development to rapidly develop projects in a quick, safe, cost-effective, and fail-proof manner. Since UAVs and UAS are becoming so commonplace, there is a need to devise a methodology for managing UAS-related projects in a manner that minimizes time and resources invested in Research and Development (R&D) as well as mitigating risks and chances of failures. The methodology presented here is referred to as the 'Top-down Design Approach' for the development of customized Autonomous UAVs.

3.1 Top-Down Design Approach for the Development of Customized UAVs

A Top-down Design Approach is a unique modular approach in which the overall system is decomposed into a number of subsystems and each system can be developed independently of the others. The three main subsystems that a customized Autonomous UAV can be divided into are:

1. the UAV itself
2. the customized application-specific payload
3. the Software Framework supporting the UAV.

Figure 2.4 shows the decomposition of the entire UAV system into these sub-systems.

Once the application and the design specifications of the autonomous UAV system have been outlined, these three sub-systems can be developed in parallel and then integrated together using a step-by-step method to quickly yield the overall system with minimal issues.

Before leaping into the development process, the different development approaches have to be carefully considered along with the project aims and constraints. The hardware

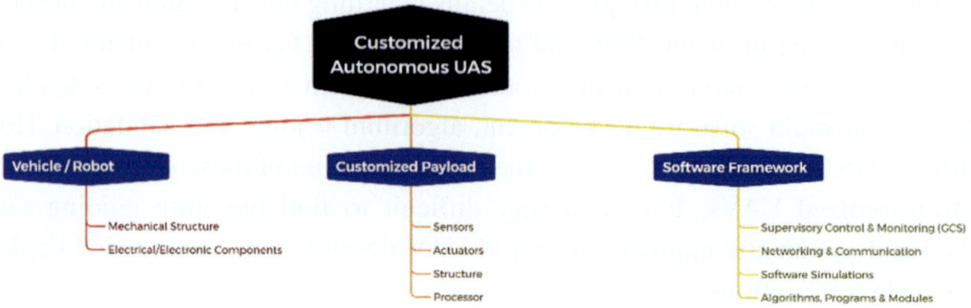

Fig. 2.4 Sub-system decomposition of a customized, autonomous UAV

development process of UAVs can follow two approaches each with its own pros and cons. One approach is to indigenously design and manufacture one or more of the UAV components; which is complex, risky, and time-consuming but worthy of high research and commercialization potential. The second approach is to use easily available, fail-proof COTS products which can significantly speed up the hardware development process. However, COTS products are expensive and may offer some limitations for more advanced custom applications. The choice between these two approaches has to be made after thoughtful consideration of these pros and cons.

Similarly, before developing the software framework of the UAV, one has to decide how and where most of the heavy computations and processing are to be carried out [2]. With Ground-based Processing, most of the processing is offloaded to a ground-based local server machine or computer, and suitable commands are then communicated to the drone. This approach is suitable for applications where extensive computation is required and small, light-weight UAVs are used which cannot carry a heavy and expensive onboard computer. However, communication delays between the ground station and the UAV pose a serious problem and the responsiveness and agility of the drone are degraded. On the other hand, with On-board Processing, most of the processing and computation is carried out on the UAV's onboard companion computer. These computers, apart from being expensive, increase the weight of the drone but they allow the UAV to actively perceive its environment, process information, make quick decisions and react to the environment almost spontaneously. The choice between these two approaches is heavily dependent on the LOA that the UAV is being designed to achieve [51]. The higher the LOA, the greater the need for onboard processing in order to make the UAV as self-sufficient as possible in terms of decision-making.

4 Design of Autonomous UAVs

4.1 Consideration of Design Specifications

The design and hardware development stage begins with clearly listing down the design specifications of the customized Multi-rotor UAV. Some of the specifications that can be considered are illustrated in Fig. 2.5 and also discussed as follows [41]:

- **Payload**—This is the maximum expected weight that the drone is required to carry excluding its structure and basic electronics. Payload refers to any additional mass fixed to the drone that, if removed, will not affect the drone's ability to fly and maneuver in the air.
- **Flight Time**—This refers to the maximum estimated time for which the drone can operate and remain airborne. It should be sufficient enough for each application.
- **Range**—This refers to the maximum distance that the drone can reach and safely operate away from the base station or Ground Control Station (GCS).
- **Operating Altitude**—This refers to the height above the ground at which the drone can fly and safely operate.
- **Operation Conditions**—These are the environmental and surrounding conditions in which the drone is expected to operate such as indoor/outdoor environments, presence of obstacles, lighting conditions (suitable for cameras), GPS-permitted or GPS-denied environments, etc.
- **The Application**—This refers to the UAV application for which it is being built such as surveillance, search-and-rescue, agriculture, etc.

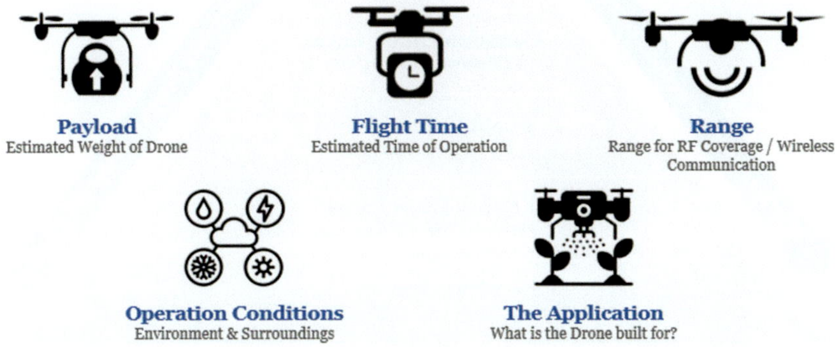

Fig. 2.5 Common design specifications considered for the development of UAVs

4.2 Typical Hardware Architecture of Multi-rotors

The next step after identifying the design specifications is selecting the required functional hardware components for the vehicle assembly. The hardware architecture of most Multi-rotor UAVs is generally divided into the following five main layers as illustrated in Fig. 2.6:

1. **Power Distribution Layer (PDL)**

This layer works like the UAV's blood circulatory system and aims to transfer the right amount of power from a power source (usually a battery) to each and every electronic component based on their prescribed power ratings. The PDL generally consists of components such as batteries, power modules, Power Distribution Boards (PDB), Universal Battery Eliminator Circuit (UBEC), Battery Management System (BMS)/Battery Monitors, Power Switches, Circuit Breakers, etc.

2. **Drive-Train Layer (DTL)**

This layer works like the skeleton and limbs of the drone and includes all the components that enable the UAV to fly and move about in a 3D environment. The main components

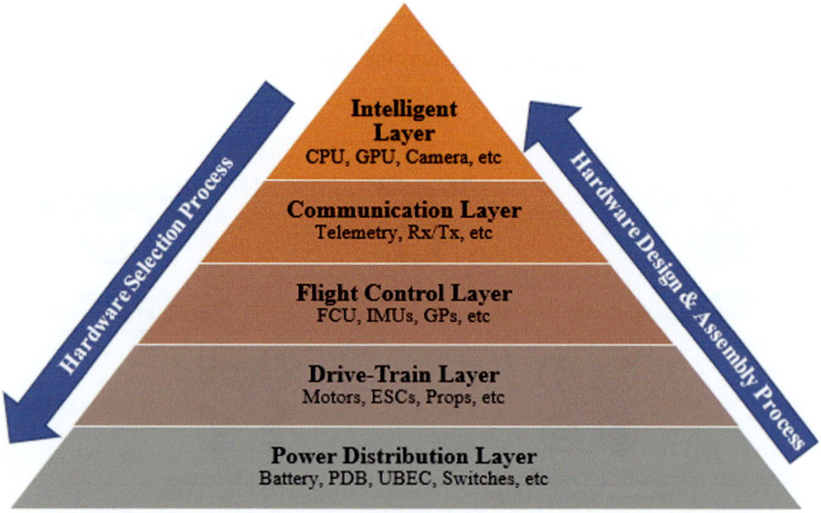

Fig. 2.6 Typical hardware architecture of multi-rotor UAVs

of this layer are the Airframe (or drone body), Motors (typically Brushless DC or BLDC motors), Electronic Speed Controllers (ESCs), and Propellers.

3. **Flight Control Layer (FCL)**

This layer is the heart of the UAV. It includes a Flight Controller Unit (FCU) which provides coordinated commands to control the drive-train components and allow the UAV system to maneuver in the desired direction. The FCU utilizes various built-in and external sensors such as Inertial Measurement Units (IMUs), GPS, and Compass to gain self-awareness of its current state (position and orientation) in space.

4. **Communication Layer (COML)**

When working with UAVs, users like to receive flight information and sensor data from a remote location (a Ground Station or Base Station) as well as maintain a degree of control over the drone's movement while it is airborne. Several communication devices can be used to achieve this including Radio Transmitters and Receivers, Radio/WiFi Telemetry, 4G/5G, GSM Networks, IoT and Cloud Platforms, etc.

5. **Advanced Computation/Intelligence Layer**

This layer is additionally included in most semi or fully autonomous UAVs. It includes an onboard companion computer that functions as the UAV's 'brain'. This layer helps to run computationally intensive programs and algorithms for more advanced autonomous applications such as image processing, Simultaneous Localization and Mapping (SLAM), Stereo Vision processing, Sensor Fusion, etc.

4.3 Hardware Development of Multi-rotor UAVs

The success of a UAV's performance, efficiency, and robustness depends greatly on the appropriate selection of components for each of the layers described in Sect. 4.2. The hardware selection process starts from the top of the pyramid illustrated in Fig. 2.6 and moves downwards in descending order while the hardware assembly process usually starts from the bottom of the pyramid and moves upwards in ascending order. This section will briefly go over some important details that should be considered during the hardware selection process. It is also noteworthy to remember that the inter-compatibility of the components between the layers should be determined and verified before progressing further. The steps for the Hardware Development process are outlined in Fig. 2.7 as well as in the following sections.

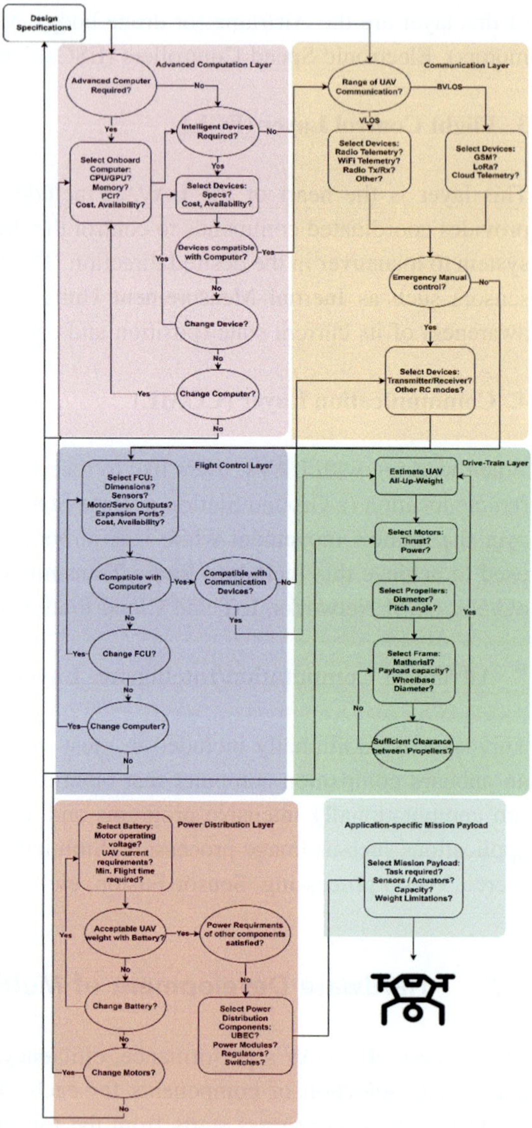

Fig. 2.7 Summary of the UAV hardware selection process

4.3.1 Selection of Advanced Computation/Intelligence Layer Components

Since a customized autonomous UAV is usually designed for practical deployment in a real-world application, it makes sense to consider its LOA and intelligent capabilities before anything else. Hence, this layer is considered first in order to fully understand the UAV's ability to process information and make intelligent decisions [51]. A suitable companion computer must be selected for this purpose that is capable of handling

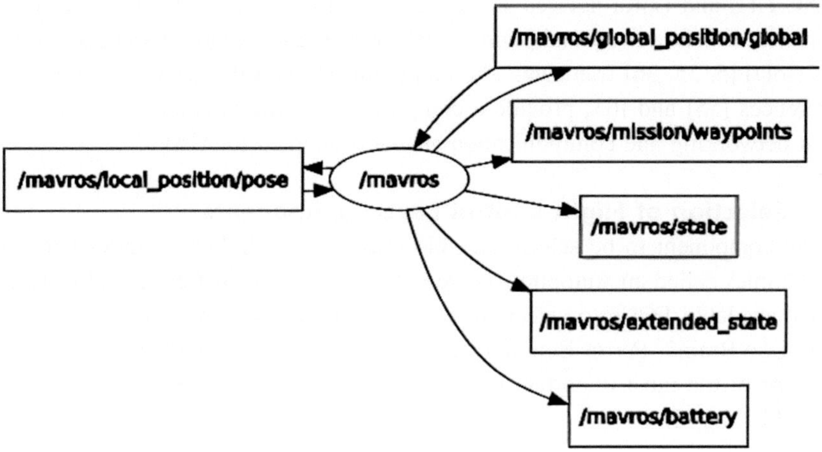

Fig. 2.8 MAVROS package or drones

the intensive computation required according to the custom application of the UAV. Some factors to be considered while selecting a suitable computer for a UAV are [2]: the amount of computation power desired, the memory and storage capacity of the computer, availability of GPIO, UART, I^2C, SPI ports, external peripheral interfacing options, CPU/GPU resources and capabilities, supported Operating Systems (OS) and amount of community support available. Some of the most popular companion computers to choose from include Raspberry Pi, Nvidia Jetson Nano, Nvidia TX1/TX2, ODroid, LYCHEE, etc. [7].

Intelligence is also highly reliant upon environmental perception and self-awareness. Therefore, suitable cameras (monocular, stereo-vision, depth, multi-spectral, etc.) and other exteroceptive and interoceptive sensors, should also be considered according to the application at hand [14, 24]. It is also necessary to ensure that these sensing devices are compatible with the chosen onboard computer.

4.3.2 Selection of Communication Layer Components

Communication with the UAV is essential and of the utmost importance especially during the development phase. For the Communication layer, the selection criteria for components are based strongly upon the desired range of communication decided in the design specifications in Sect. 4.1. The most popular and cheapest option for wireless communication between the GCS and airborne UAV is Radio or WiFi Telemetry which is suitable for low-range communication within the Visual Line of Sight (VLOS).

For drones that are being customized for Beyond Visual Line of Sight (BVLOS) operation, GSM modules may be used for transferring data from one location to another in the form of text messages using a SIM card. However, this option is not considered suitable when large packets of data have to be transmitted. Other methods for BVLOS communication are discussed in [23, 47].

4G/5G LTE and LoRa devices are the latest techniques used for long-range communication with low power consumption. These devices use Cloud Services and Internet of Drones (IoD) [5, 35, 36] frameworks to communicate with the Ground Station.

References [55] and [63] provide a comprehensive and detailed survey regarding the different networking and communication channels adopted for UAVs.

4.3.3 Selection of Flight Control Layer Components

The main component to be selected at this stage is the Flight Controller Unit or FCU, also sometimes called an Autopilot. There is an inexhaustive list of several popular, commercially available FCUs to select from such as Pixhawk, Pixracer, CUBE, APM 2.x, Navio+, Erle Brain2, Parrot Bebop, etc. [9]. Selecting the right Flight controller for the UAV is one of the most important decisions to be made and depends on several factors mentioned below [6]:

- **Physical Size**: The dimensions of the Autopilot should fit in the UAV frame
- **Sensors**: The Autopilot should have some integrated sensors such as IMUs and should also have or provide support for some redundant sensors for more accurate and reliable state estimation
- **Motor/Servo Outputs**: The Autopilot should have enough output pins for your UAV of choice
- **External Expansion Ports**: The Autopilot should provide enough ports for connections such as UART, I^2C, CAN, etc. for integration of external components such as the Telemetry module, companion computers, and additional sensors
- **Economic Cost and Local Availability**: The Autopilot should be cheap and locally available in your region
- **Community Support**: The Autopilot should have enough documentation and community support on online forums to assist one's research

Since most flight controllers come with built-in IMUs, external IMUs are seldom used. However, connecting an external GPS/GNSS and compass for autonomous flight missions is usually recommended. When selecting an appropriate GPS/GNSS and Compass model, factors such as the position estimation accuracy, sensitivity (responsiveness), and refresh rate also need to be considered. A good GPS/GNSS is preferred for applications that require accurate position estimates.

Many researchers are now interested in developing customized Flight controllers for their UAVs in order to add more sensors and redundancy in smaller package sizes [38].

4.3.4 Devising Tables for Component Weight and Power Ratings

Since most of the important components required for the UAV application have already been selected, one can devise tables listing the weights and power ratings (energy requirements, operating voltages, and currents) of all the components selected so far. Examples

Table 2.1 UAV weight estimation table

Component	Quantity	Weight/part (g)	Total weight (g)
Onboard Computer (Nvidia Jetson Nano)	1	200	200
Camera (Arducam IMX219 8MP)	1	10	10
Radio Telemetry (Holybro Radio V3)	1	30	30
Radio Receiver (FS-iA6B)	1	15	15
Intel 8265AC WiFi Chip + Antennas	1	50	50
Pixhawk 2.4.8 FCU + Peripherals	1	100	100
Additional Payload	1	1000	1500
Motors + Propellers	4	180	720
ESCs	4	60	240
Frame	1	476	476
Battery + PDL Components	1	1450	1450
Total			4791

Table 2.2 UAV power consumption estimation table

Component	Quantity	Operating voltage (V)	Max. current drawn/ component (A)	Total Max. current drawn (A)
Onboard Computer	1	5	3	3
Pixhawk 2.4.8 FCU	1	5	1	1
Additional Payload	1	12	2	2
Motors	4	22.2	33.5	134
				140

of these tables can be seen in Tables 2.1 and 2.2 in Sect. 6. The table has to be updated throughout the selection process in order to assess the drone's energy and power requirements as well as to ascertain whether the selected drone configuration is likely to be successful or not.

4.3.5 Selection of Drive-Train Layer Components

The Drive-Train Layer (DTL) is the most important layer of the UAV as it determines how well the drone will be able to fly at full-load capacity. The main Drive-Train Layer components are the Airframe, Motors, ESCs, and Propellers.

The UAV Frame is the structure that houses and holds all the components together and bears the weight of the entire system. It has to be sturdy and rigid so that it does not contribute to any undesired vibrations and sources of instability in the system. It should also be strong and durable so that it can sustain enough impact during landing or other accidental events (like minor crashes) without causing damage to the other equipment. The frame should also be as lightweight as possible to minimize the weight for better performance. The configuration (specified by the number of motors and their placement such as X-, H-configuration, co-axial, etc.) and size (specified by the wheelbase diameter) of the frame have to be carefully selected after considering the size of the propellers [48]. Generally, the smaller the size of the drone, the greater the agility, maneuverability, and stability. However, smaller size drones cannot be used to carry large payloads and have limited applications.

The choice of a suitable combination of motors and propellers is dependent on the payload capacity of the drone. For this, the estimated UAV Weight Table discussed in Sect. 4.3.4 can be used to quantify the expected All-Up-Weight (AUW) or Maximum-Take-Off-Mass (MTOM) of the UAV for different choices of motors and propellers. To design a multi-rotor, n number of BLDC motors have to be mounted such that each motor-and-prop combination can provide thrust greater than at least two times the total weight of the Multi-rotor UAV at 50% throttle. This is because the Thrust-to-Weight Ratio for a Multi-rotor should preferably be greater than 2:1 for better performance and stability [34].

Most UAVs typically use BLDC Motors because they have the capacity to operate at very high speeds at sufficient torques (above 10,000 RPMs) and better responsiveness to changes in direction and acceleration because of their very low rotor inertia. The absence of brushes also makes these motors more durable and reliable with better life expectancies. Propellers are mounted on the BLDC motors. They are used to convert the rotational energy produced by the spinning of the motors into vertical thrust. This thrust is produced by the acceleration of air or airflow in the downward direction that produces aerodynamic forces. The amount of thrust that can be produced is dependent on the following factors: the diameter of the propeller, the number of blades, the pitch angle of the blades, and also on the spinning velocity.

Once the motors are decided, one can then decide on suitable ESCs to control the phase and amplitude of the DC current pulse required to control the speed and torque of the BLDC motor. The ESCs act as an intermediary between the flight controller and the motors. They accept PWM signals from the Flight controller and convert them into varying 3-phase pulses to control the speed of the BLDC motors. The choice of ESCs depends on the maximum current rating for the selected motors. Typically, ESCs should be chosen such that their maximum current rating is at least 30–50% higher than the maximum current drawn by the BLDC motors at nearly 80–100% throttle.

The correct choice of DTL components plays a major role in the long-term success of the customized autonomous UAV. Failure to choose the right configuration can result in crashes and significant damage. A detailed literature outlining the process of sizing the DTL components is provided by Biczyski et al. [12]. It is highly recommended to use available online resources such as the Copter Ecalc Calculator to verify the combination of the DTL components [42, 43].

4.3.6 Selection Criteria for Power Distribution Layer Components

The components in the Power Distribution Layer (PDL) can only be decided on once all the components in the other layers have been finalized and their power ratings (voltage and current) are noted down. The details of the selection of the most important components are described as follows:

Battery Selection

While there are a number of different options for UAV power sources, most UAV systems prefer to use LiPo Batteries for power because they are rechargeable, provide high energy density per unit weight and also because they come in small, compact packages that can easily be mounted on drones [17, 46].

The specifications and power ratings of the battery can be roughly decided while finalizing the components of the drive-train layer as it is the most power-consuming layer in the entire UAV system. The desired battery voltage has to be roughly equal to the voltage that needs to be supplied to the BLDC motors in order to spin them at the speeds required to produce the desired amount of thrust. Once this is determined, one can choose from different options of LiPo batteries available such as 3S (11.1–12.6 V), 4S (14.8–16.8 V) or, 6S (22.2–25.2 V). There may be other versions of LiPo Batteries available locally to choose from. The terms 3S/4S refer to the configuration of the batteries which is generally represented as XSYP where X is the number of cells connected in Series and Y is the number of cells connected in Parallel [64]. The typical formula for calculating the battery voltage is represented as:

$$V = V_{cell} \times N_{series}$$

where, V_{cell} is the voltage of a single cell connected in series and N_{series} is the number of cells connected in Series. The nominal voltage of a cell is 3.7 V and the maximum voltage when it is fully charged is 4.2 V.

Apart from the voltage, batteries are also labeled with a Rated Capacity in Ah (Amperes-hour) or mAh (milli-amperes-hour). They are also labeled with a C-rating (such as 25 C or 50 C). This is the amount of energy the battery can safely discharge continuously. In order to calculate what is the maximum amount of current that can be safely drawn from the battery continuously, the C-rating is multiplied by the Rated Capacity [20]. The formula is represented as follows:

$$I_{Batt} = C_{rate} \times Q_{Batt}$$

where, C_{rate} is the C-rating of the battery, Q_{Batt} is the Rated capacity of the battery in mAh or Ah and I_{Batt} is the maximum current that can be safely drawn continuously. For example, a 1000 mAh, 20C battery can provide 1000 m $\times$ 20 = 20 A continuously.

After the selection of the Drive-Train components, the amount of current required by all the motors at 100% throttle can be determined. We can assume that the maximum current that is required by the UAV can be calculated as:

$$I_{max} = (I_{motor} \times n) + I_{other}$$

where I_{motor} is the maximum current drawn by one motor at 100% throttle, n is the number of motors, I_{other} is the assumed value of the maximum current that can be drawn by all other components and I_{max} is the maximum current that can be drawn by the UAV system.

At times when I_{other} is unknown, we can assume it to be roughly equal to 10–20% of the value of I_{motor} for simplified calculations. Once I_{max} is calculated, the battery should be selected such that it has a suitable capacity and C-rating that can supply the required amount of current.

The final calculation that needs to be made before selecting the right battery is the expected flight time. In order to do so, the AUW of the UAV system needs to be determined which is the total weight of the UAV when all required components and payloads are mounted. The formula for estimating the flight time is represented as:

$$t_{flight} = Q_{Batt} \times \frac{DoD}{AAD}$$

where Q_{Batt} is the Rated capacity of the battery in mAh or Ah, DoD is the allowable depth of discharge expressed as a percentage (since it is not recommended to discharge LiPo batteries beyond 80%, this value is typically taken as 0.8) and AAD is the Average Ampere Draw of the system

The value for AAD can be calculated using the following formula:

$$AAD = AUW \times \frac{P}{V_{Batt}} = AUW \times I_u$$

where AUW is the All-Up-Weight of the system, P is the power required to lift 1 kg of the weight in the air (for ease of calculations, this value can be assumed around 170 W/kg for most drone systems or slightly less for more efficient systems), V_{Batt} is the nominal battery voltage and I_u is the amount of current required to lift 1 kg of weight in the air.

genstattu.com [20] provides a detailed article regarding the selection, installation, and maintenance of LiPo batteries in UAVs.

In order to increase the flight time, one can consider the following options:

- Use a battery with a higher rated capacity (Q_{Batt})
- Use lighter systems (decrease the weight of the system)
- Add more batteries in parallel
- Use more efficient systems.

UBEC, Power Modules, and Battery Monitors

Many devices installed in UAVs such as companion computers, flight controllers, and external sensors usually operate at voltages much lower than the battery-supplied voltage. UBEC consists of a number of switching regulators that can convert the high LiPo battery voltage into lower required voltages (3.3 V, 5 V, 12 V) to provide a reliable source of power to other low-voltage components in the UAV system. Some models of UBEC come with Battery Monitors and protection circuitry which can protect sensitive equipment from undesired power surges or damage as well as provide information regarding the health of the battery.

Power Modules also provide similar functions as the UBEC but are available only for dedicated devices such as flight controllers. They usually come with integrated features such as protection as well as current and voltage sensing to send information about the battery's health to the flight controller.

It is also common to install a Power Distribution Board (PDB) in UAVs to easily distribute the correct amount of power to a large number of peripherals throughout the drone structure.

4.4 Customized, Application-Specific Payloads

As stated in Sect. 4.1, a payload is anything additionally fixed to the UAV assembly that, if removed, will not affect the UAV's motion. The basic underlying architecture of most multi-rotor UAVs is usually the same. The worth, practicality, and creativity of UAVs are often judged by the effectiveness of the customized payloads and the ability of the UAV to employ them for real-world applications.

The choice and/or design of payloads depends on the background application of the UAV and the tasks it is expected to perform. These payloads may include but are not limited to the following components:

- **External exteroceptive sensors**—These are used for enhancing the UAV's ability to perceive the environment and collect useful data. Examples of such sensors include different types of cameras (monocular, stereo-vision, optical flow, RGB-D, tracking cameras, etc.), rangefinders (LiDARs, sonar, etc.), RADARs, thermal/Infra-red (IR) and various others [56].
- **Redundant interoceptive sensors**—These are used to estimate the UAV's own state in the environment with greater accuracy by employing Sensor Fusion or other advanced techniques such as GPS Blending, etc. Examples include barometers, external IMUs, redundant GPS/GNSS, etc.
- **Local Processing Devices**—These may be smaller processors to offload small-scale subsystem-level processing, clean up (pre-process) gathered data quickly, and store it for analysis at a later time [13].

- **Actuators**—These are used to perform desired actions to accomplish tasks based on different applications.
- **Storage**—These include storage spaces allocated for retaining objects or materials in the UAV structure such as storage tanks, compartments for storing objects for package delivery, etc. [60].

5 Software Development of Autonomous UAVs

Development of a Software Framework for Customized Multi-rotor UAVs can be a daunting task. The basic software stack can be generally divided into the following modules:

1. **Flight Controller / Navigation Controller**—responsible for moving the UAV from one point to another either with respect to the Global/World Frame (in global coordinates) or the Local Frame (relative to the UAV's home position) [7, 8, 27, 49].
2. **Environment Perception**—responsible for obtaining data about the environment and surroundings using a number of external sensors such as monocular or stereo cameras, LiDARs, etc. [14, 45].
3. **Application-specific Decision Handler**—responsible for combining the data from the above two modules for advanced decision-making such as Object Detection and Recognition, Obstacle Avoidance, SLAM, etc.
4. **Networking and Communication**—responsible for communication with remote GCS or base station over a network or through the Internet using Web Applications, GUIs, etc. [47].

For typical UAVs that use COTS FCUs, the software modules for Flight/Navigation Controllers are available in the form of ready-to-use firmware such as ArduPilot, PX4, etc. This makes the process of software development significantly simpler as one does not have to worry about low-level control of the n number of actuators (motors) of a Multi-rotor UAV to achieve the desired pose.

Some of the major challenges arising during the Software development process is the seamless integration of the various software modules while enabling a cohesive exchange of information and the testing, validating and debugging our software algorithms in the early development stage without the risk of compromising the hardware. In order to counter these challenges, we use the widely popular ROS which not only offers a myriad of tools for the development of Robotics applications but also a suitable message-passing protocol using the publish/subscribe model [53]. We can also test our algorithms in a close-to-real-world Simulation environment using the Gazebo Simulator [19].

5.1 ROS for Multi-rotor UAVs

ROS provides a MAVROS Package which is one of the most important packages for drone developers [25, 44, 52] as it allows us to command missions and obtain feedback from drones that are capable of communicating via MAVLink messages.

MAVLink is the communication protocol utilized by most Autopilot firmwares such as ArduPilot and PX4. The MAVROS package is responsible for bringing this data into the ROS environment via various communication drivers for UDP communication. Hence, MAVROS acts as a bridge between the flight controller and ROS.

Some of the important MAVROS topics are listed as follows:

- **/mavros/state**—This topic publishes the state of the flight controller and can inform us whether or not the connection between the flight controller and ROS is established.
- **/mavros/local_position/local**—This topic receives the x, y and z coordinates of the UAV in meters as well as the orientation expressed in quaternions.
- **/mavros/local_position/pose**—This topic receives messages of type geometry msgs/ PoseStamped.
- **/mavros/global_position/global**—This topic receives messages about the global position information about the UAV in latitude, longitude and altitude.

Several ROS-based libraries and APIs (Application Programming Interface) such as the open-source iq_gnc Python and C++ API [31] have also been developed to program drones based on MAVROS. Other popular APIs such as Dronekit [1] can also be integrated with MAVROS.

5.2 Simulation of ROS-Based Multi-rotors Using Gazebo

It is usually recommended to test the software algorithms in a simulation environment prior to hardware-level testing in order to remove any pertinent bugs that may arise in real-time. Software-In-The-Loop (SITL) is a way to simulate the behavior of drones by running the drone navigation controllers directly on a local PC without any hardware. For SITL, fake environmental conditions, input commands, and sensor data are obtained from the Simulator that utilizes an in-built flight dynamics model. The resulting behavior of the drone is then visualized to produce meaningful information regarding the performance of our algorithms [40, 41, 44].

The Gazebo Simulator provides a powerful 3D simulation environment to test ROS-based algorithms on autonomous robots. It also offers some already-built drone models and worlds that can be configured to simulate realistic environments.

6 Example Case: Autonomous Quadcopter for Water Transportation

Let us consider an example to demonstrate the development process discussed so far. Collection, transportation, and release of water at target areas are some basic tasks that may be performed by agricultural and firefighting drones. We consider the development of a prototype Autonomous Quadcopter for the transportation of water from one target area to another. It should be capable of identifying the target areas, navigating itself to the center of these areas, and then collecting or releasing water precisely at the correct location. This mission scenario was built for the [58] International UAV Competition that was held in Afyonkarahisar, Turkey [58].

6.1 Hardware Development

Design Specifications

The design specifications of the Autonomous Quadcopter are defined as follows:

- Maximum Take-off-Mass (MTOM)—5 kg
- Total Estimated Payload—1.5 kg
- Minimum Water Storage Capacity—500 mL
- Minimum Flight Time—10 min.
- Operating Range—2–20 km (VLOS)
- Operating Altitude—5–30 m
- Operating conditions—daylight operation in GPS-permitted regions
- Expected Tasks—Waypoint Navigation for coverage, target area identification and recognition using Image Processing, collection of water from a blue target area (intake reservoir), the release of water to a red target area (release tank), precise positioning over the target areas, return to the Home location and safe landing.

Developing the Advanced Computation Layer

We first consider the Advanced Computation/Intelligence Layer components. Since autonomous navigation, target area detection and recognition, and precise positioning are all computation-intensive tasks, an onboard companion computer is necessary. Considering various different options, Nvidia Jetson Nano is chosen as the companion computer as it provides both CPU and GPU resources which can be utilized for fast real-time target area detection and recognition as well as navigation. It is also relatively cheaper than other boards such as Nvidia TX1/TX2.

Target area detection and recognition also require a camera to be installed in the drone assembly. After searching for a number of suitable cameras compatible with Nvidia Jetson Nano, the Arducam IMX219 8MP Camera is selected with a wide-angle Field of View (FOV) lens. It is lightweight and small compared to traditional webcams and is also very economical in terms of cost. It can also provide images of a suitable resolution at nearly 30–60 fps which is acceptable for our application.

Developing the Communication Layer (COML)

Throughout the mission, the UAV is expected to remain in contact with the Ground Control Station (GCS) and transmit its status and flight information steadily. The UAV is expected to start the execution of its mission autonomously upon receiving a signal from the GCS and should abort its autonomous mission anytime the pilot or GCS deems it necessary. Since the UAV is required to operate at only a limited range from the GCS (2–20 km), the best option for enabling GCS communication is Radio Telemetry such as Holybro Radio V3. Apart from this, a radio receiver can also be installed on the Quadcopter to allow manual remote control in cases of emergencies via a radio transmitter.

Furthermore, users also prefer to establish direct, wireless communication with the onboard computer in order to form a wireless local area network. This can be accomplished by connecting both the GCS and the onboard computer to the same gateway or router via WiFi. Since Nvidia Jetson Nano does not come with a built-in WiFi module, a separate WiFi module such as Intel 8265AC chip is needed with dual 6dBi antennas to cover the operating range.

Developing the Flight Control Layer (FCL)

The main component required for this layer is the flight controller and its peripherals. A large portion of the UAV developers' community prefers Pixhawk as the best flight controller for designing autonomous UAVs. There are several versions of Pixhawk to choose from among which the cheapest is Pixhawk 2.4.8 and the latest preferred Autopilot is Pixhawk Cube Orange. Each Pixhawk FCU has a number of built-in IMUs for accurate state estimation and redundancy. It also comes with supporting peripherals such as GPS/GNSS, safety switches, etc.

Devising the Component Weight and Power Ratings Table

During the selection process, it is advised to maintain a table for listing the weights of all components as well as their power ratings. These tables are updated throughout the selection process until the most suitable components have been selected that satisfy the design specifications. Examples of these tables are provided as Tables 2.1 and 2.2.

Developing the Drive-Train Layer (DTL)

As mentioned earlier, the Drive-Train Layer is one of the most important layers of a UAV. Referring back to our design specifications, we know that the MOTM of our Quadcopter should not exceed 5 kg. Therefore, one motor of our Quadcopter should provide a thrust equal to one-fourth of the MOTM at 50% throttle, i.e., 1.25 kg. Therefore, the motor should be capable of generating at least 2.5 kg thrust while operating at full efficiency so that the Quadcopter can have a minimum thrust-to-weight ratio equal to 2:1 for stable flight. In order to satisfy these requirements, we find that Gartt ML5008 400 kV motors along with 1755 propellers (17″ length and 5.5° pitch angle) are a suitable combination as shown in Fig. 2.10. These motors weigh around 180 g each along with their propellers. Hence, four such combinations will add 720 g to our Quadcopter.

Figure 2.10 also shows that the maximum current that can be drawn by these motors is 33.5A. Hence, ESCs with a 40A current rating will be sufficient to control these motors at their operating voltage. The ESCs we selected have a total weight of 60 g each and will, therefore, add further 240 g to our Quadcopter.

Finally, the last item to be appropriately selected is the airframe over which all the other components shall be mounted. We first need to decide the size of the frame or its wheelbase diameter. Using the online calculator [42], we can determine the best frame size to be 650 mm which will provide around 27 mm of safe clearance between the propellers. We can now search for suitable lightweight, sturdy frames for this size that can safely carry at least 10 kg (twice the MOTM). The Tarot Ironman 650 carbon-fiber frame is a suitable option and it weighs only 476 g.

Developing the Power Distribution Layer (PDL)

Table 2.2 can be used to determine the PDL components. A 6S LiPo battery is chosen as it can supply 22.2 V to the motors in order to generate the required amount of thrust. In order to determine the battery capacity, we calculate the total current drawn by four motors operating at 100% throttle as well as the other power-consuming devices. According to Table 2.2, this value is equal to 140A. Using the formulas given in Sect. 4.3.6, we need to select a battery with suitable battery capacity (Q_{batt}) and C-rating (C_{rate}) such that the maximum continuous current drawn by the battery (I_{Batt}) is nearly 140 A. Therefore, after searching for available batteries at economical cost, we considered two suitable options: a 6000 mAh 25C battery that weighs 810 g and a 10000 mAh 20C battery that weighs around 1400 g. Hence, the battery with the highest capacity (10000 mAh) is selected in order to maximize the flight time as long as it does not exceed the MTOM limit specified in the design specifications.

The estimated flight time for the UAV can be found using either an online calculator [57, 64] or manually using the formulas given in Sect. 4.3.6. The estimated flight time of our UAV is found to be nearly 10 min which is also in accordance with our design specifications.

Apart from the battery, it can be seen in Table 2.2 that some components such as the onboard computer and customized additional payload require low operating voltages. A UBEC such as the Matek UBEC Duo can be used to step down and regulate the high battery voltage into the required lower voltages for these components. The Pixhawk flight controller also comes with a special Power Module to supply regulated voltage input along with voltage and current sensing data.

Verifying Suitability of Selected UAV Hardware Configuration

The suitability of the Hardware configuration finalized during the design process is verified using Xcopter Ecalc online drone calculator [42]. The results are shown in Fig. 2.11.

Developing the Customized, Application-specific Payload

According to the design specifications, the UAV is required to pick up at least 500 mL of water from an intake reservoir, store it and then release it into another water release area. It is also provided that the total weight of the additional payload (along with stored water) should not be greater than 1.5 kg. This scenario requires a custom-designed two-way mechanism for sucking and releasing water from a certain height along with a storage tank of capacity preferably greater than 500 mL as it is difficult to ascertain the amount of air in the tank or the intake/release pipes. We chose to install a water tank of roughly 1000 mL capacity along with two lightweight 12 V water pumps for water sucking and release respectively. The pumps have a maximum head of 2.5 m which is a safe altitude for our Quadcopter to hover above the water surface. A small control circuit with a dedicated local microcontroller is also developed to control the pumps after receiving commands from the companion computer. The final design for the payload is shown in Fig. 2.13.

6.2 Software Development

Flight Controller Firmware Setup

One of the basic tasks expected from our Quadcopter is autonomous Waypoint Navigation for coverage and exploration. Navigation of a Quadcopter from one pose to the next is achieved with the help of feedback Attitude and Position controllers that are very sophisticated, especially considering the fact that a Quadcopter is an underactuated system. However, Pixhawk can be used with a number of open-source firmware such as ArduPilot and PX4 which can implement these controllers readily and easily without much hassle. We choose to use the ArduPilot firmware for our Quadcopter [8].

ArduPilot provides two different flight modes which can be used for Waypoint Navigation: Auto Mode, where the FCU saves a number of specified Waypoints and attempts to cover them sequentially, and Guided Mode, where the FCU continuously receives Waypoints one after another from a companion computer. The Guided Mode is more suited for our application as the companion computer is responsible for governing the Quadcopter's navigation.

Ground Control Station for Supervisory Control and Monitoring

A Ground Control Station (GCS) software application is required to monitor the Quadcopter's flight information and data throughout the mission. The open-source Mission Planner GCS Software [10] is easily available and is recommended for use with ArduPilot. It provides all tools for effective visualization of real-time flight data as well as basic supervisory control features that the pilot can use to take over the control of the Quadcopter anytime the need is felt.

Building the Simulation Environment for Testing

A simulation environment is built for testing our software algorithms in SDF format for implementation in the Gazebo Physics Engine and simulator as shown in Fig. 2.14. The iris Quadcopter model as shown in Fig. 2.9 is used in the simulation.

Software Algorithm

The software algorithm is divided into a number of modules. One module is responsible for guiding the Quadcopter to follow a series of Waypoints while another module runs in parallel to continuously inspect the scene and detect the water reservoir which is marked in blue color in the simulation, as shown in Fig. 2.14. In order to identify the water reservoir, the algorithm should momentarily exit from the waypoint navigation mode and shift to a vision-based control mode. It needs to be emphasized that mode switching is an important phenomenon in our mission-guided control. This mode switching causes some

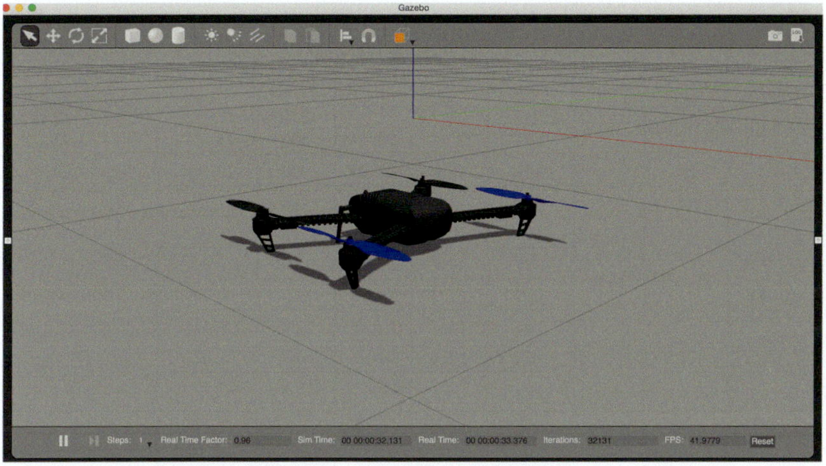

Fig. 2.9 Iris model for drone simulation

erratic behavior which will be elaborated on in subsequent paragraphs. After identifying the water reservoir, the quadcopter should descend over it while maintaining a certain height over the reservoir. It has to wait for a couple of seconds during which it is required to collect the water by initializing the sucking operation. After completing the operation, it should rise and attain the same altitude at which it was flying. The vision-based control must transfer the control back to the waypoint navigation. The drone should move along the waypoints and identify the water drop location which is marked in red color. The drone should then follow the same set of operations at the water drop location as it had conducted at the water-intake reservoir. The only difference is that due to a change in the colors of the targeted location, image filters in the image processing modules are slightly varied. The drone must continue on the same trajectory and reach the home position and land safely.

We use the Dronekit Python library to implement Autonomous Waypoint Navigation as it is extensively used in programming autonomous drones. Code for the waypoint navigation and vision-based control was developed in Python using Dronekit and OpenCV libraries. Initially the code was tested in simulation using Software in The Loop (SITL) architecture with ArduCopter as the firmware and MavProxy as the Ground Control Station (GCS). The quadcopter was simulated in Gazebo Physics Engine. Multiple ROS nodes were designed and spawned to carry out different tasks e.g., control, image processing, waypoint navigation, etc. The launch file which launches all the nodes is given below:

ML5008推力测拭								
1tem NO.	Volts (V)	Prop	Throttle	Amps (A)	Watts (W)	Thrust (g)	Efficiency (g/W)	Operating temperature(℃)
ML5008 400KV	22.2V	1555	40%	1.5	33.3	490	14.71	51℃
			45%	2.4	53.3	690	12.95	
			50%	3.5	77.7	910	11.71	
			55%	5.8	128.8	1280	9.94	
			60%	8.5	188.7	1580	8.37	
			65%	11.3	250.9	1950	7.77	
			70%	14.2	315.2	2240	7.11	
			75%	17.6	390.7	2590	6.63	
			80%	21	466.2	2830	6.07	
			100%	23.6	523.9	3100	5.92	
ML5008 400KV	22.2	1755	40%	2.1	46.6	660	14.16	68℃
			45%	3.6	79.9	930	11.64	
			50%	5.5	122.1	1260	10.32	
			55%	8.7	193.1	1680	8.70	
			60%	12.5	277.5	2110	7.60	
			65%	16.5	366.3	2610	7.13	
			70%	20.6	457.3	2980	6.52	
			75%	25.5	566.1	3290	5.81	
			80%	29.8	661.6	3640	5.50	
			100%	33.5	743.7	3880	5.22	
Notes: The test condition oftemperature is motor surface temperature in 100% throttle while the motor run 1 min. environment temperature 26℃								

Fig. 2.10 Datasheet of Gartt ML-5008 BLDC Motors (Gleagle n.d.)

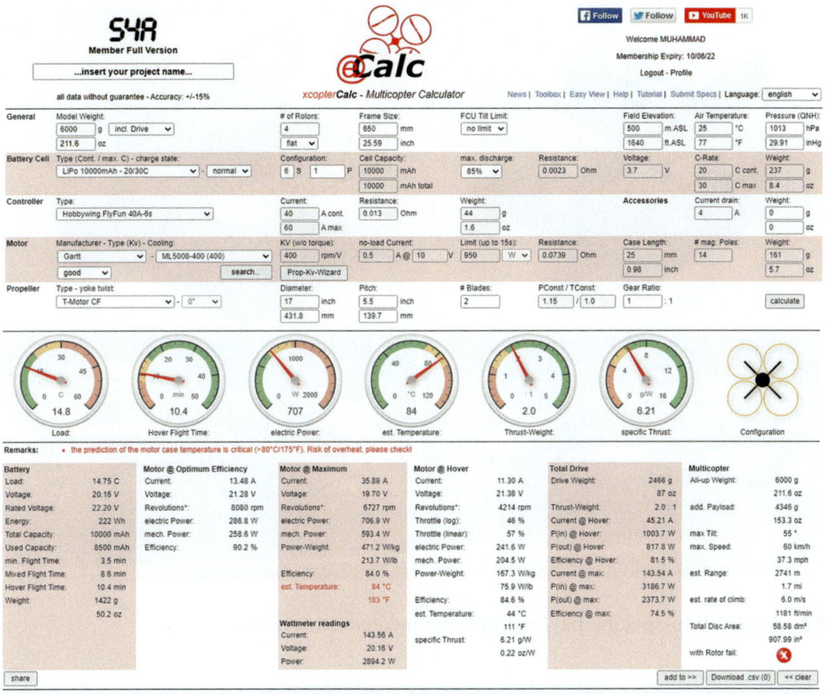

Fig. 2.11 Estimated Performance results obtained using online Ecalc for Multicopter design [43]

Fig. 2.12 Final quadcopter assembly

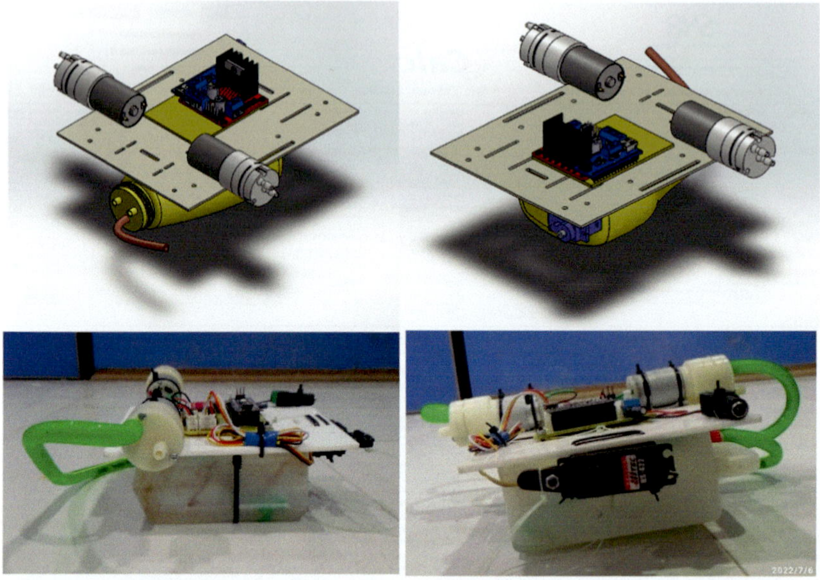

Fig. 2.13 Application-specific payload/mission mechanism system

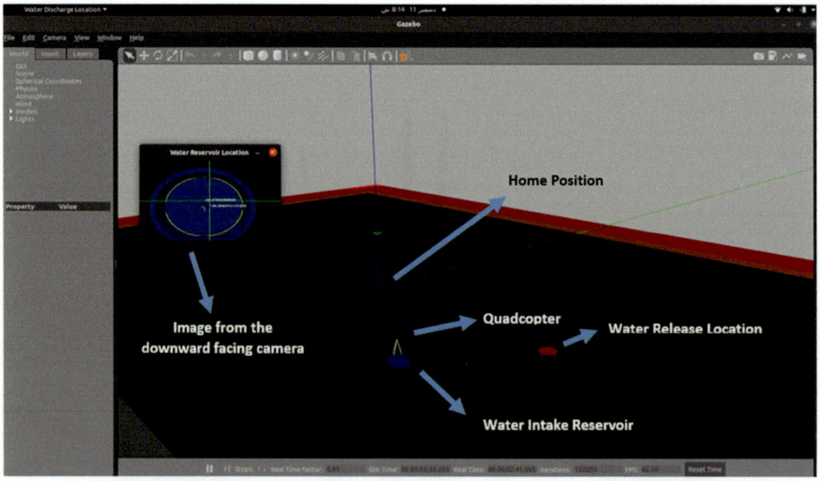

Fig. 2.14 Simulation environment in Gazebo for testing software algorithms

```
<launch>

<!--Launching all the parameters from a YAML file -->
<rosparam file="$(find
mbk_ardupilot_drone)/config/setup.yaml" />

<!--Waypoint navigation using Python script -->
<node pkg="mbk_ardupilot_drone"
type="waypoint_navigation_mission_02_for_dronekit.py"
name="Quadcopter_Navigation" output="screen" >
</node>


<!--For generating path of the quadcopter in RVIZ by
using Python script -->
<node pkg="mbk_ardupilot_drone"
type="Generate_Trajectory_In_RVIZ.py"
name="Generating_Trajectory" output="screen" >
</node>

<!-- Determining (x,y) coordinates of the Water Reser-
voir and Water Discharge Location -->

<node pkg="mbk_ardupilot_drone"
type="droneCameraCV_WaterLocation.py"
name="Water_Location_mbk" output="screen" >
</node>

<!--PID Controller for Picking and Dropping Water-->
<node pkg="mbk_ardupilot_drone"
type="WaterTracking_pidControl_ARDUPILOT_MBK.py"
name="MBK_Set_Flight_Modes_and_Tracking_PID_Controller"
output="screen" >
</node>

</launch>
```

To implement the waypoint navigation, a custom function was developed that was able to move the quadcopter to the desired location. The ***set_destination(lat, lon, alt, wp_index)*** function as shown below accepts the following arguments (Fig. 2.15):

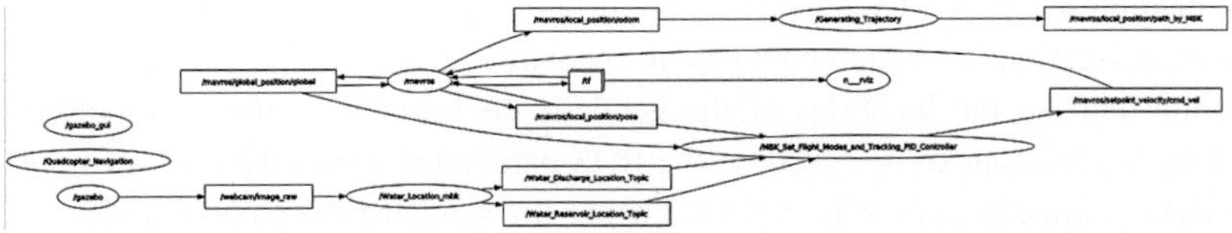

Fig. 2.15 ROS Graph showing launched ROS nodes

- lat: latitude of the desire position
- lon: longitude of the desired position
- alt: altitude at the desired position
- wp_index: is the index of the waypoint.

```
def set_destination(lat, lon, alt, wp_index):

      global vehicle
      rospy.set_param('/EXECUTING_WAYPOINT_NAVIGATION',
1)
      print("Moving to Waypoint {0}".format(wp_index))
      aLocation = LocationGlobalRelative(lat, lon,
float(alt))
      vehicle.simple_goto(aLocation)
      dist_to_wp = dist_between_global_coordinates
(vehicle.location.global_frame, aLocation)

      while dist_to_wp > 10:

dist_to_wp = dist_between_global_coordinates
(vehicle.location.global_frame, aLocation)
      print("Distance to Waypoint {0}:
{1}".format(wp_index, dist_to_wp))
      print("Reached Waypoint {0}".format(wp_index))
      rospy.set_param('/EXECUTING_WAYPOINT_NAVIGATION',
0)

time.sleep(1)
```

The ***vehicle.simple_goto(aLocation)*** from the dronekit library was used in the ***set_destination()*** function. Using this function, the drone can easily move to the desired location. The main disadvantage of ***simple_goto()*** function is that if it is interrupted during execution the UAV ends in an uncertain state. During our mission, if the drone is moving towards a way-point and suddenly the image processing modules identify the water reservoir below it (since the image processing module is getting the real-time images from the downward-facing camera) the mode has to be switched to vision-based control from waypoint navigation immediately. When the mode is switched during flight the ***simple_goto()*** function has an adverse effect on the drone flight. To circumvent this problem, the waypoint_navigation list which contains all the waypoints has to be dynamically updated. During the first lap the locations of the geodetic coordinates of the water reservoir and water drop location has to be saved on the ROS parameter server. The following variables on the ROS parameter server are updated using the standard ROS commands.

```
#setup.yaml file
Lap_Count: 1
Water_Reservoir_Location_Detected_Lap_02: 0
Water_Discharge_Location_Detected_Lap_02: 0
Water_Reservoir_Location_Detected_Lap_01: 0
Water_Discharge_Location_Detected_Lap_01: 0
Water_Sucked_by_Syringes : 0
Water_Released_by_Syringes : 0
Water_Reservoir_Location_Latitude : 0.0
Water_Reservoir_Location_Longitude : 0.0
Water_Reservoir_Location_Altitude : 0.0
Water_Discharge_Location_Latitude : 0.0
Water_Discharge_Location_Longitude : 0.0
Water_Discharge_Location_Altitude : 0.0
Water_Discharge_Location_Saved : 0
Water_Reservoir_Location_Saved : 0
EXECUTING_WAYPOINT_NAVIGATION : 0
Current_Waypoint_Index_Lap_01 : 0
Waypoint_Index_After_which_BLUE_was_detected: 0
Waypoint_Index_After_which_RED_was_detected: 0
```

Once the geodetic coordinates for both locations have been saved on the ROS parameter server, the waypoint navigation list (waypoints_lap_01) in Python has to be updated for the second lap. In the second lap the waypoint navigation has two additional points. The UAV will move on the designated waypoints as well as it will definitely fly over the water reservoir and water drop location. When the UAV will fly over the water reservoir waypoint, the control will automatically shift to vision-based control. The UAV will rectify its position i.e., place itself at the center of the reservoir by using two PID controllers, one for the x-axis and another for the y-axis. After performing the desired operation on the blue tank, it will rise and attain the same altitude. At this instant, the mode will switch back to waypoint navigation and the UAV will continue its flight toward the next waypoint. By utilizing this mechanism the ***vehicle.simple_goto(aLocation)*** is not interrupted during its execution.

```
waypoints_lap_01=[[ 24.7944000, 67.1352048,5,1],
[ 24.794442070,67.13542070,5,2],
[24.79477870,67.13535640,5,3],
[24.79475190,67.13510290,5,4],
[ 24.7944000, 67.1352048,5,5] ]
```

Code for inserting the geodetic coordinates for water reservoir and water discharge location in the way-points_lap_02 list is given below:

```
if Water_Reservoir_Location_Saved_Lap_01 and
lap_counter==1:
if ENTRY_FLAG_01==1:
Lati-
tude=rospy.get_param('/Water_Reservoir_Location_Latitude')
Longi-
tude=rospy.get_param('/Water_Reservoir_Location_Longitude')
Altitude=5
#rospy.get_param('/Water_Reservoir_Location_Altitude')
in-
dex=rospy.get_param('/Waypoint_Index_After_which_BLUE_w
as_detected')
way-
points_lap_02.insert(index,[Latitude,Longitude,Altitude
,index])
print('I am inserting the BLUE''s GPS location in the
waypoints_lap_02 at index:',index)




ENTRY_FLAG_01=0


if Water_Discharge_Location_Saved_Lap_01 and
lap_counter==1:
if ENTRY_FLAG_02==1:
Lati-
tude=rospy.get_param('/Water_Discharge_Location_Latitud
e')
Longi-
tude=rospy.get_param('/Water_Discharge_Location_Longitu
de')
Altitude=5
#rospy.get_param('/Water_Discharge_Location_Altitude')
in-
dex=rospy.get_param('/Waypoint_Index_After_which_RED_wa
s_detected')
way-
points_lap_02.insert(index+1,[Latitude,Longitude,Altitu
de,index+1]) #If we use index the RED location will be
inserted before waypoint 3 which will be incorrect
print('I am inserting the RED''s GPS location in the
waypoints_lap_02 at index',index)
ENTRY_FLAG_02=0
```

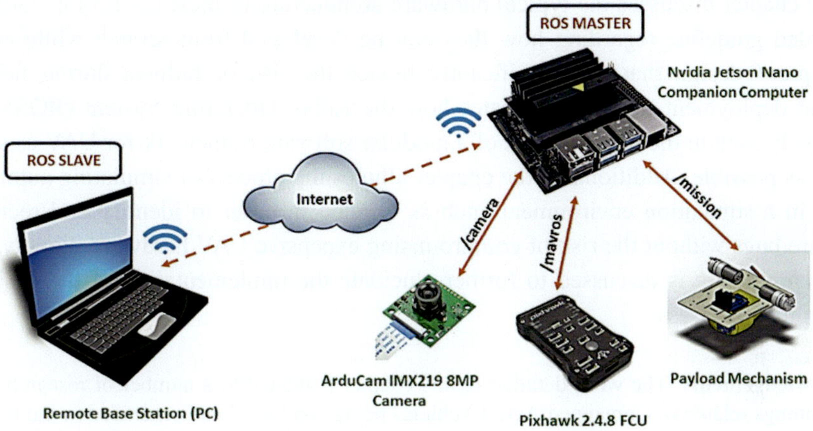

Fig. 2.16 Software-Hardware Integration of UAV system using ROS

The vehicle is instructed to return to the Home Position and land safely once all the waypoints have been reached and the water collection and release operations have been completed.

Hardware-Software Integration

When implementing our algorithms using SITL, we can migrate our software framework to the actual hardware. It is important to test each subsystem individually and resolve any issues arising due to the hardware before engaging in all-out field testing. Safety precautions for both the user and the drone need to be ensured in case of unprecedented events.

During the Hardware-Software Integration, a ROS network is built that allows all the different peripherals to communicate with one another using ROS. Just like we have our Windows Operating System that helps us connect and utilize different peripherals such as keyboard, mouse, printer, etc., the Robot Operating System or ROS allows us to utilize the functionalities of various hardware peripherals and ease communication between them via the common ROS protocol. The ROS network we developed is shown in Fig. 2.16.

7 Conclusion

The chapter provides a broad and comprehensive overview of the development process of Autonomous Multirotor UAVs which are becoming increasingly commonplace in a variety of industries and applications. The main aims of composing this chapter are to assist and encourage more individuals, with lesser prior professional experience, to develop UAVs for solving real-world problems in a safe, fail-proof, reliable and cost-effective way.

The chapter discusses the typical hardware architecture of most multirotors along with a detailed guideline regarding how they can be developed from scratch while ensuring a fail-proof design that can significantly reduce the risk of failures during field testing and deployment. It also elaborates how the Robot Operating System (ROS) can be effectively used in the development of a modular software framework for UAV as simplistically as possible. Additionally, the chapter alludes the process of simulating autonomous UAVs in a simulation environment such as Gazebo in order to identify and rectify any software bugs without the risk of compromising expensive UAV hardware. Finally, a simple example case is discussed to further elucidate the implementation of the design and development process.

Acknowledgements The work detailed in this chapter is related to a number of research projects and trainings related to Unmanned Aerial Vehicles being conducted and organized in the Intelligent Mobile Robotics (IMR) Lab in Karachi Institute of Economics and Technology (KIET), Pakistan, funded by the National Center of GIS and Space Applications (NCGSA), Pakistan.

References

1. 3D Robotics, Welcome to DroneKit-Python's Documentation!' Accessed 28 February 2023, https://dronekit-python.readthedocs.io/en/latest/
2. F. Ahmed, M. Jenihhin, A survey on UAV computing platforms: a hardware reliability perspective. Sensors **22**(16) (2022)
3. M.A. Akhloufi, A. Couturier, N.A. Castro, unmanned aerial vehicles for wildland fires: sensing, perception, cooperation and assistance. Drones **5**(1) (2021). https://doi.org/10.3390/drones5010015
4. M.A. Alanezi, A. Mohammad, Y.A. Sha'aban, H.R.E.H. Bouchekara, M.S. Shahriar, Autoencoder learning-based UAV communications for livestock management. Drones **6**(10) (2022). https://doi.org/10.3390/drones6100276
5. H. Alsharif, M.A. Khan, E.T. Michailidis, D. Vouyioukas, A review on software-based and hardware-based authentication mechanisms for the internet of drones. https://doi.org/10.3390/drones
6. ardupilot.org. n.d.-a, Choosing an autopilot — copter documentation. Accessed 9 January 2023, https://ardupilot.org/copter/docs/common-choosing-a-flight-controller.html
7. ardupilot.org. n.d.-b, Companion computers — dev documentation. Accessed 9 January 2023, https://ardupilot.org/dev/docs/companion-computers.html#companion-computers
8. ardupilot.org. n.d.-c, Copter attitude control — dev documentation. Accessed 11 January 2023, https://ardupilot.org/dev/docs/apmcopter-programming-attitude-control-2.html.
9. ardupilot.org. n.d.-d. Flight controller (Autopilot) hardware | PX4 user guide. Accessed 9 January 2023, https://docs.px4.io/main/en/flight_controller/
10. ardupilot.org. n.d.-e, Mission planner home — mission planner documentation. Accessed 11 January 2023, https://ardupilot.org/planner/
11. K. Baskaran n.d, Get started on building a VTOL drone for autonomous flight.

12. M. Biczyski, R. Sehab, J.F. Whidborne, G. Krebs, P. Luk, Multirotor sizing methodology with flight time estimation.J. Adv. Transp. (2020). https://doi.org/10.1155/2020/9689604
13. J. Cabahug, H. Eslamiat, Failure detection in quadcopter UAVs using K-means clustering. Sensors **22**(16) (2022). https://doi.org/10.3390/s22166037
14. M.P. Christiansen, M.S. Laursen, R.N. Jørgensen, S. Skovsen, R. Gislum, Designing and testing a UAV mapping system for agricultural field surveying. Sensors (Switzerland) **17**(12) (2017). https://doi.org/10.3390/s17122703
15. Collins Dictionary. n.d., Ornithopter. Accessed 11 January 2023, https://www.collinsdictionary.com/zh/dictionary/english/ornithopter
16. T. Cozzens, SPH engineering's UgCS software now supports velos UAV helicopter - GPS World : GPS World (2021). Accessed 11 January 2023, https://www.gpsworld.com/sph-engineerings-ugcs-software-now-supports-velos-uav-helicopter/
17. S. Elouarouar, H. Medromi, Multi-rotors unmanned aerial vehicles power supply and energy management, in *E3S Web of Conferences,* vol. 336, p. 00068. https://doi.org/10.1051/e3sconf/202233600068
18. Flir, Teledyne. n.d., Black Hornet PRS
19. gazebosim.org. n.d., Gazebo. Accessed 11 January 2023 (https://gazebosim.org/home).
20. genstattu.com. n.d., Lipo battery guide. Accessed 10 January 2023, https://genstattu.com/bw/
21. Gleagle. n.d., Gartt Ml 5008 330 kv brushless motor for multicopter - AliExpress. Accessed 27 February 2023, https://www.aliexpress.com/i/32752596266.html
22. J. Gustave, J. Chahal, A. Belbachir, functional architecture using ROS for autonomous UAVs, in*ICINCO 2020 - Proceedings of the 17th International Conference on Informatics in Control, Automation and Robotics* (SciTePress, 2020), pp. 506–512
23. R.J.A.L. Hartley, I.L. Henderson, C.L. Jackson, BVLOS unmanned aircraft operations in forest environments. Drones **6**(7) (2022)
24. L. Hassan-Esfahani, A. Torres-Rua, A. Jensen, M. McKee, Assessment of surface soil moisture using high-resolution multi-spectral imagery and artificial neural networks. Remote Sens. **7**(3), 2627–2646 (2015). https://doi.org/10.3390/rs70302627
25. W. Hönig, N. Ayanian, Flying multiple UAVs Using ROS, in *Studies in Computational Intelligence*, vol. 707 (Springer, 2017), pp. 83–118
26. C.E. Hrabia, A. Hessler, Y. Xu, J. Seibert, J. Brehmer, S. Albayrak, Efffeu project: towards mission-guided application of drones in safety and security environments. Sensors (Switzerland) 19(4) (2019). https://doi.org/10.3390/s19040973
27. A. Idalene, K. Boukhdir, H. Medromi, UAV control architecture: review. Int. J. Adv. Comput. Sci. Appl. **10**(11), 652–657 (2019). https://doi.org/10.14569/IJACSA.2019.0101186
28. IFAC, World drone competition (2023). Accessed 28 February 2023, https://www.ifac2023.org/program/world-drone-competition/).
29. I. Singh, Delivery drone severely damaged in test flight over Scotland (2022). https://Dronedj.Com/2022/12/20/Delivery-Drone-Crash-Incident-Scotland/
30. J.C. Milliman, TEAM public affairs, Pax River. 2000. UAVs test crash in simulator during latency testing in latency testing, https://Www.Navair.Navy.Mil/Node/3231
31. E. Johnson,. 'GitHub - intelligent-quads/Iq_gnc: ROS package for GNC of ardupilot drones (2022). Accessed 11 January 2023, https://github.com/Intelligent-Quads/iq_gnc
32. A. Kamilaris, F.X. Prenafeta-Boldú,Disaster monitoring using unmanned aerial vehicles and deep learning (2018)
33. T. Kekec, B.C. Ustundag, A. Guney, A. Yildirim, M. Unel, A modular software architecture for UAVs (2013)

34. D. Kotarski, M. Krznar, P. Piljek, N. Simunic, Experimental identification and characterization of multirotor UAV propulsion, in *Journal of Physics: Conference Series*, vol. 870 (Institute of Physics Publishing, 2017)
35. A. Koubaa, B. Qureshi, DroneTrack: cloud-based real-time object tracking using unmanned aerial vehicles over the internet. IEEE Access **6**, 13810–13824 (2018). https://doi.org/10.1109/ACCESS.2018.2811762
36. A. Koubâa, B. Qureshi, M.F. Sriti, A. Allouch, Y. Javed, M. Alajlan, O. Cheikhrouhou, M. Khalgui, E. Tovar, Dronemap planner: a service-oriented cloud-based management system for the internet-of-drones. Ad Hoc Netw. **86**, 46–62 (2019). https://doi.org/10.1016/J.ADHOC.2018.09.013
37. X. Liu, Y. Yin, Y. Su, R. Ming, A multi-UCAV cooperative decision-making method based on an MAPPO algorithm for beyond-visual-range air combat. Aerospace **9**(10) (2022). https://doi.org/10.3390/aerospace9100563
38. J. Mcbride, Flight control system for small high-performance UAVs flight control system for small high-performance UAVs (2010)
39. M. McNabb, DRONEII: Tech talk - Unraveling 5 levels of drone autonomy – DRONELIFE (2019). Accessed 11 January 2023, https://dronelife.com/2019/03/11/droneii-tech-talk-unraveling-5-levels-of-drone-autonomy/
40. J. Meyer, A. Sendobry, S. Kohlbrecher, U. Klingauf, O. von Stryk (2012) Comprehensive simulation of quadrotor UAVs using ROS and Gazebo
41. S.A.H. Mohsan, M.A. Khan, F. Noor, I. Ullah, M.H. Alsharif, Towards the unmanned aerial vehicles (UAVs): a comprehensive review. Drones 6(6) (2022)
42. M. Müller, ECalc - XcopterCalc - the most reliable multicopter calculator on the web (2022). Accessed 9 January 2023. https://www.ecalc.ch/xcoptercalc.php
43. M. Müller, ECalc - XcopterDesigner - design your multicopter, multirotor or UAV. Accessed 9 January 2023. https://www.ecalc.ch/xcopterdesigner.htm
44. C. Papachristos, M. Kamel, M. Popović, S. Khattak, A. Bircher, H. Oleynikova, T. Dang, F. Mascarich, K. Alexis, R. Siegwart, Autonomous exploration and inspection path planning for aerial robots using the robot operating system, in *Studies in Computational Intelligence*, vol. 778 (Springer, 2019), pp. 67–111
45. N. Patrinopoulou, I. Daramouskas, D. Meimetis, V. Lappas, V. Kostopoulos, A multi-agent system using decentralized decision-making techniques for area surveillance and intruder monitoring. Drones **6**(11), 357 (2022). https://doi.org/10.3390/drones6110357
46. K.-L. Pham, J. Leuchter, R. Bystricky, M. Andrle, N.N. Pham, Van Thuan Pham, The study of electrical energy power supply system for UAVs based on the energy storage technology. Aerospace **9**(9) (2022). https://doi.org/10.3390/aerospace9090500
47. E. Politi, I. Panagiotopoulos, I. Varlamis, G. Dimitrakopoulos, A survey of UAS technologies to enable beyond visual line of sight (BVLOS) operations, in*International Conference on Vehicle Technology and Intelligent Transport Systems, VEHITS - Proceedings*. Vols 2021-April (Science and Technology Publications, LDA), pp. 505–12
48. px4.io. n.d.-a, Airframes reference I PX4 user guide'. Accessed 9 January 2023, https://docs.px4.io/main/en/airframes/airframe_reference.html#copter
49. px4.io. n.d.-b, PX4 architectural overview I PX4 user guide. Accessed 11 January 2023. https://docs.px4.io/main/en/concept/architecture.html
50. P. Radoglou-Grammatikis, P. Sarigiannidis, T. Lagkas, I. Moscholios, A Compilation of UAV applications for precision agriculture.Comput. Netw. **172**. https://doi.org/10.1016/j.comnet.2020.107148
51. A. Renault, A model for assessing UAV system architectures, in *Procedia Computer Scienc,* vol. 61 (Elsevier 2015), pp 160–67

52. A. Rodriguez-Ramos, A.A.-F.H. Bavle, J. Rodriguez-Vazquez, L.L.M. Fernandez-Cortizas, R.A. Suarez Fernandez, A. Rodelgo, C. Santos, M. Molina, L. Merino, F. Caballero, P. Campoy (2021) Autonomous aerial robot for high-speed search and intercept applications
53. ros.org. n.d., ROS: Home'. Accessed 11 January 2023. https://www.ros.org/
54. Seafarer Chapter, Student unmanned aerial systems competition (2022). Accessed 28 February 2023. https://suas-competition.org/
55. A. Sharma, P. Vanjani, N. Paliwal, C.M. Wijerathna, D.N. Basnayaka, K. Jayakody, H.-C. Wang, P. Muthuchidambaranathan, Communication and networking technologies for UAVs: a survey. J. Netw. Comput. Appl. **168**, 102739 (2020). https://doi.org/10.36227/techrxiv.12912449.v1
56. M. Sivakumar, T.Y.J. Naga Malleswari, A literature survey of unmanned aerial vehicle usage for civil applications. J Aerosp Technol Manage 13
57. B. Szyk, M. Koperska, A. Benn, Drone flight time calculator (2022). Accessed 9 January 2023. https://www.omnicalculator.com/other/drone-flight-time
58. Teknofest, International UAV competition (2022). Accessed 28 February 2023, https://teknofest.org/en/competitions/competition/45
59. Tubitak, Teknofest, Teknofest 2022 rules booklet (2022)
60. USAID, and PEPFAR, Unmanned aerial vehicles landscape analysis: applications in the development context (2017)
61. A. Wheeler, Drones reconsidered in age of COVID-19 | Engineering.Com (2020). Accessed 11 January 2023 (https://www.engineering.com/story/drones-reconsidered-in-age-of-covid-19).
62. K.W. Williams,A summary of unmanned aircraft accident/incident data: human factors implications (2004)
63. Y. Zeng, R. Zhang, T.J. Lim, Wireless communications with unmanned aerial vehicles: opportunities and challenges (2016). https://doi.org/10.1109/MCOM.2016.7470933
64. A. Zolotkov, Drone LiPo battery calculator • electrical, RF and electronics calculators • online unit converters (2020). Accessed 9 January 2023, https://www.translatorscafe.com/unit-converter/en-US/calculator/multicopter lipo battery/

Robust and Intelligent Kalman Filter Residue-Based Trajectory Tracking and Fault-Tolerant Control of Quadrotor Using Bayes Classifier Fusion

Rajamani Doraiswami, Lahouari Cheded, and Marius Brinkmann

Abstract

A robust Kalman filter (KF)-based feedback controller is proposed using an internal model (IM)-based approach for the accurate tracking of the desired trajectory despite unknown disturbances and measurement noise using Bayesian classifier fusion of model-based and machine-based learning. The states of the system comprise the position coordinates, the yaw, pitch, and roll, and their derivatives. The system's input is a nonlinear function of the roll, pitch, and yaw, and its output is its position. The quadrotor is a nonlinear system (Wiener-type) formed of a linear dynamic system cascaded with a static nonlinear function. The input and output are corrupted by an unknown disturbance and measurement noise. Using a machine learning approach, the true input is estimated from the accessible corrupted input. The nonlinear system is better approximated using a linear parameter-varying (LPV) model described by a piecewise-linear Box-Jenkins (BJ) model, than with a conventional (Taylor series-based) linearized

R. Doraiswami
Department of Electrical and Computer Engineering, University of New Brunswick, Fredericton, New Brunswick, Canada
e-mail: dorai@unb.ca

L. Cheded (✉)
Independent Researcher and Consultant, Cheshire, UK
e-mail: cheded@kfupm.edu.sa; cheded@ieee.org

Formerly at Systems Engineering Department, KFUPM, Dhahran 31261, Saudi Arabia

M. Brinkmann
Technical University of Munich, Munich, Germany
e-mail: Marius.brinkmann@tum.de

M. Abdelkader and A. Koubaa (eds.), *Unmanned Aerial Vehicles Applications: Challenges and Trends*, Synthesis Lectures on Intelligent Technologies,
https://doi.org/10.1007/978-3-031-32037-8_3

model. The system and the KF are identified using emulator-generated data by minimizing the KF residual (the error between the state and its estimate) and using a classifier fusion scheme so that the identified models are accurate, consistent, and reliable. The proposed scheme was successfully evaluated and exhibits scalability as it can manage multiple quadrotors.

Keywords

Autonomous vehicles • Quadrotor • Linear parameter-varying systems • Machine learning/Predictive analytics • Bayesian classifier • Kalman filter • Identification • Robust control • Internal model-based control • Fault accommodation • Nonlinear (Wiener) systems • Fault-tolerant systems

1 Introduction

Interest in unmanned autonomous vehicles (UAVs) has been continuously growing for several years, from both the research community as well as from industry, thanks to their multiple applications that were facilitated by their cost-effective development and rapid deployment across many areas. Their applications cover a wide spectrum of applications, including surveillance and intelligence gathering, rescue missions, and reconnaissance services for various important operations including transport, monitoring, and communications. Examples of such operations may include (a) delivery of postal and vital medical items, to far-away and difficult-to-access regions, such as remote islands and high mountains visited by hikers, tourists, etc., (b) pipeline monitoring for leak detection, as well as (c) serving as communication platforms for information transmission and relay. In this last application, they present an advantage over their satellite-based brethren in that they offer attractive advantages such as simple construction, high-speed and low-latency communication capability that turn them into reliable wireless channels for information transmission. This communication capability can also impart to them a very useful function of acting as carriers of reflected intelligent surfaces that are currently used for wireless localization purposes. It should however be pointed out that their limited energy supply does reduce their flight distances and can reduce their operating range, especially when flying alone. With their ever-improving development, UAVs can also be used in a cluster form to improve their communication capabilities and to allow them to offer multitasking capabilities in their cluster communication (or multiple-UAV) network. This will equip them with some resilience against network attacks, which is an emergent capability of interaction and cooperation, and will also enable them to carry out distributed operations, thereby improving their communication reliability and their collective intelligence, and will also allow them to rapidly complete the targeted tasks thanks to the gained increase in the data transmission volume. UAVs also provide the needed capability of plugging the communication hole created by the insufficient coverage of traditional cellular wireless networks, thus widening their application area. The current efforts in miniaturization and energy harvesting may further mitigate the current challenges faced by UAVs because of their limited energy supply which may impact their operational endurance. However, the multiple advantages of UAVs outweigh their current challenges, and the current intense research is fuelling their development, and widening their applications into more diverse areas. As part of this effort to extend the area of application of UAVs, we propose in this work, a novel scheme for accurately tracking a desired trajectory using a single quadrotor. Our work proposes the judicious combination of a robust feedforward-feedback controller and a novel theory based on the residue of a Kalman Filter (KF) which, together, show how the quadrotor, driven by the KF residual, becomes an autonomous unmanned tracker of the desired trajectory.

The dynamic model of the quadrotor is described by a second-order nonlinear Euler–Lagrange equation, which generalizes the Newton's laws of motion to that of a rigid body, relating the (a) position x, y and z of the center of gravity of the quadrotor, (b) the Euler angles, namely the yaw ψ, pitch θ and roll ϕ, , and c) their derivatives, to the second derivatives of the position coordinates x, y, and z. The inputs are static nonlinear functions of ϕ, θ, and ψ while the position coordinates x, y, and z are the outputs.

Aerial vehicles such as quadrotor are invariably subject to model uncertainties and are affected by unknown exogenous stochastic and deterministic disturbances and random measurement noise. These disturbances include wind gusts updrafts, aerodynamic drags, gravitational load, Coriolis, and centripetal forces [1–5]. They fly at different altitudes and are subject to varying wind gusts that may cause accelerations or decelerations of (a) the output, i.e., the position coordinates x, y, and z, and (b) the input variables, i.e. the roll, pitch, and yaw. As a result, the quadrotor is pushed away from the desired trajectory. The autonomous flight of a quadrotor needs precise position and attitude information for control and stabilization. It carries sensors including (a) three accelerometers measuring linear accelerations of x, y, and z, (b) gyroscopes for measuring angular accelerations of ψ, θ and ϕ, (c) three-axis magnetometer for attitude determination, d) for measuring the main rotors angular velocity (rpm), and e) an ultrasonic, or barometric altitude sensor for take-off and landing. The sensor measurements are generally corrupted by noise [6]. The measurement of the roll ϕ, pitch θ, and yaw ψ and the positions x, y and z are corrupted by unknown stochastic disturbance and measurement noise. Both the input and output are corrupted by unknown disturbances and measurement noise. A model whose input and output are both corrupted with noise, is termed as an errors-in variables one [7].

The measurement output is, in general, an additive combination of the signal, disturbance and measurement noise. The output error is defined as the difference between the output and the signal estimate and is assumed to be bounded. Further, it is assumed that the disturbance is a stationary Gaussian stochastic process, and the measurement noise is a zero-mean white Gaussian noise process. The signal, the stochastic disturbance, and the deterministic disturbance, are each modeled as an output of an asymptotically stable linear time-invariant system driven by a corrupted input, a Gaussian zero-mean white noise process, and a unit impulse respectively. It is assumed that the signal, the stochastic disturbances, and the measurement noise are all uncorrelated with each other. The quadrotor model covers a wide class of physical systems where inputs and outputs are both corrupted by stochastic disturbances and measurement noise, and thus would be represented by models termed error-in-variable dynamic models. Estimation of unknown inputs and disturbances has been a research topic of great interest. It finds applications in robust control and fault diagnosis. Here, it is used in a novel way in that it avoids the conventional use of the statistics of the unknown disturbances, and measurement noise.

This chapter is organized as follows. After the introduction in Sect. 1, Sect. 2 lays down the necessary and wide background material which describes the various tools used for the design of the proposed scheme, including some introductory concepts of learning, the linear parameter-varying (LPV) approach to approximate the nonlinear quadrotor model, the powerful internal model principle and its use in the design of the feedforward-feedback controller that is at the heart of the proposed trajectory-tracking scheme, the general Box-Jenkins model of the system, the use of the accuracy-enriching concept of emulators in identification, leading on to a highlight of some interesting practical applications, including UAVs, of this set of tools, and ending with a list of contributions offered by this study.

Section 3 describes the kinematics of translation and rotation; position and orientation; velocities, the moment of inertia. Section 4 describes the dynamic model derivation using the Lagrangian method, the dynamics of translational motion and dynamics of rotational motion, and the linear parameter-varying approach to approximate the nonlinear quadrotor model. Section 5 gives exogenous and endogenous models for the disturbances and measurement noise, and shows how to extract the true input from the corrupted one. Section 6 develops the piecewise-linear dynamics model of rotation and translation in continuous time and their discrete-time equivalents. The discrete-time signal model is derived. Section 7 develops the internal model approach to designing the feedback and feedforward controller for accurate tracking of the desired trajectory and the closed-loop signal model is obtained. The system model formed of the closed-loop signal model and disturbance, termed Box-Jenkins model, is obtained. Section 8 describes the KF for the system and its key properties. Section 9 describes the emulator-based identification of the signal, estimation of signal and the output error and identification of the KF for the signal model, and covers controller-related aspects such as status monitoring, fault isolation, condition-based maintenance, controller adaptation, fault tolerance. Sections 10, 11 and 12 show respectively the results of the evaluation of the proposed trajectory-tracking scheme on a simulated quadrotor, the KF-based soft sensor used to estimate the needed derivatives of the translational and rotational variables used in this study, and the fault tolerance capability of the proposed scheme using the key properties of the KF. Section 13 ends the chapter with some conclusions.

2 Background

In the following sub-sections, the necessary background material is provided on the tools and underlying theory used to support the design of our proposed trajectory tracking scheme. Also, a brief description of the main application (pipeline leak detection) that motivated this work is given, and the main theoretical contributions made in this study are highlighted.

2.1 Machine Learning, Deep Learning and Predictive Analytics

These three approaches all rely on the powerful tools of learning system characteristics and/or dynamics, and statistical prediction. They can either supplement and improve the performance of theoretical models when these are available, or simply replace them when these models are either too complicated to use or not available at all. A brief description of these approaches follows.

Machine Learning (ML) is an emerging and futuristic technology that stands as the starting point to create automated innovations with intelligence. It relies on the learning of patterns and trends that has occurred in a period. Machine Learning is derived from AI (Artificial Intelligence) that provides a system with the ability to learn from its experiences and improve without human intervention [8, 21]. It makes use of data for training so that it can develop computer learning by itself. Machine Learning Algorithms are used for training models, using the following approaches:

- Supervised Learning if the data set is labelled,
- Unsupervised Learning, if the dataset is not labeled, and techniques like Dimensionality reduction and Clustering are used
- Reinforcement Learning (algorithm in which the model learn from its every action through the use of a critic and reward/penalty signals) for the development of machine learning solution for applications such as fault defection and isolation

Deep learning (DL) is a subset of machine learning. It provides a powerful set of artificial neural networks tools, with multiple hidden layers, notably the convolution neural networks (CNNs) type, used to perform complex analysis of large amounts of data by processing them through a set of multiple hidden layers to extract key features of the data processed from the discrepancy between the source and the target data under varying operating conditions [4], thereby ensuring an accurate tracking of the desired trajectory.

Predictive analytics (PA) is the use of data, statistical algorithms and ML techniques to identify the likelihood of future outcomes based on historical data. The goal is to go beyond knowing what has happened to provide a best assessment of what will happen in the future.

In this study, it is the latter approach (i.e. PA) which has been used here as is briefly explained next. As the accessible input is corrupted by unknown disturbance and measurement noise, PA is used to extract the *true input* from the corrupted input. It does so by exploiting the smoothness of the true input and the jaggedness of the stochastic disturbance and measurement noise. The PA-estimated input thus replaces the inaccessible true input. Moreover, the resulting model is thus free from the input disturbance and measurement noise and the system model relates the true input and the output corrupted by unknown disturbance and measurement noise. More details on this PA technique can be found in [21].

2.2 Linear Parameter-Varying Model

Most physical systems contain nonlinearities such as saturation, backlash, rate limiters and dead zones: the aircraft, drone and robotic manipulators are inherently nonlinear. The scheduling variables are measured in real time and may be selected so that they can track the degree of variations in the dynamics of the system. The scheduling variables include exogenous signals such as the set point, internal variables such as the velocity and power, environment variables such as the altitude, temperature, pressure, and air speed. In the last decades, both analysis and synthesis techniques for linear parameter varying (LPV) systems have received a lot of attention from the control community and the LPV paradigm has become a standard formalism in systems and control, for analysis, synthesis of controllers and system identification. As the analysis, design, estimation, identification, and control of nonlinear system is not mathematically intractable, a linear parameter varying (LPV) model is used to approximate the nonlinear system [9]. The resulting piecewise-linear dynamic models help develop computationally simple, efficient, and robust schemes for identification, design of KF, fault detection and isolation.

The model of the quadrotor is a second-order unstable nonlinear system of the type of Wiener-Hammerstein model, i.e., a cascade combination of a linear dynamic system and static nonlinear function [10]. The static nonlinearities are functions of roll, pitch, and yaw. Since they are estimated using predictive analytics approach, the static nonlinearities are therefore pre-computed at each of the operating points indicated by the scheduling variables. The pre-computed static nonlinearities are treated as known constants in the neighborhoods of the operating points, similar in spirit to the computed torque approach [11]. The resulting piecewise linear dynamic model is driven by the estimated true input, and not by the corrupted one, and has only its output corrupted by a stochastic disturbance and measurement noise.

2.3 Internal Model Principle

In many applications, and for the purpose of tracking a constant reference waveform, controllers such as the proportional integral (PI), proportional integral and derivative (PID), proportional derivative (PD), and Linear Quadratic Regulator (LQR) type controllers have been widely used. An efficient way of handling such a controller designed to track a given reference signal despite the presence of disturbances and perturbations in the plant is to use the powerful internal model (IM) principle [12, 13]. The principle states that the output will track the given reference *if and only if* the structure of the controller used contains a) an internal model of the reference to be tracked and the disturbance signals and is driven by the tracking error and b) the closed-loop system formed of the plant and the internal model is asymptotically stable. For nonlinear quadrotor models, adaptive controllers are employed [14].

The IM principle can only be used if the reference and disturbance waveforms are both deterministic with rational spectra and the model that generated the deterministic reference, and the disturbance are known. The mean value of the disturbance can be rejected by IM if its model is known. If the mean is continuous, then using Weierstrass approximation theorem, it is approximated by a polynomial function such as a constant, ramp, or cubic function or other. It cannot be applied if disturbances are unknown deterministic or stochastic and the measurement noise random [21].

2.4 Feedback-Feed Forward Controller

As the internal model-based state-feedback robust optimal controller for the signal model is obtained from the parameterization of all stabilizing controllers. In the class of stabilizing controllers, the state feedback controller is the most robust if the states are available. Thanks to KF, the signal and its model are derived, thus allowing the implementation of the state feedback controller. Since the estimated signal is not corrupted by the Gaussian disturbance and measurement noise, we may then use an optimal linear quadratic regulator approach in the class of all-stabilizing state-feedback controllers. Here again, the important role of the KF in allowing the use of the state-feedback control design approach is highlighted. The feedforward controller is designed to track the desired reference using the feedforward version of the reference signal so that the signal tracks the reference input asymptotically. It is pro-active thanks to its predictive nature; it overcomes sluggish dynamics and delays without compromising the stability of the system. A combination of feedback and feedforward controllers can improve the performance significantly compared to using merely a feedback controller alone [12].

2.5 Box-Jenkins Model

The system is represented here by an augmented model of the signal and the disturbances, and, as such, is termed a Box-Jenkins (BJ) model. It covers a wider class of physical systems than do conventional models such as the Auto-Regressive and Moving Average (ARMA), Moving Average (MA), and Auto-Regressive (AR) ones, which are all special cases of the Box-Jenkins model [16, 17].

The system relating the estimated input and the output is observable but not controllable, that is, it is not a minimal realization. The system transfer function has pole-zero cancellation as the denominator of the disturbance model is common to both the numerator and the denominator of the system model.

2.6 Emulator-Based Two-Stage Identification Method

The quadrotor as well as the stochastic and deterministic disturbance models are both subject to perturbations. To account for these perturbations, *emulators* are connected in cascade to the output, the input or both during the identification phase. An emulator is a transfer function block which emulates the variations in the system, signal, and disturbance models. An emulator takes the form of a static gain or an all-pass filter to induce gain or phase variations. The emulator parameters are perturbed to mimic various normal and abnormal, or faulty, operating scenarios resulting from variations in these subsystems. The emulator-generated data is employed in the identification [16]. In view of the key properties of the KF, it is the residual model of the KF, and not the system model, which is identified in our work, thus lending our proposed scheme a novelty that sets it apart from all other conventional approaches. The key features it enjoys are highlighted below:

(a) the identified models of the system, the signal, the associated KFs, and the disturbances are all accurate, and robust,
(b) to enable the development of a fault diagnosis scheme for the system and signal models,
(c) to ensure that the controller is robust,
(d) The emulator-generated data is crucial to our classifier fusion approach. The model-based and the model-free approaches are presented with the same emulator-generated

data so that the fusion of the various approaches is superior to that using only a single approach.

The theoretical justification underlying our approach is expressed in the form of *if and only if* conditions for the accuracy of the system model, the KF, the signal model, and the disturbance model. The identification objective of ensuring that the residual is a zero-mean white noise process will ensure not only that the system model is accurately identified but also that the Kalman gain is optimal, thereby avoiding the need to specify the covariance of the disturbance and the measurement noise and to use the Riccati equation to solve for the Kalman gain.

A set made of the system, the signal model and the associated KFs for the system and the signal models is identified from the emulator-perturbed data using the direct least-squares-based approach. An optimal robust identified model and the KF are selected using the H_∞-measure of deviations between the perturbed identified models. It is shown that the proposed identification scheme based on emulator parameter-perturbed experiments to generate likely model perturbations is superior to the conventional approach based on using an approach with bounds that are either conservative, optimistic or simply inexistent, instead of the true bound of the perturbed plant models. Further, for the implementation of the fault tolerance system, influence vectors which are the partial derivative of the optimal Kalman filter residual with respect to each of the emulator parameters are estimated.

The residuals of the KF for the system and the signal models are employed to determine when the system is faulty. A fault may arise due to perturbation of the disturbance model, or the subsystem associated with signal model such as the actuator, the sensor or other. The disturbances can negatively affect the system performance. It is not in general easy to determine whether the fault is the result of variations in the disturbance or the subsystem model. Thanks to the availability of two KFs, a simple solution is to analyze the residuals of both KFs as the residual of the KF for the system captures the variations in both the system and disturbance models, while that of the KF for the signal, and captures only the variations in the signal model. This is crucial for reducing false alarm, and all its concomitant risks and costs, resulting from variations in the disturbance and not in the signal model.

2.7 Practical Applications

The above tools enjoy various practical applications as highlighted below, including an application to UAVs which constitutes the main topic of this chapter, as well as in other related ones dealing with fault monitoring and soft sensor design, which, though used in this study, can also be used independently in other applications.

2.7.1 Unmanned Aerial Vehicles (UAVs)

Unmanned (or autonomous) aerial vehicles such as drones can track accurately the desired trajectory in three-dimensional space and are widely used in exploring and mapping, transporting, manipulating, and assembling objects. The motivation for this work stems primarily from the need to detect accurately and reliably leakage in pipelines running in a dangerous and/or environmentally unfriendly area and collecting data from sensors placed at some suitable locations along the pipeline, to detect possible leaks in the pipeline. The proposed feedforward-feedback controller scheme is evaluated on a simulated model of a drone moving in a horizontal plane to track accurately the desired trajectory despite wind gusts and model perturbations. The performance of the proposed controller is compared with that of the conventional combined controller.

2.7.2 Fault Isolation, Fault Tolerance, Condition-Based Monitoring, and Soft Sensor

If the performance degradation is high or the controller adaptation fails to meet the acceptable bound, the presence of a fault is asserted [17]. The faulty subsystem is then isolated using the influence vector and the fault size is estimated from the emulator parameter variations. The faulty subsystem is either repaired or replaced. The residuals of the KF are employed to monitor the status of the overall system and to detect and isolate faults in the signal and disturbance-based models and the sensors. The proposed scheme provides a sound framework for redeveloping condition-based maintenance systems as well. Some key aspects of a typical such system are discussed below.

If performance degradation is high or the controller adaptation fails to meet the acceptable bound, the presence of a fault is asserted [17]. The faulty subsystem is then isolated using the influence vector and the fault size is estimated from the emulator parameter variations. The faulty subsystem is either repaired or replaced. The residuals of the KF are employed to monitor the status of the overall system and to detect and isolate faults in the signal and disturbance-based models and the sensors. The proposed scheme provides a sound framework for developing fault-tolerant systems and condition-based maintenance systems as well, as discussed next.

Fault Tolerance: Fault tolerance is a crucial requirement for aerial vehicles [16, 17]. The KF, which forms the fault-tolerant system, monitors the performance and diagnoses faults thanks to the emulators. Influence vectors, which maps the KF residual to the subsystem parameter variations to the plant input–output data, are estimated from the emulator-perturbed experiments. Incipient faults are detected, and the fault is isolated using the influence vectors. If there is a performance degradation, then the controller is redesigned (adapted), or the plant is re-identified. The fault-tolerant autonomous controller is developed using properties of the residuals when there is a model-mismatch.

Condition-based monitoring: Stable closed-loop system performance degradation may result from changes in its operating condition, malfunctioning of its components, variations in its noise and disturbance statistics, etc. In a high-performance control system, it is imperative that the closed-loop stability, performance, and the system's health condition be all monitored to initiate quick remedial actions, when needed. For instance, if the performance degradation stems from a plant model variation, then the plant is re-identified and the robust controller redesigned or if a fault is detected, it is then isolated and either the fault is accommodated, or the system is shut down for repair.

KF-based soft sensor: The sensor measurements are generally corrupted by noise [6]. The measurement of the roll ϕ, pitch θ, and yaw ψ and the positions x, y and z are corrupted by unknown stochastic disturbance and measurement noise.

The KF-based, derivative-free soft sensor replaces the conventional hardware sensor [18]. The soft sensor is used to estimate the first, second and third derivatives of the translational variable and of the rotational variable. As such, the soft sensor effectively and in a differentiation-free manner, replaces the hardware one in estimating the required derivatives [19, 20].

2.8 Main Contributions

The proposed scheme is an extension of our previous work on quadrotor [19] with the novel feature described below:

A. ***Fusion of model-based and Predictive Analytics (Machine Learning)-based approach***:
- The input is corrupted by an unknown stochastic disturbance and measurement noise. In order to separate the true input from the corrupted one, the smoothness of the true input and the randomness of the corrupted input are exploited using two-stage frequency domain identification. The corrupted output is replaced by the estimated true input
- Extracting features automatically from the discrepancy between the source and the target data under variable operating conditions thereby ensures that the desired trajectory is accurately tracked [4, 21].

B. ***Linear parameter-varying model***

Most of the papers on quadrotors use nonlinear models whereas the proposed scheme uses a piecewise linear-dynamic model approach, which provides a better approximation to the nonlinear system than the conventional Taylor series-based one. For a large class of physical systems, a linear parameter-varying model to cover the entire operating region is employed as it is mathematically tractable, and there is a wealth of readily available and powerful analysis and design tools to use for such linearized models. For example, state-feedback controller is based on the internal model principle which for a linear system gives necessary and sufficient whereas a nonlinear controller such as adaptive one provides only sufficient condition. Conventional approaches based on observer, nonlinear filters, other nonlinear device cannot handle stochastic disturbance and measurement noise or gives only sufficient condition. The KF on the other hand gives necessary and sufficient condition for estimation of signal from corrupted output, detection and fault isolation, development of fault tolerance, and soft sensor to estimate the signal and its derivatives.

C. ***Two-stage identification***

In system identification, the criterion for determining whether the identified model has captured completely the dynamic behavior of the system is that the residual (error between the output and its estimate obtained using the identified model) is a zero-mean white noise process. If the equation error is a zero-mean white noise process, the estimate will be unbiased and efficient; otherwise, it will be biased. In view of the above-stated key properties, the residual model structure of the KF is employed, instead of the linear regression model of the system whose equation error is a colored noise process. The identification objective of requiring that the residual be a zero-mean white noise process, will ensure not only that the system model is accurately identified but also that the Kalman gain is optimal, thereby avoiding the need to specify the covariance of the disturbance and the measurement noise and the use of the Riccati equation to solve for the Kalman gain. The system and the associated KF are identified using the emulator-perturbed data covering all likely operating scenarios so that the identified models are accurate, consistent, and robust to model perturbation so that the identified robust model can be used to design a robust controller. During identification, a vector, termed *influence vector,* plays a crucial role in the fault isolation process. The influence vector relates the variation of residual to the variation of the emulator associated with a subsystem.

In stage I, the system model and the associated KF are identified: the signal buried in the output corrupted by disturbance and measurement noise, and the signal models are estimated; the output error and the disturbance model are estimated; KF for the identified signal model is identified. In stage 2, using the key properties of the Kalman filter, the robust signal model and its associated Kalman filter are obtained as explained in Sect. 2.6 on the Emulator-based two-stage identification method.

D. ***State feedback and feed forward controller design***

A combined feedback and feed forward controller based on the internal model principle and the key properties of the KF is proposed so that the drone can track the desired trajectory accurately despite the presence of the deterministic and stochastic disturbance and measurement noise without causing any *lane departure*, unlike with the use of the conventional controller where the vehicle will be subject to random deviation around the desired trajectory. The feed forward controller is designed to track the desired reference using the feed forward version of the reference signal so that signal tracks the reference input asymptotically. Thanks to its predictive nature, it is pro-active. It also overcomes sluggish dynamics and delays without compromising the stability of the system.

E. ***Status monitoring; and fault tolerance***

The two KFs, one associated with the system and the other with the signal model, are employed to develop a fault-tolerant, condition-based maintenance system, and a disturbance-tolerant (or disturbance-accommodating) system.

F. ***Soft sensor***

The KF-based, derivative-free soft sensor replaces the conventional hardware sensor. The soft sensor is used to estimate the first, second and third derivatives of the translational variable and of the rotational variable. As such, the soft sensor effectively and in a differentiation-free manner, replaces the hardware one in estimating the required derivatives.

G. ***Theoretical justification***

The key properties of the Kalman filter forms the backbone of the proposed scheme. The identification of the system model, the Kalman filter, the signal model and the disturbance model are accurate since theoretical justifications, in the form of both necessary and sufficient conditions for the proposed scheme, are given. The conventional scheme generally ensures only sufficiency condition as stated in Sect. 2.6 on *Emulator-based two-stage identification method*.

3 Kinematics: Translation and Rotation

The position is described in the Cartesian coordinate system with the coordinates axes denoted x, y, and z, which denote the position vector of the center of mass of the quadrotor relative to a fixed inertial frame F_G:

$$\xi = (x, y, z) \in R^3 \tag{1}$$

where R is the field of real numbers and $\in$ denotes 'Belong to'. The position for the body-fixed frame F_B is:

$$\xi_B = (x_B, y_B, z_B) \in R^3 \tag{2}$$

The orientation of $\boldsymbol{\xi}_B$ with respect to $\boldsymbol{\xi}$ is defined by the Euler angles representing the three elementary rotations, namely the roll ϕ, the pitch θ and the yaw ψ. The orientation of the rotorcraft is expressed by:

$$\boldsymbol{\eta} = (\phi, \theta, \psi) \in R^3 \tag{3}$$

The orientation $\boldsymbol{\eta}$ in the body frame F_B is:

$$\boldsymbol{\eta}_B = (\phi_B, \theta_B, \psi_B) \in R^3 \tag{4}$$

The generalized coordinate of the quadrotor $\boldsymbol{q}$, termed *signal* $\boldsymbol{s}$, is a vector formed of $\boldsymbol{\xi}$ and $\boldsymbol{\eta}$:

$$\boldsymbol{q} = (q_1, q_2, ..q_6) = (\phi, \theta, \psi, x, y, z) \in R^6. \tag{5}$$

4 Dynamic Model

The following sub-sections give the details of the derivation of the dynamical model of the quadrotor.

4.1 Dynamics of Translational Motion

The translation motion along the x, y and z axis is [12]:

$$\begin{bmatrix} \ddot{x} \\ \ddot{y} \\ \ddot{z} \end{bmatrix} = \begin{bmatrix} f_{th}(\sin\phi \sin\psi + \cos\phi \cos\psi \sin\theta)/m \\ f_{th}(-\cos\psi \sin\phi + \sin\psi \sin\theta \cos\phi)/m \\ f_{th}(\cos\theta \cos\phi)/m \end{bmatrix} - g\begin{bmatrix} 0 \\ 0 \\ 1 \end{bmatrix} - \begin{bmatrix} k_x\dot{x} \\ k_y\dot{y} \\ k_z\dot{z} \end{bmatrix} \tag{6}$$

k_x,k_y and k_z are aero-dynamical drag forces in, x, y, and z directions, and $g = 9.81\,\text{m}/\sec^2$ is the gravitational constant and m is the mass of the quadrotor.

4.2 Dynamics of Rotational Motion

The input torque to the motor is $\boldsymbol{\tau} = \left(\tau_\psi, \tau_\theta, \tau_\phi\right) \in R^3$:$\tau_\psi$, τ_θ and τ_ϕ are respectively the total yawing, pitching and rolling torque. The expression relating the second derivative of the Euler angles, the torque, and the Carioles forces $\boldsymbol{V}(\boldsymbol{\eta}, \dot{\boldsymbol{\eta}}) = \left[\dot{\psi}\dot{\theta}\cos\theta \;\; -\dot{\psi}\dot{\phi}\cos\theta \;\; \dot{\phi}\dot{\theta}\cos\theta\right]^T \in R^3$:

The Coriolis force includes the gyroscopic load torque and the centripetal forces [1–3].

$$\begin{bmatrix} \ddot{\phi} \\ \ddot{\theta} \\ \ddot{\psi} \end{bmatrix} = \frac{1}{\cos^2\theta} \begin{bmatrix} 1 & 0 & \sin\theta \\ 0 & \cos^2\theta & 0 \\ \sin\theta & 0 & 1 \end{bmatrix} \left(\begin{bmatrix} \tau_\phi \\ \tau_\theta \\ \tau_\psi \end{bmatrix} - \begin{bmatrix} \dot{\psi}\dot{\theta}\cos\theta \\ -\dot{\psi}\dot{\phi}\cos\theta \\ \dot{\phi}\dot{\theta}\cos\theta \end{bmatrix} \right) \tag{7}$$

where $f_1(\theta) = 1/\cos^2\theta; f_2(\theta) = \sin\theta/\cos^2\theta$;

Remarks There are 3 translations (x, y, z) along the coordinates x, y and z axes, and 3 rotational motions about each of the coordinate axes. Euler angles (ϕ, θ, ψ) and their time derivatives do not depend on the translational dynamics of (x, y, z) whereas the translation dynamics of $(\ddot{x}, \ddot{y}, \ddot{z})$ depend on the Euler angle (ϕ, θ, ψ) but not on their time derivatives. There are four control inputs (ϕ, θ, ψ) and f_{th} that drives the quadrotor along three translational components along the x, y and z axes.

5 Disturbance and Measurement Noise

The input, and the output are subject to *unknown* stochastic and deterministic disturbances and measurement noise. The sensors located on the quadrotor include the attitude sensor to provide the measurements of the attitude (yaw, pitch, and roll), three-axis rate gyro, the three-axis accelerometer to measure the angular and linear velocities, a global positioning system (GPS) to determine the location and a camera [10, 13].

Machine Learning Approach: The inaccessible true input $\boldsymbol{u}_o(k)$, termed as *true input*, buried in the accessible input $\boldsymbol{u}(k)$ corrupted by the unknown stochastic disturbance is extracted using Machine Learning exploiting the smoothness of $\boldsymbol{u}_o(k)$ and jaggedness of $\boldsymbol{u}(k)$.

5.1 State-Space Models of the Disturbances

Input disturbance: The zero-mean stochastic disturbance $\boldsymbol{d}_{dru}(k)$ is modeled as the output of $\left(\boldsymbol{A}_{dr\eta}, \boldsymbol{B}_{dr\eta}, \boldsymbol{C}_{dr\eta}\right)$ driven by a zero-mean white Gaussian white noise process $\boldsymbol{d}_r(k) \in R^3$;

$$\begin{aligned} \boldsymbol{x}_{dr\eta}(k+1) &= \boldsymbol{A}_{dr\eta}\boldsymbol{x}_{dr\eta}(k) + \boldsymbol{B}_{dr\eta}\boldsymbol{u}_{d\eta}(k) \\ \boldsymbol{d}_{u\eta}(k) &= \boldsymbol{C}_{dr\eta}\boldsymbol{x}_{dr\eta}(k) + \boldsymbol{v}_\eta(k) \end{aligned} \tag{8}$$

Output disturbance: The output disturbance $\boldsymbol{d}_{d\xi}(k)$ includes in addition to the zero-mean stochastic disturbance $\left(\boldsymbol{A}_{dr\xi}, \boldsymbol{B}_{dr\xi}, \boldsymbol{C}_{dr\xi}\right)$, a deterministic disturbance

$(\boldsymbol{A}_{dd\xi}, \boldsymbol{B}_{dd\xi}, \boldsymbol{C}_{dd\xi})$ to mimic the behavior of wind gusts. The output disturbance model $(\boldsymbol{A}_{d\xi}, \boldsymbol{B}_{d\xi}, \boldsymbol{C}_{d\xi})$ is driven by a white Gaussian white noise process $\boldsymbol{d}_{r\xi}(k) \in R^3$ and Kronecker Delta function denoted $\boldsymbol{\delta}(k) \in R^3$.

$$\begin{cases} \boldsymbol{x}_{d\xi}(k+1) = \boldsymbol{A}_{d\xi}\boldsymbol{x}_{d\xi}(k) + \boldsymbol{B}_{d\xi}\boldsymbol{u}_{d\xi}(k) \\ \boldsymbol{d}_{d\xi}(k) = \boldsymbol{C}_{d\xi}\boldsymbol{x}_{d\xi}(k) + \boldsymbol{v}_{\xi}(k) \end{cases} \tag{9}$$

where $\boldsymbol{A}_{d\xi} = \begin{bmatrix} \boldsymbol{A}_{dr\xi} & 0 \\ 0 & \boldsymbol{A}_{dd\xi} \end{bmatrix} \in R^{n_{d\xi}xn_{d\xi}}$, $\boldsymbol{B}_{d\xi} = \begin{bmatrix} \boldsymbol{B}_{dr\xi} & 0 \\ 0 & \boldsymbol{B}_{dd\xi} \end{bmatrix} \in R^{n_{d\xi}x6}$,$\boldsymbol{u}_{d\xi}(k) = \begin{bmatrix} \boldsymbol{d}_{r\xi} \\ \boldsymbol{\delta} \end{bmatrix} \in R^6$

The corrupted input $\boldsymbol{u} \in R^6$ is a sum of the true input $\boldsymbol{u}_0 \in R^6$, the stochastic disturbance $\boldsymbol{d}_{r\eta}(k)$ and measurement noise $\boldsymbol{v}_{\eta}$. Similarly the corrupted output $\boldsymbol{y}$ is a sum of the true output $\boldsymbol{y}_0$, the stochastic disturbance $\boldsymbol{d}_{r\xi}(k)$ deterministic disturbance $\boldsymbol{d}_{dd}(k)$ and measurement noise $\boldsymbol{v}_{\xi}$:

$$\begin{aligned} \boldsymbol{u}(k) &= \boldsymbol{u}_0(k) + \boldsymbol{d}_{u\eta}(k) + \boldsymbol{v}_{\eta}(k) \\ \boldsymbol{y}(k) &= \boldsymbol{y}_0(k) + \boldsymbol{d}_{d\xi}(k) + \boldsymbol{v}_{\xi}(k). \end{aligned} \tag{10}$$

5.2 The Inaccessible Input Signals

The inaccessible input $\boldsymbol{u}_0(k)$ given in (10) is identified using a Machine Learning approach with, $\boldsymbol{u}_0(k)$ being a smooth input such as a sum of sinusoids while the noisy input error $\boldsymbol{\vartheta}_u(k)$ exhibits random fluctuation.

$$\hat{\boldsymbol{u}}(k) = \left[\hat{\phi}(k)\ \hat{\theta}(k)\ \hat{\psi}(k)\right]^{\mathrm{T}} \in R^3 \tag{11}$$

The estimated input $\left(\hat{\phi}, \hat{\theta}, \hat{\psi}\right)$ replaces the input (ϕ, θ, ψ) at each operating point $\boldsymbol{\xi}_i$,$i = 1, 2, 3, .., L$ thereby avoiding the use of corrupted input at all operating points in the rest of this paper. The resulting model relating the estimated input and the output is s linear dynamic model. Further, the proposed computed torques approach based on deep learning helps handle the errors-in-variables model like the model-based approach [16–18]. As the true input is extracted using *Machine Learning,* only the output is corrupted by disturbance and measurement noise.

5.3 Nonlinear State-Space Model

The dynamic model of the translation (6) and the dynamic model of the rotation (7) are described using a multiple-input multiple-output (MIMO) nonlinear state-space model. The states, the inputs and the outputs of the system are:

$$\boldsymbol{x}(k) = \left[\phi\ \dot{\phi}\ \theta\ \dot{\theta}\ \psi\ \dot{\psi}\ z\ \dot{z}\ y\ \dot{y}\ x\ \dot{x}\right]^T \in R^{12};\ \boldsymbol{u} = \left[\tau_\phi\ \tau_\theta\ \tau_\psi\ f_{th}\right] \in R^4;$$

$$\begin{cases} \boldsymbol{x}(k+1) = \boldsymbol{f}(\boldsymbol{x}(k), \boldsymbol{u}(k), \boldsymbol{u}_w(k)) \\ \boldsymbol{y}(k) = \boldsymbol{g}(\boldsymbol{x}(k), v(k)) \end{cases} \tag{12}$$

where $\boldsymbol{x}(k)$ is the state, $\boldsymbol{u}(k)$ is the control input,$\boldsymbol{y}(k) \in R^6$ is an output vector formed of all measured (accessible) outputs, and $v(k) \in R^6$ is a measurement noise; The input $\boldsymbol{u}(k)$ and the output $\boldsymbol{y}(k)$ are corrupted by disturbance and measurement noise, $\boldsymbol{u}_w(k) \in R^6$ is an inaccessible input that generates the stochastic and deterministic disturbances, $\boldsymbol{f} \in R^{12}$, $\boldsymbol{g} \in R^6$ are nonlinear functions.

6 Piecewise-Linear Dynamic Model

6.1 Linear Parameter-Varying Model

In recent years, many complex, high-order, and nonlinear physical systems including aircrafts, robot manipulators, ships, mechatronic systems, multi-machine power systems, power electronics devices, and automotive systems have been successfully modeled using the Linear Parameter-Varying (LPV) framework [3–5] for developing status including fault diagnosis, fault tolerance, and condition-based maintenance. LPV methods help develop less conservative optimization-based solutions with low computational burden for complex nonlinear systems. For a large class of physical systems, a linearized model covering the entire operating region, however, may not capture the behavior of the system over a wide range of operating regimes resulting from variations in the input and parameters of the subsystems.

Subfigure A of Fig. 1 shows the operating region. The trajectory of the operating region (in solid line) as a function of the scheduling variable. The input segments and their corresponding output ones are indicated. The scheduling variables denoted by $\{\xi_i\}_{i=1}^{L} \in R^L$ denote L operating points selected to cover the entire operating regime. Based on this, a piecewise linear dynamic model that best approximates the nonlinear system is derived [16–20].

The operating conditions change not only due to disturbances and variations in the states but also due to changes in the parameters of the system, and environmental variables, production quantities, quality of raw material etc. The scheduling variables, related

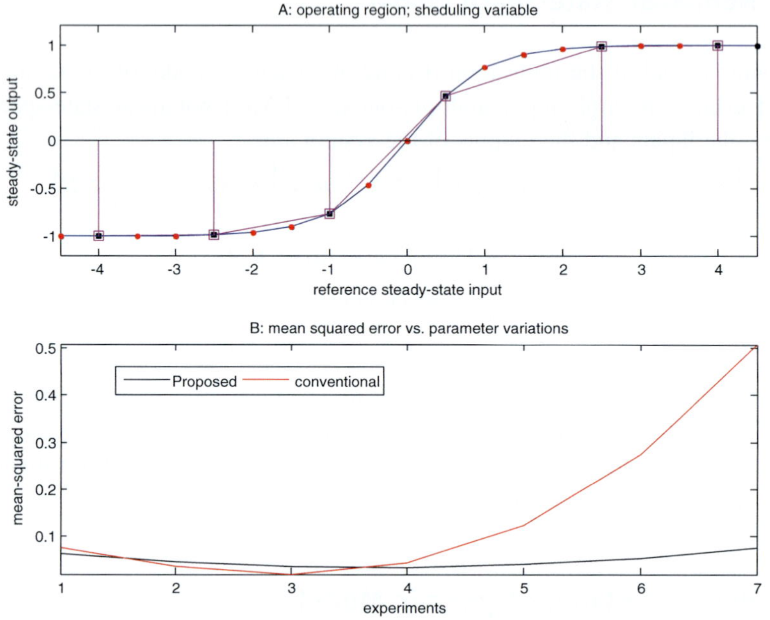

Fig. 1 Piecewise-linearization of system (Top) and the mean-square error deviation for the proposed LPV model and the conventional Linearized model (Bottom)

to the operating points used, include steady-state reference input, state variables, velocity and power, and external variables such as the altitude, temperature, pressure, and air speed. A finite number of scheduling variables is selected to cover the entire operating regime, including the linear and nonlinear ones. The operating points along the trajectory are shown as dots. The input segments and their corresponding output ones are indicated.

Subfigure A of Fig. 1, where both axes are in cm, shows a generic Input/Output function of a nonlinear system and its linearized version obtained by the LPV modeling approach based on various operating points.

Subfigure B of Fig. 1 compares the performance of the proposed LPV-based scheme with the conventional one, using mean-squared identification errors as the metric. Here, the x-axis is unitless as it shows the experiment number number, with each experiment is carried out by varying the system parameters, whereas the unit y-axis in cm^2 as it shows measures of a squared error on a variable measured in cm.

The proposed scheme is superior to those of conventional schemes. Note that the mean-squared errors for the proposed scheme are relatively larger for experiments indicated by 2, 3 and 4 and smaller for the rest, than their corresponding values in the conventional scheme. The conventional method employs the model identified by performing a single experiment indicated by 3, using the rotational and translational dynamics of the quadrotor, which belong to class of Wiener-Hammerstein model. The piecewise linear dynamic models covering the operating region $\xi_i,\ i = 1 : L$ are developed using the LPV approach.

6.2 Piecewise Linear Dynamic Box-Jenkins Model

The piecewise linear (MIMO) Box-Jenkins (BJ) model $(\boldsymbol{A}, \boldsymbol{B}, \boldsymbol{C})$ is obtained using the LPV approach which approximates the nonlinear system at an operating point and is given by:

$$\begin{aligned} \boldsymbol{x}(k+1) &= \boldsymbol{A}\boldsymbol{x}(k) + \boldsymbol{B}\boldsymbol{u}(k) + \boldsymbol{E}_w\boldsymbol{u}_w(k) \\ \boldsymbol{y}(k) &= \boldsymbol{C}\boldsymbol{x}(k) + \boldsymbol{v}(k) \end{aligned} \tag{13}$$

where $\boldsymbol{A} \in R^{nxn}$, $\boldsymbol{B} \in R^{nxp}$, $\boldsymbol{C} \in R^{qxn}$ and $\boldsymbol{E}_w \in R^{nxp}$; $(\boldsymbol{A}, \boldsymbol{B}, \boldsymbol{C})$ is a linearized model at an operating point ξ_i,$\boldsymbol{x}(k) \in R^n$ is the state, $\boldsymbol{u}(k) \in R^p$ is a user-defined input that is accessible, $\boldsymbol{u}_w(k) \in R^q$ is a zero-mean Gaussian white noise process that is inaccessible,$\boldsymbol{v}(k) \in R^q$ is a measurement noise that is a zero-mean Gaussian white noise process. The BJ model is an augmented model of the signal and disturbance and is given by:
$\boldsymbol{A} = \begin{bmatrix} \boldsymbol{A}_s & 0 \\ 0 & \boldsymbol{A}_w \end{bmatrix}$; $\boldsymbol{B} = \begin{bmatrix} \boldsymbol{B}_s \\ 0 \end{bmatrix}$; $\boldsymbol{E}_w = \begin{bmatrix} 0 \\ \boldsymbol{B}_w \end{bmatrix}$;$\boldsymbol{C} = \begin{bmatrix} \boldsymbol{C}_s & \boldsymbol{C}_w \end{bmatrix}$ $(\boldsymbol{A}_w, \boldsymbol{B}_w, \boldsymbol{C}_w)$ is termed the signal model as it represents a noise-free and disturbance-free model.

The auto-regressive and moving-average (ARMA), moving-average (MA), and auto-regressive (AR) models are all special cases of the Box-Jenkins model. This model covers several imporatant practical systems including, among others, autonomous vehicles such as autonomously-guided drones flying at different altitudes and subject to varying wind gusts that may accelerate or decelerate, push, or pull the drone away from the desired trajectory. These devices are currently enjoying an ever-increasing field of important applications. Note here that, for such devices, an ARMA model will not be suitable to use as it cannot model a disturbance whose denominator is different from that of the signal model. The transfer functions of the signal and disturbance models are respectively denoted by $\boldsymbol{G}_s(z)$ and $\boldsymbol{G}_w(z)$. The state-space model $(\boldsymbol{A}_s, \boldsymbol{B}_s, \boldsymbol{C}_s)$ of the signal $\boldsymbol{s}(k) \in R^q$, is given by:

$$\begin{aligned} \boldsymbol{x}_s(k+1) &= \boldsymbol{A}_s\boldsymbol{x}_s(k) + \boldsymbol{B}_s\boldsymbol{u}(k) \\ \boldsymbol{s}(k) &= \boldsymbol{C}_s\boldsymbol{x}_s(k) \end{aligned} \tag{14}$$

The state-space model $(\boldsymbol{A}_w, \boldsymbol{B}_w, \boldsymbol{C}_w)$ of the disturbance $\boldsymbol{w}(k) \in R^q$, is given by:

$$\begin{aligned} \boldsymbol{x}_w(k+1) &= \boldsymbol{A}_w\boldsymbol{x}_w(k) + \boldsymbol{B}_w\boldsymbol{u}_w(k) \\ \boldsymbol{w}(k) &= \boldsymbol{C}_w\boldsymbol{x}_w(k) \end{aligned} \tag{15}$$

where $\boldsymbol{x}_s \in R^{n_s} \boldsymbol{x}_w \in R^{n_w}, \boldsymbol{A}_s \in R^{n_s x n_s}$, $\boldsymbol{A}_w \in R^{n_w x n_w}, \boldsymbol{B}_s \in R^{n_s x p}$, $\boldsymbol{B}_w \in R^{n_w x p}$, $\boldsymbol{C}_w \in R^{q x n_w}, \boldsymbol{u}_w \in R^p; n_s$ and n_w are respectively the orders of the signal $\boldsymbol{s}(k)$ and disturbance models.

Assumptions

- The signal model $(\boldsymbol{A}_s, \boldsymbol{B}_s, \boldsymbol{C}_s)$ and disturbance model $(\boldsymbol{A}_w, \boldsymbol{B}_w, \boldsymbol{C}_w)$ are controllable and observable
- The disturbance model $(\boldsymbol{A}_w, \boldsymbol{B}_w, \boldsymbol{C}_w)$ is asymptotically stable
- The input $\boldsymbol{u}_w(k)$ driving the disturbance model is a zero-mean white Gaussian noise process
- The input $\boldsymbol{u}(k)$ is accessible, deterministic, and smooth such as the constants, sinusoids, and their combinations
- The measurement noise $\boldsymbol{v}(k)$ is a zero-mean white Gaussian noise process
- The inputs $\boldsymbol{u}(k)$ and $\boldsymbol{u}_w(k)$, and measurement noise $\boldsymbol{v}(k)$ are uncorrelated and hence the signal $\boldsymbol{s}(k)$,the disturbance $\boldsymbol{w}(k)$ and measurement noise $\boldsymbol{v}(k)$ are uncorrelated

The piecewise linear dynamic models captures the local behavior of the nonlinear model, in the neighborhood $\boldsymbol{\xi}_{i-1} < \boldsymbol{\xi}_i < \boldsymbol{\xi}_{i+1}$ of the operating point. For notational convenience, the operating points ξ_i , $i = 1, 2, ..., L$ are not explicitly indicated. The measurement output $\boldsymbol{y}(k)$ is assumed to be an additive combination of the signal $\boldsymbol{s}(k)$, the stochastic disturbance $\boldsymbol{w}(k)$ and the measurement noise $\boldsymbol{v}(k)$. The measurement model relating the signal, the disturbance and the measurement noise is given by:

$$\begin{aligned} \boldsymbol{y}(z) &= \boldsymbol{s}(z) + \boldsymbol{\vartheta}(z) \\ \boldsymbol{\vartheta}(z) &= \boldsymbol{w}(z) + \boldsymbol{v}(z) \end{aligned} \tag{16}$$

$\boldsymbol{s}(k) \in R^q$, $\boldsymbol{v}(k) \in R^q$; $\boldsymbol{w}(k) \in R^q$, $\boldsymbol{\vartheta}(z) \in R^q$, $\boldsymbol{y}(k) \in R^q$.

7 Feedback and Feedforward Controller Design

The desired trajectory denoted $\boldsymbol{\xi}_d = (x_d, y_d, z_d) \in R^3$ is specified for the translation in the (x, y, z) plane. The four desired control inputs $(\phi_d, \theta_d, \psi_d)$ and f_{th} drives the quadrotor along three translational components along the x, y, and z axes [9, 29].

$$\phi_d = \arcsin\left(\frac{u_{xd}\sin\psi_d - u_{yd}\cos\psi_d}{\sqrt{u_{xd}^2 + u_{yd}^2 + (u_{zd} + g)^2}}\right); \theta_d = \arctan\left(\frac{u_{xd}\cos\psi_d + u_{yd}\sin\psi_d}{u_{zd} + g}\right);$$

$$f_{thd} = \sqrt{u_{xd}^2 + u_{yd}^2 + (u_{zd} + g)^2} \tag{17}$$

$$\phi_d = \arcsin\left(\frac{u_{xd}\sin\psi_d - u_{yd}\cos\psi_d}{\sqrt{u_{xd}^2 + u_{yd}^2 + (u_{zd}+g)^2}}\right);\ \theta_d = \arctan\left(\frac{u_{xd}\cos\psi_d + u_{yd}\sin\psi_d}{u_{zd}+g}\right);$$

$$f_{thd} = \sqrt{u_{xd}^2 + u_{yd}^2 + (u_{zd}+g)^2}$$

Equation (17) is termed as inverse kinematics.

The desired trajectory $\boldsymbol{q}_d$ for q is obtained from the desired translation $\boldsymbol{\xi}_d$ and $(x, y, z, \phi_d, \theta_d, f_{thd})$ from (5) as

$$\boldsymbol{q}_d = (x_d, y_d, z_d, \phi_d, \theta_d, f_{thd}) \in R^6. \tag{18}$$

7.1 Feedback Controller: An Internal Model-Based Design Approach

Since the signal model is unstable, the objective is to design a feedback controller so that the closed-loop system formed of the signal model and the controller is asymptotically stable, and so that its output tracks a given specified trajectory. In the literature, several approaches to controller design such as proportional, proportional derivative, proportional integral and derivative, model reference adaptive controller, and internal model-based controller are well known It has been shown in that the output of a linear time-invariant system will track a desired reference waveform despite the disturbance if and only if controller includes internal models of the reference and the disturbance waveform driven by the error between the output and the desired reference waveforms.

A feedback controller is thus designed using the internal model principle to track the desired reference despite the perturbation in the system. The internal model that generates the reference input $\boldsymbol{q}_d(k)$ is

$$\begin{aligned} \boldsymbol{x}_I(k+1) &= \boldsymbol{A}_I\boldsymbol{x}_I(k) + \boldsymbol{B}_I\boldsymbol{e}(k) \\ \boldsymbol{y}_I(k) &= \boldsymbol{C}_I\boldsymbol{x}_I(k) \end{aligned} \tag{19}$$

where $\boldsymbol{A}_I \in R^{n_I x n_I}\boldsymbol{B}_I \in R^{n_I x 6}$; $\boldsymbol{C}_I = \boldsymbol{I} \in R^{n_I x n_I}$ is an identity matrix; $\boldsymbol{e}(k) = \boldsymbol{q}_d(k) - \boldsymbol{y}(k) \in R^6$

A feedback controller is thus designed using the internal model principle to track the desired reference despite the perturbation in the system.

7.2 Feedforward Controller Design

Even in the presence of model perturbations, the feed forward controller can mitigate the effect of the output error on the performance of the combined controller. The feed forward controller quickly rejects the output error without waiting for the deviation in the output to occur, hence its anticipatory action. The reference input is injected into the feed-forward path to reject the effects of the output error on the output, thereby ensuring that the output is generated by the reference input and not by the output error [16, 17]. The feed forward controller is given by:

$$\boldsymbol{u}_{ff}(z) = \boldsymbol{H}_{ff}\,\boldsymbol{q}_d(z) \tag{20}$$

The expression for $\boldsymbol{H}_{ff} \in R^{6x6}$ is given in the later section.

7.3 Combined Feed-Forward-Feedback Controller

$$\begin{aligned} \boldsymbol{u}_s &= \boldsymbol{u}_{ff} - \boldsymbol{F}_I \boldsymbol{x}_I \\ \boldsymbol{u}_p &= \boldsymbol{u}_s - \boldsymbol{F}_s \boldsymbol{x}_p \end{aligned} \tag{21}$$

where $\boldsymbol{x}_{Fi}$ is the state of the internal model $\left(\boldsymbol{A}_I, \boldsymbol{B}_I, \boldsymbol{C}_{I_i}\right)$; $\boldsymbol{u}_p$ is the input that drives the plant model; $\boldsymbol{F}_I$ and $\boldsymbol{F}_p$ are state feedback gains.

The state-feedback controller $\boldsymbol{u}_{fb} = \boldsymbol{F}_s \boldsymbol{x}_p$ is an optimal linear quadratic regulator obtained from the parameterization of all stabilizing controllers to ensure robustness to model perturbations [12].

Remarks The quadrotor model is nonlinear. Using the LPV framework, the system is approximated by piecewise linear dynamic models. An internal model-based controller is employed for the piecewise linear dynamic models at each operating point. The asymptotic tracking holds independently of all variations in the plant and the stabilizer parameters if the closed system is asymptotically stable, and the internal model parameters are not perturbed. For nonlinear quadrotor models, nonlinear controllers such as adaptive controllers are employed [3, 9].

7.4 Closed-Loop Signal Model Derivation

The augmented closed-loop model $(\boldsymbol{A}_{cl}, \boldsymbol{B}_{cl}, \boldsymbol{C}_{cl})$ is formed of the plant model, and the internal model $(\boldsymbol{A}_I, \boldsymbol{B}_I, \boldsymbol{C}_I)$ of the desired reference $\boldsymbol{q}_d(k)$:

$$\begin{cases} \boldsymbol{x}_{cl}(k+1) = \boldsymbol{A}_{cl}\boldsymbol{x}_{cl}(k) + \boldsymbol{B}_{cl}\boldsymbol{q}_d(k) \\ \boldsymbol{q}(k) = \boldsymbol{C}_{cl}\,\boldsymbol{x}_{cl}(k) \end{cases} \tag{22}$$

where $\boldsymbol{x}_{cl}(k) = \begin{bmatrix} \boldsymbol{x}_s \\ \boldsymbol{x}_I \end{bmatrix} \in R^{n_{cl}}, \boldsymbol{A}_{cl} = \begin{bmatrix} \boldsymbol{A}_s - \boldsymbol{B}_s\boldsymbol{F}_s & -\boldsymbol{B}_s\boldsymbol{F}_I \\ -\boldsymbol{B}_I\boldsymbol{C}_s & \boldsymbol{A}_I \end{bmatrix} \in R^{n_{cl}xn_{ci}}, \boldsymbol{B}_{cl} = \begin{bmatrix} \boldsymbol{B}_{cl}\boldsymbol{H}_{ff} \\ \boldsymbol{B}_I \end{bmatrix} \in R^{n_{cl}x6}$, $\boldsymbol{C}_{cl} = \begin{bmatrix} \boldsymbol{C}_I & \boldsymbol{C}_s \end{bmatrix} \in R^{6xn_{cl}}$, $n_{cl} = n_s + n_I$.

The transfer function of the closed-loop signal model $\boldsymbol{G}_{cl}(z) = \frac{N_{cl}(z)}{D_{cl}(z)}$ is:

$$\boldsymbol{q}(z) = \boldsymbol{G}_{cl}(z)\boldsymbol{q}_d(z) \tag{23}$$

The tracking error input $\boldsymbol{e}(k)$ drives the internal model, $(\boldsymbol{A}_I, \boldsymbol{B}_I, \boldsymbol{C}_I)$ to generate the class of all desired reference waveforms $\boldsymbol{q}_d(k)$.

Assumptions It is assumed that the signal $\boldsymbol{s}(k)$, the stochastic disturbance $\boldsymbol{d}_{r\xi}(k)$, the deterministic disturbance $\boldsymbol{d}_{d\xi}(k)$ and the measurement noise $\boldsymbol{v}(k)$ are uncorrelated, and that $\boldsymbol{G}_s(z)$ is a minimum-phase transfer matrix so that the feedforward gain remains finite at all frequencies.

The stochastic disturbance model $\left(\boldsymbol{A}_{dr\xi}, \boldsymbol{B}_{dr\xi}, \boldsymbol{C}_{dr\xi}\right)$ is asymptotically stable.

The signal model $\boldsymbol{G}_s(z)$ assumed to be a minimum-phase one, therefore has all of its zeros and poles within the unit circle, i.e. $\boldsymbol{G}_s(z)$ and its inverse $\boldsymbol{G}_s^{-1}(z)$ are both stable.

7.5 Box-Jenkins Model of the System

The BJ model $(\boldsymbol{A}, \boldsymbol{B}, \boldsymbol{C})$ relating the corrupted output $\boldsymbol{y}(k)$ and the input $\boldsymbol{q}_d(k)$, is termed *system model*, and the augmented model, shown in (24), is formed of the internal model-based closed loop control system $(\boldsymbol{A}_{cl}, \boldsymbol{B}_{cl}, \boldsymbol{C}_{cl})$

$$\begin{aligned} \boldsymbol{x}(k+1) &= \boldsymbol{A}\boldsymbol{x}(k) + \boldsymbol{B}\boldsymbol{q}_d(k) + \boldsymbol{E}_{dr}\boldsymbol{w}(k) + \boldsymbol{E}_{dd}\boldsymbol{\delta}(k) \\ \boldsymbol{y}(k) &= \boldsymbol{C}\boldsymbol{x}(k) + \boldsymbol{v}(k) \end{aligned} \tag{24}$$

$\boldsymbol{x} = \begin{bmatrix} \boldsymbol{x}_s \\ \boldsymbol{x}_{dr\xi} \\ \boldsymbol{x}_{dd\xi} \end{bmatrix} \in R^n; \boldsymbol{A} = \begin{bmatrix} \boldsymbol{A}_{cl} & -\boldsymbol{B}_{cl}\boldsymbol{C}_{dr\xi} & -\boldsymbol{B}_{cl}\boldsymbol{C}_{dd\xi} \\ 0 & \boldsymbol{A}_{dr\xi} & 0 \\ 0 & 0 & \boldsymbol{A}_{dd\xi} \end{bmatrix} \in R^{nxn}; \boldsymbol{B} = \begin{bmatrix} \boldsymbol{B}_{cl} \\ 0 \\ 0 \end{bmatrix} \in R^{nx6}; \boldsymbol{E}_{dr} = \begin{bmatrix} 0 \\ \boldsymbol{B}_{dr\xi} \\ 0 \end{bmatrix} \in R^{nx6}; \boldsymbol{E}_{dd} = \begin{bmatrix} 0 \\ 0 \\ \boldsymbol{B}_{dd\xi} \end{bmatrix}; \boldsymbol{y}(k) \in R^6.$

The output $\boldsymbol{y}(z)$ in terms of the closed-loop signal model, the stochastic and the deterministic disturbance given in (10) is:

$$\boldsymbol{y}(z) = \boldsymbol{q}(z) + \boldsymbol{y}_{dr\xi}(x) + \boldsymbol{y}_{dd\xi}(z). \tag{25}$$

8 Kalman Filter Design

The KF associated with the true system model (fault-free unperturbed model) $(\boldsymbol{A}_0, \boldsymbol{B}_0, \boldsymbol{C}_0)$ given in s re-written:

$$\begin{aligned}\hat{\boldsymbol{x}}(k+1) &= (\boldsymbol{A}_0 - \boldsymbol{K}_0\boldsymbol{C}_0)\hat{\boldsymbol{x}}(k) + \boldsymbol{B}_0\boldsymbol{q}_d(k) + \boldsymbol{K}_0\boldsymbol{y}(k) \\ \hat{\boldsymbol{y}}(k) &= \boldsymbol{C}_0\hat{\boldsymbol{x}}(k)\end{aligned} \tag{26}$$

$$\boldsymbol{e}_{kf}(k) = \boldsymbol{y}(k) - \hat{\boldsymbol{y}}(k) \tag{27}$$

where $\hat{\boldsymbol{x}}(k) = \left[\hat{\boldsymbol{x}}_{cl}(k)\ \hat{\boldsymbol{x}}_{dr\xi}(k)\ \hat{\boldsymbol{x}}_{dd\xi}(k)\right]^T \in R^n$ and $\hat{\boldsymbol{y}}(k)$ are respectively the best estimate of the state $\boldsymbol{x}(k)$, and of the output $\boldsymbol{y}(k)$; $\boldsymbol{e}_{kf}(k)$ is the residual or the innovation sequence; the optimal Kalman gain $\boldsymbol{K}_0 \in \mathbb{R}^{nx6}$ ensures the asymptotic stability of the KF, i.e. $(\boldsymbol{A}_0 - \boldsymbol{K}_0\boldsymbol{C}_0)$ is strictly Hurwitz having all its eigenvalues strictly inside the unit circle. $(\boldsymbol{A}_0, \boldsymbol{B}_0, \boldsymbol{C}_0)$ is the identified system model embodied in the KF.

8.1 *The Residual Model of the* KF

The residual model of the KF relates the residual $\boldsymbol{e}_{kf}(z)$ to the system, the input $\boldsymbol{q}_d(z)$ and output $\boldsymbol{y}(z)$ is:

$$\boldsymbol{e}_{kf}(z) = \frac{\overline{\boldsymbol{D}}(z)}{\overline{F}(z)}\boldsymbol{y}(z) - \frac{\overline{\boldsymbol{N}}(z)}{\overline{F}(z)}\boldsymbol{q}_d(z) \tag{28}$$

where $\frac{\overline{\boldsymbol{D}}(z)}{\overline{F}(z)}$ and $\frac{\overline{\boldsymbol{N}}(z)}{\overline{F}(z)}$ relates respectively $\boldsymbol{y}(z)$ to $\boldsymbol{e}_{kf}(z)$ and $\boldsymbol{q}_d(z)$ to $\boldsymbol{e}_{kf}(z)$; $\overline{F}(z) = \det(zI - (\boldsymbol{A}_0 - \boldsymbol{K}_0\boldsymbol{C}_0))$ is termed here as the Kalman polynomial.

8.2 Identification of Signal Model, Disturbance Model, Estimates of Signal, and Output Error

8.2.1 Signal Model

The signal model $\hat{\boldsymbol{G}}_s(z)$ or its state space representation $(\boldsymbol{A}_s, \boldsymbol{B}_s, \boldsymbol{C}_s)$ and the signal $\boldsymbol{s}(z)$ are [16]:

$$\begin{aligned}\hat{\boldsymbol{G}}_s(z) &= \frac{\overline{\boldsymbol{N}}(z)}{\overline{\boldsymbol{D}}(z)} \\ \hat{\boldsymbol{s}}(z) &= \hat{\boldsymbol{G}}_s(z)\boldsymbol{u}(z).\end{aligned} \tag{29}$$

8.2.2 Disturbance Model

Here the output error is defined as the difference between the output and the estimate of the signal.

If the matrix $\overline{\boldsymbol{D}}(z)$ is invertible, the output error $\hat{\boldsymbol{\vartheta}}(z) = \boldsymbol{d}(z) + \boldsymbol{v}(z) = \boldsymbol{y}(z) - \hat{\boldsymbol{s}}(z)$ is given by [16]:

$$\begin{aligned} F^{-1}(z)\overline{\boldsymbol{D}}(z)\hat{\boldsymbol{\vartheta}}(k) &= \boldsymbol{e}_{kf}(z) \\ \hat{\boldsymbol{\vartheta}}(z) &= \boldsymbol{G}_d(z)\boldsymbol{e}_{kf}(z) \end{aligned} \tag{30}$$

where $\boldsymbol{G}_d(z) = F(z)\overline{\boldsymbol{D}}^{-1}(z)$ is the inverse of the whitening filter $F^{-1}(z)\overline{\boldsymbol{D}}(z)$, and is termed the output error model, as it generates the output error $\hat{\boldsymbol{\vartheta}}(k)$ when excited by the KF residual. The KF whitens the output error.

$\hat{\boldsymbol{G}}_d(z) = \frac{F(z)}{\overline{D}(z)}$ is termed the disturbance model $\overline{D}(z) = D_s(z)D_{dr\xi}(z)D_{dd\xi}(z)$. The disturbance model is the inverse of the whitening filter.

8.3 KF for the Signal Model

Similar to the KF for the system model, the KF $(\boldsymbol{A}_{0cl}, [\boldsymbol{B}_{0cl}\ \boldsymbol{K}_{0cl}], \boldsymbol{C}_{0cl})$ associated with the closed loop fault-free signal model $(\boldsymbol{A}_{0cl}, \boldsymbol{B}_{0cl}, \boldsymbol{C}_{0cl})$

$$\begin{cases} \hat{\boldsymbol{x}}_{cl}(k+1) = (\boldsymbol{A}_{0cl} - \boldsymbol{K}_{0cli}\boldsymbol{C}_{0cli})\hat{\boldsymbol{x}}_{cl}(k) + \boldsymbol{B}_{0cl}\boldsymbol{q}_d(k) + \boldsymbol{K}_{0cl}\hat{\boldsymbol{q}}(k) \\ \hat{\boldsymbol{q}}(k) = \boldsymbol{C}_{0cl}\hat{\boldsymbol{x}}_{cl}(k) \\ \boldsymbol{e}_{ks}(k) = \boldsymbol{q}(k) - \hat{\boldsymbol{q}}(k) \end{cases} \tag{31}$$

where $\hat{\boldsymbol{x}}_{cl}(k) \in R^{n_{cl}}$ and $\hat{\boldsymbol{q}}(k)$ are respectively the best estimate of the state $\boldsymbol{x}_{cl}(k)$, and of the output $\boldsymbol{q}(k)$, $\boldsymbol{e}_{ks}(k)$ is the residual or the innovation sequence. The optimal Kalman gain $\boldsymbol{K}_{0cl} \in R^{n_{cl}x6}$ ensures the asymptotic stability of the KF, i.e. $(\boldsymbol{A}_{0s} - \boldsymbol{K}_{0s}\boldsymbol{C}_{0s})$ is strictly Hurwitz having all its eigenvalues strictly inside the unit circle. The identified signal model $(\boldsymbol{A}_{0cl}, \boldsymbol{B}_{0cl}, \boldsymbol{C}_{0cl})$ is embodied in the KF. The residual model is given by:

$$\boldsymbol{e}_{kf}(z) = \frac{\overline{\boldsymbol{D}}(z)}{\overline{F}(z)}\boldsymbol{y}(z) - \frac{\overline{\boldsymbol{N}}(z)}{\overline{F}(z)}\boldsymbol{q}_d(z). \tag{32}$$

9 Emulator-Based Identification Method

The fault-free unperturbed Box-Jenkins model of the system $(\boldsymbol{A}_0, \boldsymbol{B}_0, \boldsymbol{C}_0)$ and the associated KF $(\boldsymbol{A}_0, [\boldsymbol{B}_0\ \boldsymbol{K}_0], \boldsymbol{C}_0)$ shown in (26) are identified by minimizing the residual of the KF given in from the following optimization problem:

$$\min_{\overline{D},\overline{N},\overline{F}}\left\{\left(e_{kf}(k)\right)^T e_{kf}(k)\right\},\ \text{such that.}\ e_{kf}(z)=\frac{\overline{D}(z)}{\overline{F}(z)}y(z)-\frac{\overline{N}(z)}{\overline{F}(z)}q_d(z). \tag{33}$$

The input and the output data are the emulator-generated data. The input is selected to be the same for all emulator perturbed experiments.

9.1 Signal Model

The identified transfer functions $\frac{\overline{D}(z)}{\overline{F}(z)}$ and $\frac{\overline{N}(z)}{\overline{F}_i(z)}$ are used to obtain the closed-loop signal model. The signal $s(z)$ and the disturbance $w(z)$ are assumed to be the outputs of linear time-invariant systems, namely a signal model $G_s(z) = D_s^{-1}(z)N_s(z)$ driven by inputs $u(z)$;

$$s(z) = G_s(z)u(z) \tag{34}$$

Thanks to the emulator-based two-stage identification of the system and its associated Kalman filter, the key properties of the KF are exploited for developing the state feedback controller, fault detection, fault isolation, condition-based maintenance, controller adaptation and fault tolerant autonomous vehicle.

9.2 Disturbance Model

If the matrix $\overline{D}(z)$ is invertible, the output error $\vartheta(z) = d(z) + v(z)$ is given by:

$$\begin{aligned}\hat{\vartheta}(k) &= y(k) - \hat{s}(k) = d(z) + v(z)\\ \hat{\vartheta}(z) &= F(z)\overline{D}^{-1}(z)e(z)\end{aligned} \tag{35}$$

where $F(z)\overline{D}^{-1}(z) = \left(I - C^0\left(zI - A^0 + K^0 C^0\right)^{-1}K^0\right)^{-1}$ is termed the disturbance model estimate which generates the output error when excited by the residual. The disturbance model $G_d(z) = F(z)\overline{D}^{-1}(z)$ is the inverse of the whitening filter $F^{-1}(z)\overline{D}(z)$, and is termed the output error model, as it generates the output error $\hat{\vartheta}(k)$ when excited by the KF residual. The KF whitens the output error.

9.3 KF-Based State Feedback and Feedforward Controller

The Kalman filter-based state-feedback control strategy generates a class of all stabilizing controllers if and only the system is both stabilizable and detectable [16, 20]. Since the signal model is both controllable and observable, the state feedback strategy is derived using the linear quadratic optimization approach that relies on the estimated states of the Kalman filter of the signal model. The state feedback strategy is obtained using the separation principle governing the estimation of the states, and the implementation of state feedback strategy [15, 16]. First, the state feedback strategy is derived using the linear quadratic optimization approach assuming that the states of the signal model $\left(\boldsymbol{A}_s^0, \boldsymbol{B}_s^0, \boldsymbol{C}_s^0\right)$ are available. Then the states of the signal model are substituted by those estimated by the associated Kalman filter. The augmented state-space model of the signal and the internal models becomes:

$$\begin{bmatrix} \boldsymbol{x}_s(k+1) \\ \boldsymbol{x}_I(k+1) \end{bmatrix} = \begin{bmatrix} \boldsymbol{A}_s^0 & \boldsymbol{0} \\ -\boldsymbol{B}_{im}\boldsymbol{C}_s^0 & \boldsymbol{A}_I \end{bmatrix} \begin{bmatrix} \boldsymbol{x}_s(k) \\ \boldsymbol{x}_I(k) \end{bmatrix} + \begin{bmatrix} \boldsymbol{B}_s^0 & \boldsymbol{0} \\ \boldsymbol{0} & \boldsymbol{B}_I \end{bmatrix} \begin{bmatrix} \boldsymbol{u}(k) \\ \boldsymbol{r}(k) \end{bmatrix} \tag{36}$$

The optimal state-feedback controller for the augmented system is determined using a linear quadratic cost function, assuming the states of the signal model are available, as in [1]. The assumed states of the signal model are substituted by those estimated by the associated Kalman filter. The feedback control input to the signal model becomes:

$$\boldsymbol{u}_{fb}(k) = -\boldsymbol{F}_s\hat{\boldsymbol{x}}_s(k) - \boldsymbol{F}_{im}\boldsymbol{x}_I(k) \tag{37}$$

9.4 Feedforward Controller

Even in the presence of model perturbations, the feedforward controller can mitigate the effect of the output error on the performance of the combined controller. The feedforward controller quickly rejects the output error without waiting for the deviation in the output to occur, hence its anticipatory action. The output error and the reference input are injected into the feedforward path to reject the effects of the output error on the output, thereby ensuring that the output is generated by the reference input and not by the output error. As neither the output error nor its entry point in the system is known, it is first estimated using the key properties of the Kalman filter, and then injected into the signal model using the feedforward controller. The feedforward controller of the reference, denoted by the steady-state gain $\boldsymbol{H}_r$, is the inverse signal model evaluated at the poles of the reference model. Thus:

$$\boldsymbol{H}_r = \boldsymbol{G}_s^{-1}(z_r) \tag{38}$$

where z_r denotes the poles of the reference model $\boldsymbol{G}_r(z)$. Even when there are modeling errors, the feedforward control can often reduce the effect of the measured disturbance on the output in a way that is better than that achievable by feedback control alone. The feedforward controller $\boldsymbol{u}_{ffr}(k)$ of the reference input is given by:

$$\boldsymbol{u}_{ffr}(z) = \boldsymbol{H}_r(z_r)\boldsymbol{r}(z) \tag{39}$$

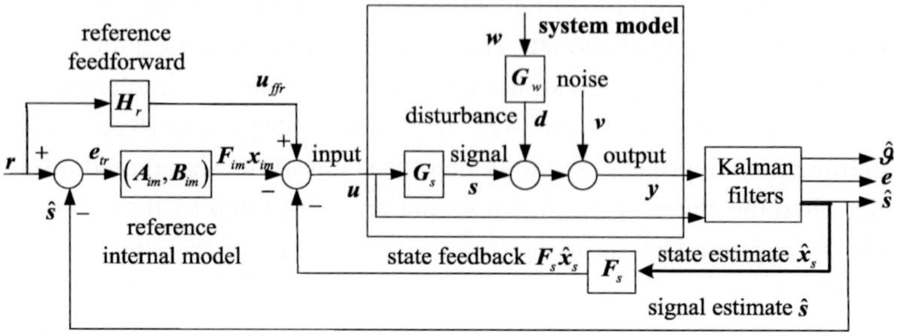

Fig. 2 Block diagram of the feedforward-feedback controller

A block diagram of the feedforward-feedback controller implemented using the Kalman filter and internal model is shown in Fig. 2. The system and signal models; Kalman filters associated with the system and the signal models; feedforward-feedback controller; the signal, the disturbance, and the measurement noise; residual, estimates of the signal and the output error of the Kalman filter are all explicitly shown in Fig. 2.

The signal $\hat{s}(z)$, shown in Fig. 2 in its shorter version of $\hat{s}$, is employed instead of the noisy output $\boldsymbol{y}(z)$ for implementing the state-feedback controller as shown in Fig. 2. This figure shows the implementation of the closed-loop controller, the role of KF of the system, and the signal models in the inplementation of the proposed trajectory-tracking system.

Remarks The KF residual of the system model $\boldsymbol{e}_{kf}(k)$ and that of the signal model $\boldsymbol{e}_s(k)$ or $\boldsymbol{e}_s(k)$ are employed to monitor the status of the overall system and to detect and isolate faults in the signal and disturbance models and the sensors. The proposed scheme provides a sound framework for developing fault-tolerant systems and condition-based maintenance systems as well. An efficient scheme to monitor the status of the system may be implemented. First the status of the system is monitored by analyzing the residual of the KF of the system model. If there is a variation, then the residual of the KF of the signal model is analyzed to ascertain whether a fault has occurred.

9.5 Fault Isolation

If the performance degradation is high or the controller adaptation fails to meet the acceptable bound, the presence of a fault is asserted. The faulty subsystem is then isolated using the influence vector. The faulty subsystem is either repaired or replaced. The residuals $\boldsymbol{e}_{kf}(k)$ and $\boldsymbol{e}_s(k)$ of the KFs are employed to monitor the status of the overall system and to detect and isolate faults in the signal and disturbance-based models and the sensors.

The proposed scheme provides a sound framework for developing fault-tolerant systems and condition-based maintenance systems as well.

9.6 Condition-Based Maintenance

Stable closed-loop system performance degradation may result from changes in its operating condition, malfunctioning of its components, variations in its noise and disturbance statistics, etc. In a high-performance control system, it is imperative that the closed-loop stability, performance, and the system's heath condition be all monitored to initiate quick remedial actions, when needed. For instance, if the performance degradation stems from a plant model variation, then the plant is re-identified and the robust controller redesigned or if a fault is detected, it is then isolated and either the fault is accommodated, or the system is shut down for repair.

9.7 Fault Tolerance

If there is performance degradation, then the controller is redesigned (adapted) or the plant identified, thereby "nipping the incipient faults in the bud" and preventing them from growing into serious faults capable of damaging the system, and possibly causing all the concomitant nefarious consequences, including system breakdown, loss of productivity and danger to the operating personnel. The role of the fault-tolerant system consists of effectively and reliably performing the following functions: (a) condition monitoring, (b) detection of performance degradation, (c) controller adaptation, and (d) detection and isolation of faults.

10 Evaluation of the Proposed Scheme on a Quadrotor

The proposed feedforward-feedback controller scheme is evaluated on a simulated model of a drone moving in a horizontal plane to track accurately the desired trajectory despite wind gusts and model perturbations. The performance of the proposed controller is compared with that of the conventional combined controller. The input and the output of the quadrotor are both corrupted by exogenous and endogenous stochastic disturbance and measurement noise.

10.1 Estimation of the True Input Using Machine Learning

The true input, denoted by $u_0(k)$ is corrupted by stochastic disturbance and measurement noise $\vartheta_u(k)$ given in (10). The true input extracted from the corrupted input $u(k) = u_0(k) + \vartheta_u(k)$ using the Machine Learning approach. The inaccessible true input signal $u_0(k)$ is assumed to be a smooth input such as for example a constant or a sinusoid while the noisy output error $\vartheta_u(k)$ exhibits random fluctuations. The inaccessible signal $u_0(k)$ is extracted from the corrupted input $u(k)$ using the frequency-domain approach by exploiting its spectral characteristics such as the smooth waveform $u_0(k)$ has line-spectra, and the noisy signal $\vartheta_u(k)$ has wildly-varying spectra..

10.2 The Closed-Loop Signal, Disturbance Models

The Box-Jenkins model $(\boldsymbol{A}, \boldsymbol{B}, \boldsymbol{C})$ relating the corrupted output $\boldsymbol{y}(k)$ and the input, termed *system model*, the augmented model formed of internal model based closed loop control system $(\boldsymbol{A}_{cl}, \boldsymbol{B}_{cl}, \boldsymbol{C}_{cl})$ the combined stochastic disturbance model and the deterministic disturbance model. The closed-loop signal model $(\boldsymbol{A}_{cl}, \boldsymbol{B}_{cl}, \boldsymbol{C}_{cl})$ is:

$$\boldsymbol{A}_{cl} = \begin{bmatrix} -0.5248 & -2.035 & 1.0531 & -8906 \\ 0.2376 & -0.0153 & 0.5265 & -0.4453 \\ 0 & 0 & 1.9319 & -1 \\ 0 & 0 & 1 & 0 \end{bmatrix}, \boldsymbol{B}_{cl} = \begin{bmatrix} 1 \\ 0.5 \\ 1 \\ 0 \end{bmatrix}, \tag{40}$$

$$\boldsymbol{C}_{cl} = \begin{bmatrix} 0 \; 1 \; 0 \; 0 \end{bmatrix}$$

The stochastic disturbance model $(\boldsymbol{A}_{dr}, \boldsymbol{B}_{dr}, \boldsymbol{C}_{dr})$ is given by:

$$\boldsymbol{A}_{d\xi} = \begin{bmatrix} 0.3960 & -0.8025 \\ 1 & 0 \end{bmatrix}, \boldsymbol{B}_{d\xi} = \begin{bmatrix} 1 \\ 0 \end{bmatrix}, \boldsymbol{C}_{d\xi} = [1.460 \; 0] \tag{41}$$

The deterministic disturbance model $\left(\boldsymbol{A}_{dd\xi}, \boldsymbol{B}_{dd\xi}, \boldsymbol{C}_{dd\xi}\right)$ is given by:

$$A_{dd\xi} = 1, \; B_{dd\xi} = 1, \; C_{dd\xi} = 1. \tag{42}$$

The desired translation trajectory in $(x.y, z) \in R^3$ is:

$$\begin{bmatrix} x_d(k) \\ y_d(k) \\ z_d(k) \end{bmatrix} = \begin{bmatrix} \cos(\pi k/12) \\ \sin(\pi k/12) \\ 1 \end{bmatrix} \tag{43}$$

10.3 The Extraction of the Signal Using a KF

The proposed KF and the internal model-based combined state-feedback and feed-forward shown in Fig. 3, is evaluated using the BJ model. Using the inverse kinematics, the desired control input $(\phi_d, \theta_d, \psi_d)$ is determined.

Remarks The internal model includes only the reference signal as the deterministic disturbance is unknown, and the internal model cannot handle the stochastic waveforms. Hence the combined effect of the unknown deterministic disturbance, stochastic disturbance, and measurement noise, continues to be present in the output, which is exhibited respectively as an offset and random fluctuations in the outputs, as shown in subfigures A, B and C of Fig. 3a and b. The subfigures D, E, and F show that the KF estimates the true signals from the corrupted output, which are accurately estimated. It can be deduced that the objective of the quadrotor is to accurately track the desired circular trajectory in the (x, y) plane and remain stationary along the z-axis at a desired fixed position despite the disturbance and measurement noise corrupting the measurement output. The importance of estimating the signal accurately despite disturbances such as wind gusts that may accelerate or decelerate, push, or pull the quadrotor away from the desired trajectory is emphasized.

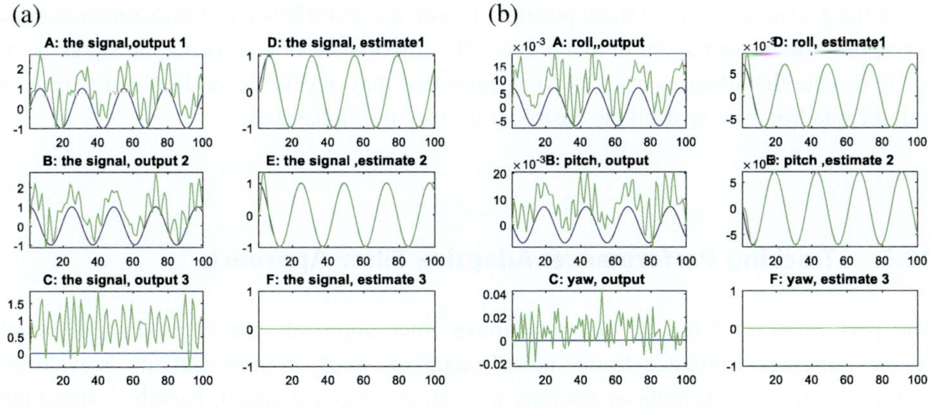

Fig. 3 **a** Positions: outputs, and estimates. **b** Euler angles: outputs, and estimates

10.4 Tracking Performance: KF-Aided Tracking of Signals and Their Corrupted Versions

Figure 3 a and b, below, show the signals, which are the desired outputs (x_d, y_d, z_d), and the desired Euler angles $(\phi_d, \theta_d, \psi_d)$, and the corresponding corrupted output measurement. In Fig. 3 a, where all x, and y, -axes are in cm, displays Subfigures A, B and C which show the desired positions (x_d, y_d, z_d), and Subfigures D, E and F which show their KF estimates. Similarly, Fig. 3 b, where all x-axes are in cm and all y-axes in radians, shows, in Subfigures A, B and C, the desired outputs (Euler angles) $(\phi_d, \theta_d, \psi_d)$, and, in Subfigures D, E and F, their KF estimates. As is clearly shown, the KF estimates of the signals are accurate and are devoid of any random fluctuations.

Remarks The internal model includes only the reference signal as the deterministic disturbance is unknown, and the internal model cannot handle the stochastic waveforms. Hence the combined effect of the unknown deterministic disturbance, stochastic disturbance, and measurement noise, continues to be present in the output as explained in the Introduction, which is exhibited respectively as an offset and random fluctuations in the outputs, as shown in subfigures A, B and C of Fig. 3a and b. Subfigures D, E, and F show that the KF estimates the true signals from the corrupted output, which are accurately estimated**.** It can be deduced that the objective of the quadrotor is to track accurately the desired trajectory in three dimension,(x, y, z): circular trajectory in the (x, y) plane while remaining stationary along the z-axis at a desired fixed position despite the disturbance and measurement noise corrupting the measurement output is met. The importance of estimating the signal accurately despite disturbances such as wind gusts that may accelerate or decelerate, push, or pull the quadrotor away from the desired trajectory is emphasized.

10.5 Tracking Performance: Adaptive Filter Approach

The performance of the KF-based adaptive filter approach, shown in Fig. 4, which ensures robust performance despite model perturbations given in its sections, is evaluated with an illustrative example of tracking a common practical signal, namely a sinusoidal input,$r(k) = \sin(\omega k)$ as shown below:

Figure 4, where the x-axes are in radians and the y-axes in cm, shows the performance of the state-feedback controller using the online KF-based adaptive filter under model perturbations. Subfigures A, B, and C show the nominal unperturbed signal, the KF residual, which is negligibly small, and the KF estimate, respectively. Subfigures D, E and F show their perturbed counterparts, namely the perturbed signal, the resulting KF residual, which is non-negligible as expected, and the accurately recovered KF estimate, namely output of the adaptive filter.

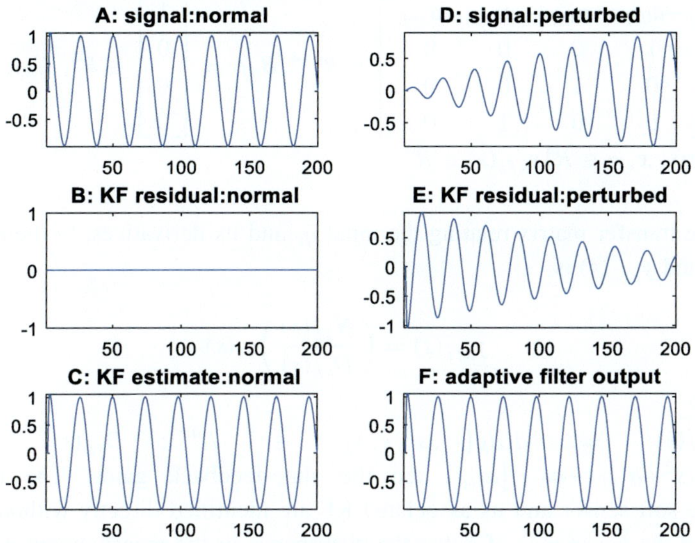

Fig. 4 Normal and perturbed signals, and adaptive filter output

Remarks

It is important to use the Kalman filter and the internal model-based controller so that the quadrotor tracks the desired trajectory accurately without random deviations about the true trajectory and prevent deterministic disturbances pushing it away.

11 KF-Based Soft Sensor Design

The KF-based, derivative-free soft sensor replaces the conventional hardware sensor. The estimates of the generalized coordinate q_i, its first derivative $\dot{q}_i$, second derivative $\ddot{q}_i$, and third derivative $\dddot{q}_i$ are all estimated from the output y_i corrupted by the disturbance and measurement noise. The soft sensor is used to estimate the first, second, and the third derivatives of the translational variable ξ_i and of the rotational variable η_i. As such, the soft sensor then replaces the hardware one in estimating the required derivatives, in an effective and differentiation-free manner. The state- space model $\left(\boldsymbol{A}_{sfi}, \boldsymbol{B}_{sfi}, \boldsymbol{C}_{sfi}\right)$ is given by:

$$\begin{cases} \dot{\boldsymbol{x}}_{sfi}(t) = \boldsymbol{A}_{sfi}\boldsymbol{x}_{sfi}(t) + \boldsymbol{B}_{sfi}q_i(t) \\ \boldsymbol{y}_{sfi}(t) = \boldsymbol{C}_{sfi}\boldsymbol{x}_{sfi}(t) \end{cases} \tag{44}$$

$$A_{sfi} = \begin{bmatrix} -a_{0sf} & -a_{1sf} & -a_{2sf} & -a_{3sf} \\ 1 & 0 & 0 & 0 \\ 0 & 1 & 0 & 0 \\ 0 & 0 & 1 & 0 \end{bmatrix} \in R^{4x4}, B_{sfi} = \begin{bmatrix} 1 \\ 0 \\ 0 \\ 0 \end{bmatrix} \in R^4,\ C_{sfi} = I \in R^{4x4}$$

Identity matrix; $x_{sfi} \in R^4 y_{sfi}(s) \in R^4$.

Lemma The transfer matrix relating the input q_i and its derivatives, to the output of the sensor, denoted $y_{sfi}(s)$ is:

$$y_{sfi}(s) = \left(\frac{N_{sf}(s)}{D_{sf}(s)}\right) q_i(s) \tag{45}$$

where $N_{sf}(s) = diag(s^3, s^2, s, 1) \in R^{4x4}$; $D_{sf}(s) = 1 + a_{3sf}s^3 + a_{2sf}s^2 + a_{1sf}s + a_{0sf}$; $\{a_{isf}\}$ are the state-feedback gains. The discrete-time model of the soft sensor and its associated KF are identified directly without an a-priori knowledge of the covariance of either the disturbance or the measurement noise, thereby avoiding the need to specify these two covariances in order to solve the Riccati equations to find the Kalman gain [17–19].

11.1 KF-Based Soft Sensor

The identified KF-based soft sensor is given by

$$\begin{cases} \hat{x}_{kfsi}(k+1) = A_{kfsi}\hat{x}_{kfsi}(k) + B_{sfi}q_i(k) + K_{ksfi}y_{sfi}(k) \\ \hat{y}_{kfsi}(k) = C_{sfi}\hat{x}_{ksfi}(k), \end{cases} \tag{46}$$

where $A_{kfsi} = A_{sfi} - K_{ksfi}C_{sfi}$.

12 Fault Tolerant System: An Illustrative Example

The KF forms the backbone for developing the fault-tolerant system. In so doing, it monitors the system performance and accurately diagnoses faults (if any) thanks to the use of the powerful emulators. To illustrate the fault-tolerance capability of the proposed approach, an illustrative example is treated next.

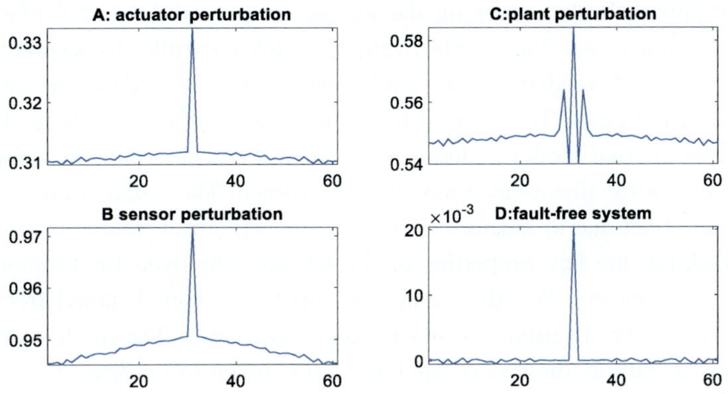

Fig. 5 Residual autocorrelation functions of different perturbed sub-systems

12.1 Illustrative Example

The residual of the KF monitors the condition of the system, detects the incipient fault and helps isolate faulty subsystems, namely the actuator, the plant, and the sensors. The non-whiteness of the KF is an indicator of an incipient fault while a zero-mean white noise indicates that the system is healthy.

Figure 5, where all y-axes are in cm^3 and all x-axes in cm, displays Subfigures A, B, and C which show, respectively, the autocorrelations, of the residual when the actuator, the sensor, and the plant are respectively perturbed. The first 3 autocorrelations (sub-figures A, B and C) are not zero-mean white noise processes as there are fault-indicator terms in each of these 3 cases. In contrast, Subfigure D shows that the autocorrelation is a Dirac function (indicative of the presence of a zero-mean white noise process) which implies that there are no perturbations, and the system is structurally healthy.

13 Conclusion

In this paper, a novel approach of accurately tracking a desired known trajectory with the help of a quadrotor guided by an autonomous KF has been proposed using a fusion of the Kalman filter and machine learning. The success of this approach was shown to rely on the new KF residue–based theory, complemented by the powerful LVP modeling technique, robust emulator-based identification and estimation schemes, and a Bayes' classifier fusion of the results obtained from the KFs and a machine learning/predictive analysis of the corrupted signal that portrays the trajectory to be tracked. It was also shown that our approach achieves its objective without any recourse to the use of the statistics of the unknown stochastic disturbances and measurement noise, as is conventionally done. It is noteworthy to point out that, besides using the powerful KF in an unconventional way, our proposed approach also relies on the use of the powerful internal model principle to efficiently design a combined feedforward-feedback controller to accurately guide the quadrotor along the desired trajectory while combatting the undesirable and disturbing effects such as wind gusts. The internal model is shown to successfully track the desired reference as the internal model includes the modes of the deterministic reference model only, since those of the disturbance model are unknown. The tracking error is noisy as it is subject to an offset due to a deterministic disturbance. To overcome the offset-induced inaccurate tracking, the key properties of the KF are employed for estimating first the deterministic disturbance, then the signal and finally the signal model from the output corrupted by unknown disturbances and measurement noise. The models of the system, and the signal as well as their associated KFs, are shown to be directly identified using only the residuals of their associated KFs. As a by-product, a novel KF-based soft sensor has also been developed to replace the error- and maintenance-prone hardware sensor to estimate the required and important position and rotations angles and their derivatives, in a differentiation-free manner. Though our new scheme has been designed and demonstrated here using a known desired trajectory, its power and application can be extended to unknown desired trajectories. Moreover, as our approach is also scalable, it can also be extended to a swarm of quadrotors, thus leading to a networked KF-based trajectory tracking system capable of dealing simultaneously with multiple trajectories. Work is currently underway to achieve both of these extended goals.

Acknowledgements The first author acknowledges the support of the Fulbright Scholar Programs, Canada; the invitation of Florida Polytechnic University, Lakeland, Florida, and Mr. Mustapha Achoubane, the Associate Director and the graduate students; the University of New Brunswick, Fredericton, New Brunswick, Canada.

References

1. R. Mahony, V. Kumar, P. Corke, Modeling, estimation, and control of quadrotor. IEEE Robot. Autom. Mag. (2012). (Sept 2012)
2. M. Fanni, D. Rashad, A. Khalifa, A new 6-DOF quadrotor manipulation system: design, kinematics, and control. IEEE/ASME Trans. Mechatron. **22**(3), 1315–1326 (2017)
3. N. Wang, S. Lev, M. Joo Er, W.-H. Chen, Fast and accurate trajectory tracking control of an autonomous surface vehicles with modelled dynamics and disturbances. IEEE Trans. Intell. Veh. **1**(3), 230–243 (2016)
4. S.D. Fabiyi, Review of unsupervised artificial neural networks with applications. Int. J. Comput. Appl. (0975–8887) **181**(40), (2019). (February 2019)
5. S. Viktor et al., Simulation of wind effect on a drone flight. ARPN J. Eng. Appl. Sci. **10**(4), (2015). March, ISSN 1819-6608
6. G. Heredia, A. Ollero, Sensor faults in small helicopter UAVs using observer/Kalman filter identification. Hindawi Publishing Corporation, Article ID 174618, 20 (2011). https://doi.org/10.1155/2011/174618

7. T. Soderstrom, System identification for the errors in-variables problem. Trans. Inst. Meas. Control. **34**(7), 780–792 (2006)
8. D. Abbott, *Applied Predictive Analytics: Principles and Techniques for the Professional Data Analyst* (Wiley, 2014)
9. Guest Editorial, *Emerging trends in LPV-Based Control of Intelligent Automotive Systems*, vol. 14, Issue 18. Institute of Engineering and Technology (IET) Control Theory & Applications (2020). (17 Dec 2020)
10. L. Li, X. Ren, Identification of nonlinear Wiener-Hammerstein systems by a novel adaptive algorithm based on cost function framework. ISA Trans. **80**, 146–159 (2018)
11. F. Lewis, D.C. Dawson, C.T. Abdallah, *Robot Manipulator Control: Theory and Practice* (Marcel Dekker Inc., New York, 2004)
12. G.C. Goodwin, S.F. Graeb, M.E. Salgado, *Control System Design* (Prentice Hall, New Jersey, USA, 2001)
13. B. Francis, M. Wonham, The internal model principle of control theory. Automatica **12**, 457–465 (1976)
14. B. Whitehead, S. Bieniawski, Model reference adaptive control of a quadrotor UAV, in *AIAA Guidance, Navigation, and Control Conference.* Published Online: 26 Jun 2012 (2012). https://doi.org/10.2514/6.2010-8148
15. Z. Guo, J. Guo, J. Zhou, J. Chang, Robust tracking for hypersonic re-entry vehicles via disturbance estimation-triggered control. IEEE Trans. Aerosp. Electron. **56**(2), 1279–1289 (2020) (and others from IEEE AIAA, IEEE Robot. Autom.)
16. L. Cheded, R. Doraiswami, A novel framework for fault diagnosis with application to process safety. Process Saf. Environ. Prot. **154**, 168–171 (2021). www.elsevier.com/locate/psep. (Journal Homepage)
17. R. Doraiswami, L. Cheded, Fault tolerance in nonlinear systems: a model-based approach with a robust soft sensor design, in *The Institution of Engineering and Technology (IET) Control Theory & Applications* (John Wiley & Sons Ltd on behalf of The Institution of Engineering and Technology, 2021), pp. 1–13
18. R. Doraiswami, L. Cheded, Robust fault-tolerant control using an accurate emulator-based identification technique. Int. J. Control 1336–5820 (2017). ISSN: 0020-7179. Journal homepage: http://www.tandfonline.com/loi/tcon20, https://doi.org/10.1080/00207179.2017.1318452
19. R. Doraiswami, L.Cheded, M. Brinkmann, Kalman-filter-based accurate trajectory tracking and fault-tolerant control of quadrotor, in *Proceedings of the 8th International Conference of Control Systems, and Robotics (CDSR'21) Niagara Falls, Canada Virtual Conference–May 23–25, 2021, Paper No. 302* (2021). https://doi.org/10.11159/cdsr21 302
20. O. Imran, S. Pradeepan, P. Sevigny, P. Caniglia, S. Rajani, B. Balaji, R. Doraiswami, Deep learning on trajectory images, in *SPIE Conference in Defense, and Commercial Sensing* (Orlando, 2022)
21. G.J. Mallory, R. Doraiswami, A Filter for on-line estimation of spectral content. IEEE Trans. Instrum. Meas. **48**(6), 1047–1055 (1999)

Rajamani Doraiswami received his B.E from Victoria Jubilee Technical Institute, Bombay, India, M.E. from the Indian Institute of Science, Bangalore, India, and Ph. D from Johns Hopkins University, Baltimore, USA all in Electrical Engineering, in 1962, 1965, and 1971 respectively. He was a professor in Brazil at the University of Santa Catharina and the University of Rio de Janeiro from 1972–1981. Since 1981, he is a professor at the University of New Brunswick, Canada, and in 2006 was awarded the title of professor emeritus.

Visiting Positions: He has held visiting positions at the ADERSA/GERBIOS, France, the Xian Jiaotong University, Xian, China, the King Fahd University of Petroleum and Minerals (KFUPM), Dhahran, Saudi Arabia, the Indian Space Research Organization, Trivandrum, India, the Delft Institute of Technology, Delfts, the Netherlands, the University of Western Australia, Perth, Australia, and the University of Applied Sciences, Furtwangen, Germany. He was a United Nations expert at the Bharat Earth Movers Limited, India.

Consultancy: He has done consultancy and collaborative research work with Brazilian utilities, Brazil, Lockheed Martin Corporation, USA, Tetra-Laval Private Limited, Singapore, Siemens Corporate Research, USA, Lizotte Consultants, Biotech Diffusion and the Institute of Clinical Research of Montreal, Biopeak Corporation, Ottawa, S5 systems, Ottawa.

Areas of interest: Areas of interest include control systems, power systems, signal processing, communication, detection and estimation, system identification, pattern classification, and filtering with applications including performance monitoring, fault diagnosis, and pattern classification of industrial processes, power utilities, biomedical systems, and devices, wind power generation, etc. Machine Learning, Deep Learning, and Predictive Analytics.

Professional Affiliation: He is a Senior Member of the IEEE, a fellow of the Indian Institute of Engineers, and a Professional Engineer in the province of New Brunswick, Canada.

Lahouari Cheded has a BS (Honors) in Physics (Univ. of Oran, DZ), an MS. in Electronic Control (Univ. of Salford, UK), and a Ph.D. in Signal processing from the University of Manchester (ex-UMIST), UK. While at UMIST, he taught Physics (1980–1981) and was a research assistant in the now nonextant Department of Instrumentation and Analytical Science (DIAS) during the period 1981–1984. He worked as a Professor at the Systems Engineering department (Control and Instrumentation Systems Engineering (CISE) section) of the King Fahd University of Petroleum and Minerals in the Kingdom of Saudi Arabia (KSA) during the period 1984–2017, at the end of which he retired to become an independent researcher, mentor and consultant in the UK.

He remains active in research with diverse research interests including digital signal processing (theory and applications) and its many interactions with Control, detection and Estimation, System Identification, fault diagnosis, leak detection in pipelines, communication, stochastic filtering, and soft computing. He has published more than 110 conference and journal papers, 5 book chapters, and one book (2010) in these areas, has 5 granted US patents, consulted on a number of research projects, and was an investigator in a number of funded research projects.

Besides teaching, course development, and research, he was also involved in undergraduate and graduate (MS and Ph.D.) students' supervision, paper reviewing for both international conferences and journals (including IEEE Tr. Sig. Proc.), and review of grant proposals for grant-awarding agencies. He was a key member of the departmental ABET team in two successive and successful accreditation periods and was the chairman of the CISE program development committee. He has also organized and taught numerous short courses in Control and Signal Processing to local and national companies in KSA and was involved in the organization of the international conference ISSPA in 2001 and other local and national workshops and symposia. In recognition of his efforts in student advising, he was elected the best advisor of his college in 1994. He is also currently involved in mentoring former junior colleagues. He was a past member, for numerous years, of the European Signal Processing Society (EURASIP) and the Institution of Engineering and Technology IET (UK), served as a member on the editorial board of the International Journal of Electrical Eng. (1994–2005), published by SAGE. He is currently a Life Senior Member of the IEEE and serves on the editorial board of the International Journal on Automation and Control Research (JACR), Open Access. ISSN:2368 6677, as well as on the organizational board of the yearly international conference on Control, Dynamic Systems and Robotics (CDSR), both during the period 2016-Present.

Marius Brinkmann is a Fulbright alumnus from Germany. He received his B.Eng. degree in Electrical Engineering from the University of Applied Sciences Meschede, Meschede, Germany. From 2018 to 2019, he studied in the USA as a Fulbright scholar and received his M.Sc. degree in Electrical Engineering at the Florida Polytechnic University, Lakeland, FL, USA. At Florida Polytechnic University, he worked as a research assistant on automotive radar signal processing algorithms.

He was a recipient of the Deutschland stipendium funded by the Federal Government of Germany, businesses, foundations, and private individuals during his bachelor's program from 2015 to 2017, the Dr. Kirchhoff award in 2017, the Fulbright Scholarship from 2018 to 2019, and the Rohde & Schwarz Innovation Award in 2021.

He received his M.Sc. degree in Electrical Engineering and Information Technology from the Technical University of Munich, Munich, Germany, in 2021, with a focus on microwave engineering and high-frequency technology. He is currently working as a development engineer at Rohde & Schwarz GmbH & Co. KG, researching microwave imaging algorithms.

Applications

Heterogeneous Drone Fleet for Radiological Inspection

Alberto Vale, Rodrigo Ventura, José Corisco, Norberto Catarino, Nuno Veiga, and Susana Sargento

Abstract

The defense against Chemical, Biological, Radiological and Nuclear (CBRN) threats is an increasing demand due to wars, terrorist attacks, disasters or simply caused by negligence (e.g., decommissioned mines, medical waste). Drones play an important role in the increased protection of critical infrastructures and public spaces in the face of imminent threats of CBRN attacks. In particular, radiological and nuclear threats, which are diffi-

A. Vale (✉)
Instituto de Plasmas e Fusão Nuclear, Instituto Superior Técnico, Av. Rovisco Pais 1, 1049-001 Lisboa, Portugal
e-mail: avale@ipfn.tecnico.ulisboa.pt

R. Ventura
Instituto de Sistemas e Robótica, Instituto Superior Técnico, Av. Rovisco Pais 1, 1049-001 Lisboa, Portugal
e-mail: rodrigo.ventura@isr.tecnico.ulisboa.pt

J. Corisco · N. Catarino
Campus Tecnológico e Nuclear, Instituto Superior Técnico, Estrada Nacional 10, 2695-066 Bobadela, Portugal
e-mail: corisco@ctn.tecnico.ulisboa.pt

N. Catarino
e-mail: norberto.catarino@ctn.tecnico.ulisboa.pt

N. Veiga
Instituto Superior Técnico, Av. Rovisco Pais 1, 1049-001 Lisboa, Portugal
e-mail: nuno.veiga@tecnico.ulisboa.pt

S. Sargento
Universidade de Aveiro, Campus Universitário de Santiago, 3810-193 Aveiro, Portugal
e-mail: susana@ua.pt

M. Abdelkader and A. Koubaa (eds.), *Unmanned Aerial Vehicles Applications: Challenges and Trends*, Synthesis Lectures on Intelligent Technologies,
https://doi.org/10.1007/978-3-031-32037-8_4

cult to intercept and release high levels of ionizing radiation. In this work, a set of drone profiles are presented, being equipped with Light Detection And Ranging (LIDAR) to perform the cartography of the scenario and support real-time navigation, when combined with Global Navigation Satellite System (GNSS) and inertial measurement units, red, green and blue (RGB) and RGB plus and depth camera (RGBD) cameras for image collection and support autonomous landing, and the radiological sensors Geiger-Muller counter (GMC) and Gamma Ray Spectrometer (GRS). These drone profiles are being used in operations divided into three classes: (1) scouting for mapping the scenario with LIDAR and GNSS returning a 3D georeferenced map, (2) monitoring using the GMC and GNSS to provide a radiological intensity map and the estimation of the potential hotspots, and (3) inspection where the GRS is used to characterize the potential radionuclides present in the scenario. Successful experiments were conducted in old mines of uranium ore. In the future, the heterogeneous drone fleet architecture can be extended to integrate chemical and biological sensors.

1 Motivation

1.1 CBRN Threats

The defense against CBRN threats is an increasing demand due to wars, terrorist attacks, disasters, or simply due to negligence. Recent projects have achieved improvements, but we are still far from safeguarding humankind against Chemical, Biological, Radiological and Nuclear (CBRN) threats.

Chemical threats involve toxic substances that affect humans and other living organisms through inhalation, contact or ingestion. Toxic substances are easily transportable, which contaminate the soil and groundwater. The use of lethal gases such as chlorine and sarin gas against civilian areas in the ongoing Syrian war [1], or the poison gas used in Tokyo subway, where six people died and hundreds are hurt [2], are examples of chemical threats in a recent past.

Biological threats occur through live microorganisms, such as bacteria or viruses, which easily install themselves in the food chain and in the air we breathe. They can spread, cause highly contagious diseases and cause epidemics. They can be used as weapons or simply result from poor maintenance of potential infrastructure for the accommodation of microorganisms. An example of this was the outbreak of the pathogenic bacterium *Legionella* [3], which occurred in Portugal in 2014 within a radius of a few kilometers north of the Tagus River. This outbreak, which caused legionnaires' disease in about 375 people, of whom twelve died, was related to the cooling towers of a nearby company [4].

Radiological and nuclear threats are associated with radioactive elements that release large amounts of energy through alpha, beta particles, X-rays, gamma rays and neutrons

[5]. Radioactive elements, when properly used, can be useful, for example, in radiotherapy in the treatment of cancerous tumors, or in agriculture with radiation to combat infestation by microorganisms. However, when exposed to high amounts and/or for long periods of time, they can cause damage to the nervous system, gastrointestinal system, bone marrow, even causing death. They can have harmful effects not only for humans with biological consequences on organisms, but also for the environment. The best-known examples are perhaps the atomic bombings of the cities of Hiroshima and Nagasaki during World War II, which killed between 150,000 and 250,000 people in the first few months [6], and the nuclear incidents which occurred in Chernobyl in 1986 and in Fukushima in 2011, with about 8000 km and 25 years of distance, whose causes are mentioned as construction errors and negligence [7]. Radiological and nuclear threats are commonly stereotyped as associated only with nuclear bombs and nuclear fission plants. However, in addition to terrorist threats, radiological and nuclear risks can be very close, as in nuclear medicine waste, deactivated mining operations or even in radioactive waste storage sites.

CBRN threats can be caused intentionally by humans, such as in weapons and terrorism, directly by negligence or technical failure as in the case of Chernobyl and the *Legionella* outbreak, or by indirect effects such as in the case of Fukushima, caused by a design flaw, underestimated the risk of high waves caused by a strong tsunami. Faced with these threats, it is necessary to act before, during and after in defense of the well-being of humanity and the planet. It is necessary to act before, in order to prevent threats from happening, during, namely in control, circumscription and in possible search and rescue operations, and after, in order to prevent them from being repeated.

1.2 Radiological and Nuclear Threats

The nuclear incidents which occurred in Chernobyl and in Fukushima several years ago, are still causes of concern. The radiation doses have increased and are carefully tracked, namely across Europe [22], and in Japan [23]. Besides these ageless and newsworthy incidents, the storage of nuclear waste and mining exploration are also silent threats, occasionally addressed by the media. The radiation doses shall be mapped and tracked along time, and, in case of leakage detection, an urgent intervention is required. There are plenty of storage sites for low to high level nuclear waste from the nuclear industry, requiring continuous management in order to prevent environmental leakage, or severe accidents in case of high level waste melting [20]. Uranium ore exploration and its legacy of old disabled mines and surface deposits of debris (uranium tailings), also require proper management and monitoring for the protection of the environment and human populations [21]. Between 1908 and 2001, in Portugal, several deposits of radioactive ore were extracted for Radium and Uranium production. These mining sites, which are predominantly in northern interior Portugal, exhibit higher levels of background radioactivity when compared with other regions. All mining and milling activities ceased in 2001 and the assets were transferred to a Portuguese public

Fig. 1 Orphan sources of radiation: photos by International Atomic Energy Agency (IAEA)

enterprise—*Empresa de Desenvolvimento Mineiro, S.A.*—that care of the management and remediation of these sites.

Other incidents related to safety and security issues associated to radioactive materials involve theft or loss of radioactive sources, unauthorized disposal, shipment of contaminated scrap metal, and the discovery of radioactive sources, commonly known as "orphan sources", out of regulatory control. Some examples are illustrated in Fig. 1.

The last, but definitely not the least, is the threat of usage of nuclear weapons, which is a subject much discussed under the Ukraine war, a conflict that is ongoing at the date of this chapter. The consequences can be catastrophic in a large extension of territory and during a long period of time. We all already have the legacy of the nuclear bombs used in the Second World War.

In summary, the radiological and nuclear threats can be divided in

- Radiological dispersal devices ("dirty bombs"),
- Nuclear power plant incidents,
- Nuclear waste storage,
- Mining sites, and
- Orphan sources.

According to the International Atomic Energy Agency (IAEA) [24], between 1993 and 2019, a total of 3686 incidents, of which 290 involved a confirmed or likely act of trafficking or malicious use, are reported in the illegal traffic database.

Motivated by the reasons presented above, this chapter is particularly focused on radiological and nuclear threats, which arise due to the emerging applications of nuclear technologies in industry, power production, medicine, and the military. North Atlantic Treaty Organization (NATO) has a Task Force working towards this objective and the National Academy of Engineering presented in 2008 "The 14 Grand Challenges for Engineering", where the prevention of nuclear terrorism was identified as one of the great challenges for the 21st century [8, 9].

1.3 Unmanned Aerial Vehicles

The Unmanned Aerial Vehicle (UAV) is an aircraft without a human pilot aboard. The UAV were originated mostly in military applications, but it is being widely spread for the purpose of personal enjoyment. The most common types of UAV are blimp like systems, fixed-wing airplanes, and rotary-wing aircraft. The last one is well known as multi-copters (or multi-rotor), such as the helicopters, quadcopters, hexacopters or octacopters, according the number of rotors. The multi-copters with four or more rotors are also known as "drones". In this chapter only multi-copters are considered for UAV and for the sake of simplicity, hereinafter, they are referred as drones.

The vertical takeoff and landing, the reasonable time duration of the flight, the possibility of customization, its payload, and the value for money, are just some of the contributions for the boom of applications and development with drones during the last decade.

Drones have assumed a prominent position in several military missions, such as intelligence, surveillance, reconnaissance and namely for strikes with onboard weapons or bombs. On the other side, medical response drones are capable of transporting medical supplies, such as blood, insulin, syringes, and wound dressings for saving lives. The search-and-rescue variants are a high-endurance drones capable of providing optical and infrared imaging to map certain areas and locate survivors [41]. Drones are, therefore, compelling tools for the present time and upcoming years and we should all push them for the best of human kind. In particular to increase the protection of critical infrastructures and public spaces in the face of imminent threats of CBRN and Explosive Materials (CBRNe) attacks.

Given the time duration of the flight, as well as the loitering capability, drones are potential tools for radiological inspection operations. Drones are able to transport lightweight GMC, and fly at low adjustable speeds and close to the potential target sources of radiation. Recently, the authors developed and tested with success a prototype of a drone with ducted propellers to perform radiological inspections, i.e., near to the radioactive sources resilient to collisions and with the best flight time [19]. In addition, drones can land automatically close to the detected hotspots of radiation activity, perform spectra gamma analysis using onboard GRS during long periods of time (seconds to minutes) and collect samples if necessary. According to the endurance of each drone, different hotposts can be visited and analyzed in a single mission. A fully equipped drone can perform all the required operations, but becomes complex, costly, risky and probably compromises the performance of the entire operation. Therefore, an heterogeneous drone fleet is an interesting option for radiological inspection of scenarios.

2 Background and Related Work

There are a number of recent projects which aim to address the issues of radiological and nuclear threats. MANUELA [25] is a project of the AREVA company, currently ORANO, a French multinational group in nuclear and renewable energy; the mobile robot is equipped with LIDAR systems and it performs a 3D reconstruction of internal scenarios with the indication of the radiation levels of hotspots. RIANA [27], also from AREVA, is a project that has developed wheeled mobile robots with the capacity to operate in nuclear scenarios to provide radiological readings and take physical samples from hazardous scenarios.

Project REWARD [26] extends the radiological survey to large outdoor environments using radiological sensors installed on cars equipped with Global Positioning System (GPS) receivers. The sensors used are multiple miniaturized solid state radiological sensors: Cadmium Zinc Telluride (CZT) for gamma radiation detection and a high efficiency neutron detector based on new semiconductor technologies. The ENCIRCLE project [29], funded by the European Union, aims to strengthen the European CBRNe industry and its capabilities.

The use of drones enables further expansion of the total area coverage for a given scenario. Systems comprised of UAV mounted with radiological sensors have already been tested as a solution to cover broader areas [30, 31], and commercial solutions already exist, such as [32, 33]. GAMMAex [28], is a project which uses drones to perform monitoring and recognition of chemical and radiological patterns in scenarios where biological, chemical, and radiological threats are identified. The application of drones has been extended to scenarios where the danger of explosive atmospheres exists, through the construction of an Explosive atmospheres (ATEX) system. This project was mainly focused on the sensory part, being open to the application of new AI techniques. The authors of this chapter performed preliminary tests by applying drone and Simultaneous Localization and Mapping (SLAM) techniques, commonly used in robotics, resulting in a 3D map of a scenario with radiological information [55].

Radiation detection systems can be divided into three classes: gas-filled detectors, scintillation detectors, and solid-state detectors [34]. Gas-filled detectors, like the GMC, are used to obtain the radiation dose rate. Scintillation detectors and solid-state detectors are used for radiation dose rate measurements and for gamma spectrometry, due to their good sensitivity and energy resolution.

Scintillation detectors may be gaseous, liquid, or solid, organic (plastics, liquids) or inorganic. Their operating principle is based on the interaction of the incident radiation with the scintillator material and the conversion of the resulting light into electrical signals by a photodetector. They are normally used for spectroscopy purposes (even for low energy radiation) [13]. Neutron detection is normally used as a complement to the gamma-ray detection and is essential for security and defense applications. Since neutron detection is characterized by a lower natural background, an intrinsic difficulty to shield neutrons, and much fewer neutron sources are commercially available, a neutron alarm can immediately trigger a response.

Most of the radiation detection systems are compatible with mobile robots, in particular with drones. The performance of the radiation detection systems are related to the following parameters: (i) exposed area of the sensor, (ii) acquisition time interval and (iii) proximity to the sources of radiation. The number of counts increase when at least one of the previous parameters increases. The gas-filled detectors, such as the Geiger-Muller counter (GMC), can acquire data in real time during a walk when installed inside a backpack, or during a journey of a ground mobile robot. When installed on a drone, the GMC is valuable when flying close to the sources of radiation (from the safety margin required for drone operation up to 3 m) and at low speeds (less than 1 m/s). Instead, for spectroscopy purposes, the acquisition time should take from several seconds to some minutes, which requires landing approaches to be usable with drone.

In 2017, the authors of this chapter developed the Mobile Application for Radiation Intensity Assessment (MARIA) [36], a solution to acquire and record georeferenced radiological intensity data based on data acquired by GNSS receivers and affordable gas-filled radiological sensors. The proposed solution was specially designed for mobile devices, such as common smartphones or tablets equipped with GPS receiver and a connection to support a GMC sensor. MARIA can be directly used by a person while walking in the scenario, or as a mobile device installed on a drone flying close to the potential sources of radiation. MARIA can also be used by anyone without expertise or experience in radiological and nuclear subjects. MARIA was developed out of a need to acquire and record georeferenced radiological intensity data. Figure 2 shows the application in sensor mode while performing an inspection operation along an old mine of uranium ore.

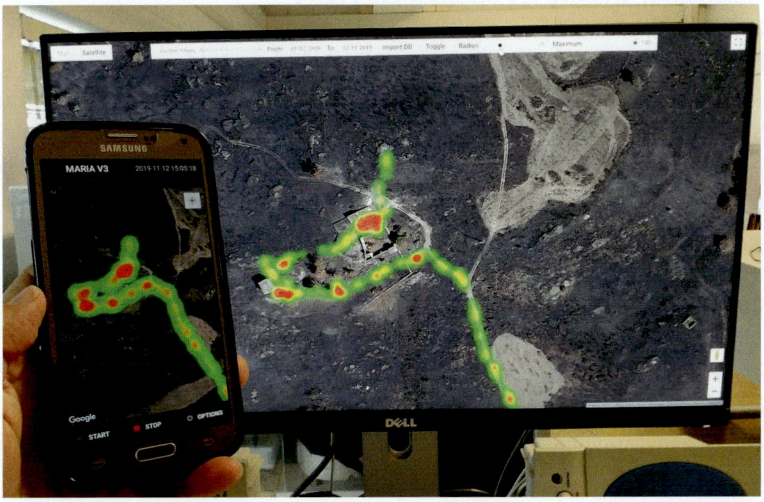

Fig. 2 Mobile Application for Radiation Intensity Assessment (MARIA) running on a mobile smartphone, with radiological map overlaid on Google Maps^{TM} being also presented on a personal computer

Fig. 3 Commercial drone (DJI Phantom 3) equipped with MARIA installed on a smartphone connected to a gas-filled GMC (left image), and the resulted radiological heatmap, where two hotspots can be identified (right image)

An algorithm is then applied to the combined location and intensity data to create a radiological heatmap of the scenario, displayed as an overlay on Google Maps^{TM}. After the first promising results with MARIA used by a person walking on the scenario, tests with a smartphone and GMC installed on a commercial drone were performed in a controlled scenario, where two sources of radiation were randomly placed. The setup, scenario and resulted radiological heatmap are illustrated in Fig. 3.

The GNSS readings combined with an Inertial measurement unit (IMU) in an Extended Kalman Filtering (EKF) algorithm may not be satisfactory for localization, navigation and mainly for map representation. Then, an upgrade with an affordable and monochrome depth camera (Structure by Occipital) was done in drone and tested in the lab, where one source of radiation was temporally placed. The 3D representation of the scenario was achieved, where the location of the source, identified as "hotspot", was estimated. The results are illustrated in Fig. 4. At that time, the tests were not performed in outdoor scenario given the technical limitation of the depth camera used.

The combination of different affordable sensors, including radiological ones, installed on commercial drone showed the value to collect a vast set of data in a short period of time, estimate the radiological heatmap and the identification of possible radiation sources. In addition, drone can perform automated operations in hard to reach areas. In 2021, an IAEA developed instrumentation and methodology for drone equipped with radiation detectors, cameras and GPS devices has been tested and validated under real conditions in the Fukushima [37]. The low cost of drones and the fact that people avoid being exposed to radiation by using them are significant advantages of this technology.

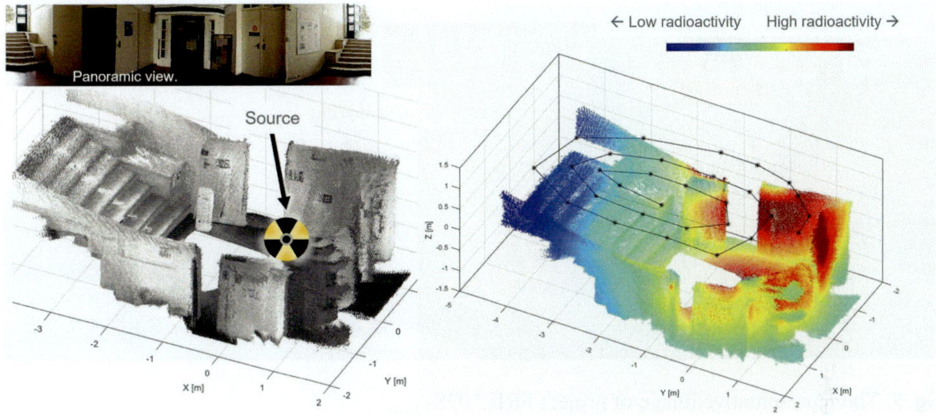

Fig. 4 Indoor scenario with a 3D reconstruction based on a depth camera (left image) and the resulted radiological heatmap where one hotspot can be identified (right image)

A drone may not be fully equipped, depending on the scenario of operation or the type of inspection, such as hotspots localization or characterization based on gamma spectroscopy. For large scenarios of operation and given the reduced endurance of commercial drone, a fleet of drone is more convenient.

The proposed solution using drones for radiological inspection bring the following added values: (i) scalable implementation from a single to multiple drones, from small to large scenarios accordingly, (ii) homogeneous or heterogeneous drone configurations (each drone profile for a specific operation), (iii) applicable to affordable drones, affordable sensors and ordinary CPU units, (iv) mission can be performed with drones in manual or in autonomous operation, (v) automatic generation of enriched 3D map output with the georeferentiation, radiation intensity, localization and characterization of identified hotspots, (vi) guidance and navigation, (vii) safe landing selection, and (viii) control and auto-landing. In addition, the solution is envisaged to be extended for the protection against CBRNe threats, and new sensing technologies will emerge.

3 Proposed Solution

Fleet of dRones for radIological inspEction, commuNication anD reScue (FRIENDS) is a project of research and development that addresses the quick characterization of the geographic extent of radiological contamination by a fleet of drones equipped with advanced technologies of sensors. In the project, the authors addressed this broad challenge by designing, developing, and validating an heterogeneous drone fleet architecture, using different drone profiles. Drone profiles can be medium/large size, able to integrate multiple sen-

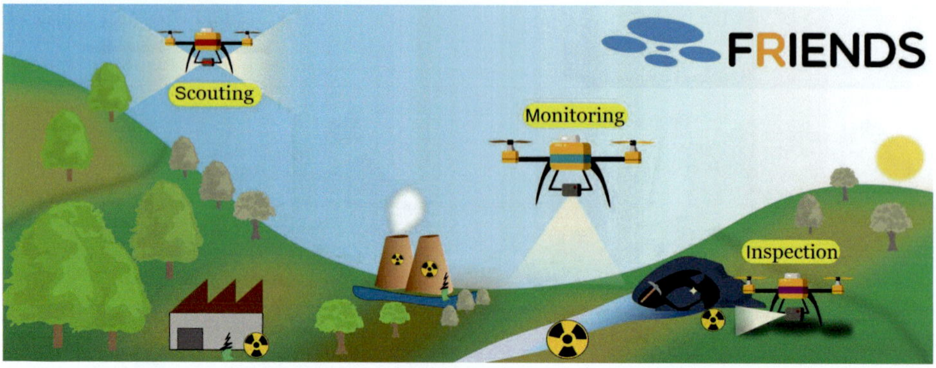

Fig. 5 The representative image of project FRIENDS

sors and an onboard computer for data acquisition and processing. The drone profiles are according to the operation.

The types of operations are: (1) scouting for mapping the scenario with LIDAR and Global Navigation Satellite System (GNSS) returning a 3D georeferenced map, (2) monitoring using the GMC and GNSS to provide a radiological intensity map and the estimation of the potential hotspots, and (3) inspection where the GRS is used to characterize the potential radionuclides present in the scenario, as illustrated in Fig. 5. All these operations consider connected drones that communicate between each other to control the mission, and that can send their sensing information in real-time.

In (1) the drone navigates teleoperated and far from the ground for maximum coverage of the scenario. The success of the overall mission depends first on exploring and characterizing the scenario. For this operation the drone is equipped with the common navigation sensors, namely GNSS and IMU included in all operations, and a scanning range finder to enable a 3D representation of the scenario. In (2) the drone navigates autonomously following trajectories optimized based on the 3D map and close to the ground to get the maximum readings of the radiation sources. The drone must be equipped with the radiological sensor GMC and a light-weight obstacle avoidance systems, such as RGBD or ultrasound pointed forward, to thoroughly sweep the selected area of interest. In (3) the drone navigates autonomously between the hotspots, identifies the safest and closest place of radiation hotspot for landing and performs auto-landing to get the gamma ray spectrum of the source (typically during a data acquisition of several minutes). The drone is equipped with a CZT and a RGBD camera pointed to the ground to find the best and safe place for landing.

Scouting, monitoring and inspection operations can be performed with a single drone or with a fleet of drones. The scouting operation is performed before the other missions. Monitoring and inspection can be performed sequentially or simultaneously with the inspection using the on live hotspots already identified by the monitoring operation. The proposed solution can be applied by teleoperating the drones, or these can be configured to work in autonomous operation. The output at the end of the entire mission is a 3D georeferenced heatmap with the localization of the hotspots and the characterization of each hotspot with the potential radionuclides identified based on the gamma spectrometry. The elapsed time taken for downloading and offline processing depends on the CPU, but is less than the duration time of the mission in an ordinary computer.

4 Drone Profiles

The drone profile can be full equipped, prepared for any operation, or partially equipped according to the target operation. A full equipped drone compromises the endurance of the operation given the extra weight of devices not used. Therefore, the proposed solution is to design the drone in two parts: (i) the drone structure and the commons devices for navigation (e.g., GNSS and IMU), detailed in Sect. 4.1 and (ii) a modular Sensorbox able to be adjusted for each type of operation, detailed in Sect. 4.2.

4.1 Drone

Drones are built from the ground-up using a carefully-selected set of components. An hexacopter configuration has been chosen as it provides better stability under heavier payloads than the more common quadcopter option. The Computer-aided design (CAD) design and one of the implemented drones are presented in the two images of Fig. 6. All the drones are assumed with this configuration.

Table 1 shows the list of selected parts and their weights. The total weight is approximately 2.7 kg. A full equipped drone is designed to have a payload of 2 kg and to carry the Sensorbox with an endurance between 15 and 20 min of flight time. This configuration assumes that the drone is simply hovering in place: A lower-bound 50% flight time estimate is provided as a conservative figure that tries to account for losses in battery capacity over time, inefficiencies in power conversion, and other factors. The radio communication is detailed in Sect. 6.11.

The legs of the frame were replaced by a systems of a spider like structure, with 6 legs, each one performing a obtuse angle (approximately 120°) with the arms of the frame that

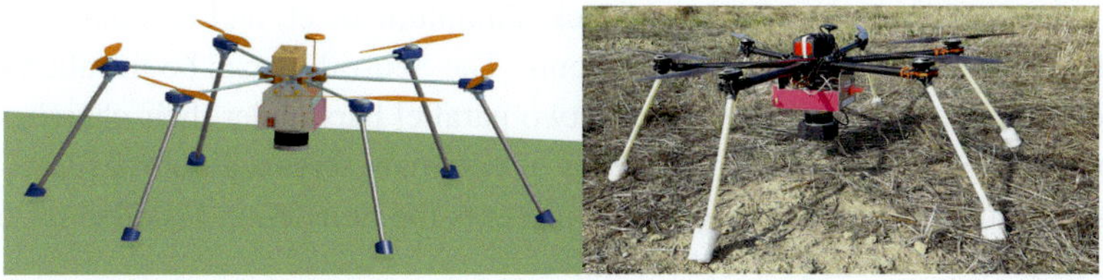

Fig. 6 Generic drone profile: CAD design (left) and the real one (right)

Table 1 List of selected drone components and the respective weights

Item	Weight (g)
Frame Tarot 680 Pro (hexacopter, landing gear and power distribution board)	780
Motors Tarot 5008-340KV x6 (3 kg max thrust per motor, i.e., a total thrust of 18 kg)	168
Electronic speed controllers (ESC) Hobbywing XRotor 40A Pro x6	50
Propellers Quanum 13" Carbon Fiber Propeller x6	18
Battery Multistar High Capacity 12000 mAh 6S 10C	1,600
Flight Controller Radiolink Pixhawk 4	16
GNSS and magnetometer	50

Fig. 7 Support specially designed to use PVC pipes as legs and printed in PLA

supports the central part of the drone. The legs are made in PVC and attached to the arms by a specific support specially designed in CAD and printed in PLA for this purpose, as illustrated in Fig. 7. The proposed legs offer a more flexible solution for landing in harsh conditions and resilient in case of damage up to 3 legs. In addition, PVC pipes are affordable and the support for the legs can be easily printed.

4.2 Sensorbox

The Sensorbox comprises its case, sensors, a computer, and the electrical power system. The case was machined out of a 2 mm thick aluminum sheet, and its main purpose is to protect and carry all the hardware components. The Sensorbox can be installed under any of the drones of the fleet. All drones have two parallel bars under their chassis where the Sensorbox slides in and is secured by means of two main screws. Figure 8 shows the CAD and a snapshot of the Sensorbox. Figure 9 presents the Sensorbox installed on one of the drones, after an experiment in a real scenario.

Sensors include a Light Detection And Ranging (LIDAR) to perform the cartography of the scenario and support real-time navigation when combined with IMU, RGB and RGBD

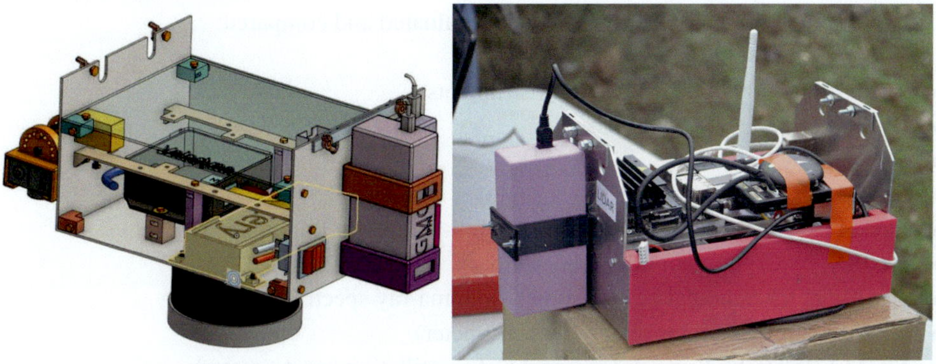

Fig. 8 Sensorbox carries the onboard computer, sensors and internal battery: CAD design (left) and the real Sensorbox, with the LIDAR sensors inside a cardboard just for support (right)

Fig. 9 Sensorbox installed on the drone after an experiment in a real scenario

cameras for image collection and autonomous landing, and the radiological sensors GMC and CZT.

The selected LIDAR is the Velodyne VLP-16, which provides a 3D scanning with 360° Horizontal Field of View (FoV), ±15° Vertical FoV along 16 fixed viewing angles (+15° to −15° with 2° increments), range up to 100 m and 830 g of weight. Every distance measurement arrives in spherical coordinates, and it additionally provides the surface reflectivity where the laser hits (with no need for calibration). It should be possible to synchronize data with precise GPS-supplied time. Connection to the VLP is done through an ethernet cable with a data rate of 100 Mb/s. The VLP-16 sensor does not have a physical power switch. The power source must be capable of providing as much as 3.0 A to accommodate the current surge that occurs during rotor spin-up when power is supplied to the sensor. In normal operation, the sensor draws approximately 8W of power. The sensor requires an input voltage between 9 and 18 V.

The following radiological sensors were evaluated and compared:

- PRM-7000/8000/9000, by Mazur Instruments
- RadEye G-10, by ThermoFIsher
- PM1405, by Radiansa
- Sparkfun-11345 with tube LND712 (geiger counter)
- Pocket Geiger radiation sensor type 5, SEN-14209, by Radiation Watch
- RasPIX, by Crytur (scintillation detectors)
- RadAngel for Schools™, by Kromek (gamma-ray spectrometer)
- GR1, by Kromek (compact CZT spectrometer)
- SIGMA 25/50, by Kromek (compact CsI scintillation spectrometer)
- MGS60-USB (ultra compact gamma CZT spectrometer)
- MGS500-USB (ultra compact gamma spectrometer)
- μSPEC: μSPEC60, μSPEC500, μSPEC1500, by BSI.

The main considerations into the selection analysis are (1) the resolution (up to 3 MeV), (2) the volume of the sensor (larger volume, better performance), (3) the number of channels, and (4) the cost. The CZT sensors have typically greater precision to identify the type of material, but low resolution. The ScI (scintillators) are the opposite.

The selected radiological sensors are: a spectrometer μSPEC500 from BSI, and two GMC, a PRM-900 and a Sparkfun-11345. The μSPEC500 from BSI provides reasonable energy resolution at 662 keV ($<$2.5%), number of channels up to 2 k, 80 g of weight, and with a price supported by the budget of the project. The basic software to acquire raw data is already included. The Sparkfun GMC is also connected to the computer via a Universal Serial Bus (USB) port masked as a serial connection. At a fixed timestamp (more than 1 s and always an integer) the GMC publishes in the serial port the counts measured in that time interval. This time stamp can be changed writing to the serial port the new desired value. This GMC was already used in MARIA [36]. Regarding the Mazur GMC counter, the operation is similar except that the publishing timestamp is fixed by the sensor to 1 min due to the company's existing firmware. During the last months of negotiation with the Mazur company, a testing firmware was already released to be included in this project. In addition, the Mazur GMC is a handheld device with a screen display that can be used standalone and to validate the activity of the identified hotspots.

Two RGB and depth camera (RGBD) cameras were tested during the project. Initially a Structure camera from Occipital was used, which only provides infrared and depth images and operates in indoor scenarios [42]. Later, the previous RGBD camera was replaced by an Intel© RealSense™ Depth Camera D415 [43], which is prepared for indoors/outdoor scenarios, with a better resolution, up to 1280×720 of depth pixels and 1920×1080 of Red, green and blue (RGB) pixels, and higher frame rate, up to 90 and 30 fps, respectively.

The Sensorbox is also prepared to support a thermal camera, namely the FLIR Duo R [44]. The camera acquires both thermal and RGB pictures and video. In picture mode, the camera

Table 2 List of selected sensors and computer and the respective weights

Item	Weight (g)
Velodyne VLP-16 (LIDAR)	830
Structure Sensor (depth camera)	55
FLIR DUO R (RGB and thermal camera), replaced by the next one	84
Intel RealSense D415 (RGB and depth camera)	72
BSI μSPEC (CZT)	80
Sparkfun 11345 (GMC)	200
Mazur PRM 9000 (GMC)	368
NVIDIA Jetson Nano (computer)	88
Battery 3C 12V	250
Cables and aluminium box	500

is able to produce an infrared image with a resolution of 160×120 pixels. It is possible to convert the pixel value directly to Celsius. The RGB image has a resolution of 1920×1080 pixels and encompasses accessible metadata regarding the camera model, calibration, and GPS position. In video mode, it is possible to stream with equal resolution on both channels, but it is necessary to use High-Definition Multimedia Interface (HDMI) connections and therefore an HDMI input on the onboard computer. An alternative is to have a converter from HDMI to USB. Otherwise, the video can only be accessed when the recording is finished. During the real experiments, the information retrieved by the thermal images were useless for radiological inspection. Nevertheless, the Sensorbox and the software is prepared to support thermal cameras, which can be relevant for other areas of CBRNe that are no radiological nor nuclear. When necessary, the RGB is extracted from the RealSesne camera.

The onboard computer is a NVIDIA Jetson Nano single-board computer [45], to which all the sensors are connected. The main specs are: Graphics Processing Unit (GPU) of 128-core Maxwell, CPU quad-core ARM A57 1.43 GHz, 4 Gigabytes of memory RAM, micro-SD storage, Gigabit Ethernet connection, 4x USB 3.0 and 1x USB 2.0. The computer runs Ubuntu 18.04 and all of the required Robot Operating System (ROS) Melodic software for data recording, including telemetry and sensor data.

All the sensors and the computer are powered by an internal battery of 12 V, and additional Direct Current (DC) converters are used to reduce the voltage according to the power specification of each device. Cables and power switches are also included.

The weights of all selected sensors and computer are presented in Table 2.

When full equipped, the total weight is approximately 2.3 kg. Since the two GMC are redundant, only one of them is used in each drone and the total weight reduces. In addition, some of the modules (sensors and the respective cables) can be removed according to the operation. Their weights are listed in Table 3.

Table 3 Weights of the various drone profiles

Drone profile	Weight (kg)
Scouting drone: only the LIDAR, computer and battery are required	1.7
Monitoring drone: only the RGB and depth camera, the GMC, the computer and the battery	1.1
Inspection drone: only the RGB and depth camera, the CZT, the computer and the battery	1

4.3 Hardware Architecture

The drone is powered by a battery, which is monitored and delivered by a Power distribution board (PDB) to the Electronic speed controllers (ESC) of the motors and all of the other electrical components. The target speed of each motor is dictated by the Autopilot, an off-the-shelf component closing the loop through an IMU and GNSS. For manual control, a Remote Control receiver is directly connected to the Autopilot. For autonomous control, a GNSS is included for waypoint navigation and an ultrasound for height measurement and control.

The Sensorbox has a single computer in charge of collecting sensor data from multiple sensors connected to it, namely the LIDAR, RGBD, GMC, and CZT. The computer can also process data during flight if necessary. There is also a WiFi module for high-level communication with the ground station. As mentioned above, the Sensorbox has its own battery, which can work independently of the drone. Besides the master power switch, there is an individual switch for the LIDAR to save energy when the device is not in use. The Sensorbox is able to issue motion commands to the drone via a data-connection between the computer and the Autopilot. By design, the manual Remote Control commands should always have higher priority than the Sensorbox commands. A ground station is able to interact with the Sensorbox via wireless communication, in particular using a USB WiFi dongle (IEEE 802.11).

5 Fleet of Drones, Operations and Sequences

A full equipped drone, or a scouting drone, or a combination of multiple drones, are able to perform the scouting operation, i.e., the reconnaissance of the scenario. Multiple drones can cover different parts of the scenario and the resulted maps, obtained by the mapping algorithm described in Sect. 6.1, are combined. Assuming that the maps are produced at the end of scouting operation, i.e., when the drones have landed, since the map is necessary to perform the monitoring and inspection, no other operations are performed simultaneously. During the scouting operation, drones are teleoperated and fly at different altitudes, typically above

5 m of height from the ground up to the maximum range of LIDAR sensor or the maximum altitude allowed by the local authority. The flight is reasonably far from the obstacles, not only for safety purpose, but also to minimize the effects of occlusion. Since the map is obtained offline but in a short period of time (from seconds to minutes depending on the dimension and complexity of the scenario and the available Central Process Unit (CPU)), the operator can observe and decide if a new scouting operation is required to complete uncover areas.

A full equipped drone, a monitoring drone, or a combination of multiple drones, are able to perform the monitoring operation sequentially or simultaneously. When using multiple monitoring drones, different regions of the map are distributed for each one. At the end of all monitoring operations the resulted heatmaps and identified hotspots, obtained by the algorithms described in Sects. 6.6 and 6.7, are combined. During the monitoring operation, the drones can be teleoperated but are intended to navigate autonomously by the algorithm described in Sect. 6.4 and using the map resulted from the scouting operation. The flights are snaking between obstacles and close to the ground, where the sources of radiation activity are expected to be located. Since the performance of the GMC decreases inversely proportional to the square distance between the sensor to the source, heights of the monitoring flights are typically above 1 m, for safety in terms of clashes, and below 3 m given the adopted GMC model. The resulted heatmap and the identified hotspots are georeferenced by the algorithm described in Sect. 6.2.

A full equipped drone, an inspection drone, or a combination of multiple drones, are able to perform the inspection operation sequential or simultaneously. The hotspots are referenced with World Geodetic System WGS-84 [46] coordinates, to be visited by the drones using its onboard GNSS receiver. The sequence of hotspots can be decided by the radiation intensity or by other criteria according the purpose of the mission. The drones fly above the obstacles between hotspots and land automatically as close as possible to the points of interest using the algorithm described in Sect. 18. Different sequence of hotspots without intersection can be distributed by different drones.

The algorithms to obtain the heatmap and to identify the hotspots can run in real time during the flight, or offline after landing. When in real time the monitoring and inspection operations can be performed simultaneously with the live map and the hotspots identified along the mission.

In teleoperation or autonomous operation, all the drones communicate between them and with a ground station, as detailed in Sect. 6.11.

In autonomous operation the drone follows its path computed in Sect. 6.4 for monitoring. For inspection, a waypoint navigation approach is used to visit the identified hotspots. The waypoint navigation and the automatic landing, or simply auto-landing, are detailed in Sect. 6.9. In this work the scouting was not designed to be implemented in autonomous operation. An algorithm inspired on boustrophedon path, as described later in Sect. 6.4 for maximizing the coverage area, can be also applied for scouting in combination with control described in Sect. 6.9. However, the scouting in autonomous operation was left for future work.

6 Algorithms

During the three operations of scouting, monitoring and inspection, different algorithms are used to achieve specific goals. Some of the algorithms are shared by different operations, and can run online or offline. The algorithms, detailed in the following subsections, are:

- Mapping to create a 3D cloud of points representing the assessed scenario;
- Georeferencing the cloud of points;
- Segmentation to identify the points associated to the ground;
- Guidance and navigation of the drone to perform the inspection operation based on the segmented cloud of points; Vertical navigation included;
- Heatmap estimation, i.e., the map with the doses of radiation;
- Hotspots identification, i.e., the coordinates of Points of Interest (PoI) where the doses of radiation are above a given threshold value;
- Best place identification for a safe auto-landing and close to the PoI;
- Setting up the radiological sensors;
- Radionuclides identification; and
- Communication between drones and between drones and the ground to provide the data and the control interaction between the drones.

6.1 Mapping

During the scout operation the drone is teleoperated to perform a reconnaissance of the scenario and to delimit the area of interest. The teleoperation can be assisted with the streaming of video images captured by an onboard RGB camera. The expected output is a 3D representation of the scenario that is going to be used by the navigation and by the auto lading during the monitoring and inspection operations, respectively. The 3D representation is a cloud of points, which is georeferenced by the algorithm described in Sect. 6.2, used to represent the estimated heatmap and the identified hotspots, as described in Sects. 6.6 and 6.7, respectively.

The mapping is accomplished by the Advanced implementation of LiDAR Odometry and Mapping (ALOAM) [47], an optimized implementation of the LiDAR Odometry and Mapping (LOAM) algorithm [48], which is mainly based on SLAM, with the LIDAR data collected during the flight. The ALOAM algorithm can work in real time during the flight, but since its output is not used for the teleoperation, the 3D map is built offline, immediately after the scouting operation. The output is a 3D cloud of points and the estimated path followed by the drone, both represented in a local referential.

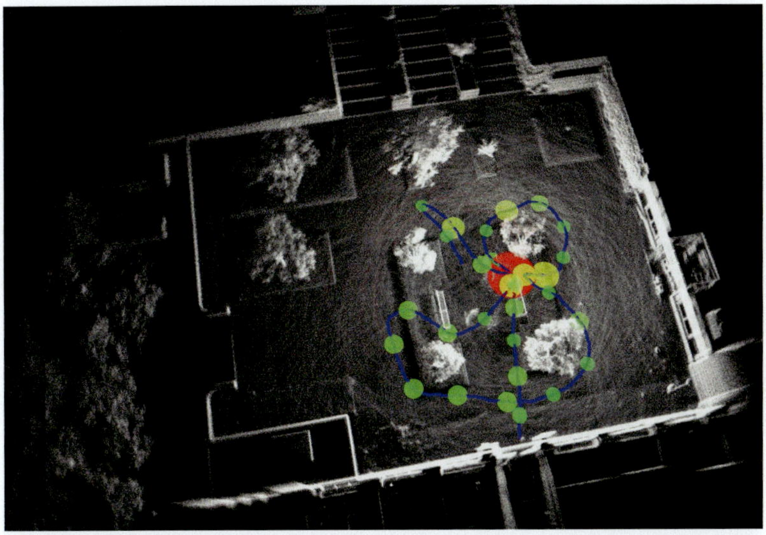

Fig. 10 Resulted map with a 3D cloud of points representing a garden in the IST campus, where a controlled source of radiation was temporarily placed in the middle

Figure 10 illustrates a map with a 3D cloud of points after running the Advanced implementation of LiDAR Odometry and Mapping (LOAM) (ALOAM) algorithm, representing a garden at the Instituto Superior Técnico (IST) campus. The raw data was acquired by the Sensorbox transported in hands. A controlled source of radiation was temporarily placed in the middle, and the counting results achieved by the GMC. The mapping algorithm is more detailed in [12].

6.2 Georeferencing

The ALOAM returns not only the 3D cloud of points but also the estimated path followed by the drone, which encompasses the positions and orientations of the Sensorbox, and hence, the LIDAR. The 3D map, as well as the estimated path, are represented in a local frame, usually the frame of the first acquisition, when the LIDAR starts the acquisition. The path can be also estimated by the GNSS, represented in the World Geodetic System WGS-84. A transformation function exists between both frames, which can be calibrated and saved at the beginning knowing the Sensorbox design and how it is installed on the drone.

Each pose returned by the ALOAM or by the GNSS receiver has a timestamp. The internal clocks of ALOAM and GNSS are assumed to be synchronized. Since the sampling frequency of the ALOAM algorithm is higher than the GNSS, the georeferencing algorithm is performed as follows: (1) for a ALOAM sample find the two closest GNSS samples, i.e., the next and previous closest samples based on the timestamps; (2) perform an interpolation

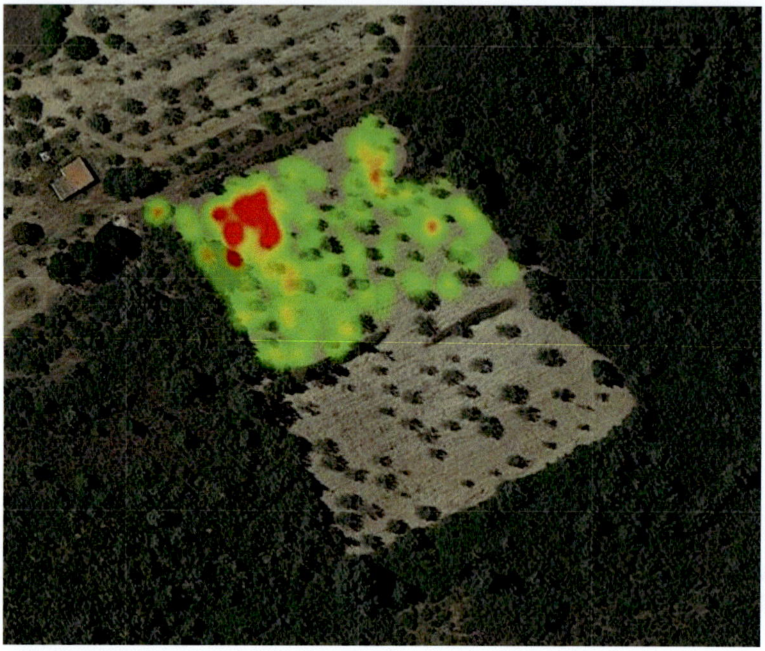

Fig. 11 Heatmap georeferenced represented using the Google Maps API

between the two closest GNSS samples taking into account each timestamp; (3) set the gepreferencing algorithm result of the ALOAM sample as the interpolation result; (4) repeat this sequence for all ALOAM samples.

The gepreferencing algorithm can be applied to the 3D cloud of points, segmented or not segmented as described later in Sect. 6.3 and to the resulted heatmap, which is explained how it is estimated in Sect. 6.6. Figure 11 illustrates an heatmap already in World Geodetic System WGS-84 and represented using the Google Maps API. The georeferencing algorithm is more detailed in [12].

6.3 Segmentation

The sources of radiation are assumed to be located at the ground or close to it. Therefore, the drone aims to fly close to the ground to get the maximum readings of the radiation sources during the monitoring operations. To keep a constant flight height, implement the path planning and navigate safely through the obstacles, the cloud of points are segmented in two classes: (i) ground and (ii) non-ground. The first represents the floor surface, or the lowest surface on the map. The second represents the obstacles above it, such as trees, bushes, rocks or building structures.

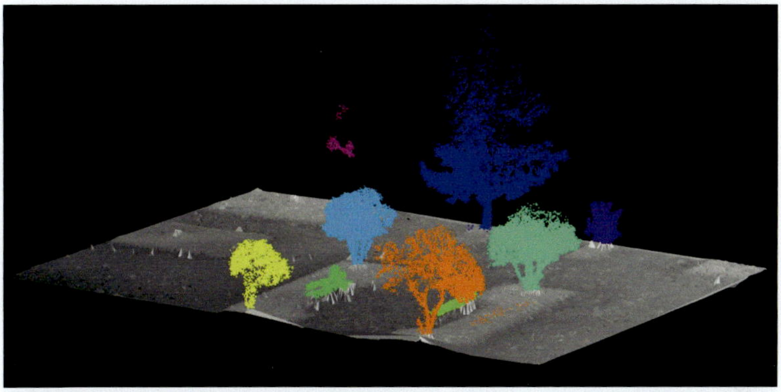

Fig. 12 Segmentation of part of the map

The segmentation algorithm operates the following way. First, the Area of Interest (AoI) is defined by the user within the whole generated 3D map. Then, the mapping algorithm selects a small sector, i.e., a vertical squared prism with base dimensions and infinite height. This sector is centered at given coordinates, and the algorithm filters all AoI points outside this prism. The point with minimum height coordinate is identified (the lowest point). A region with higher height up to a given threshold is considered and all the points inside are candidates to be classified ground. For the last step, the Random Sample Consensus (RANSAC) algorithm [52], is applied to find the plane that best fits the previous points. The plane is not horizontal and, typically, has a slope. The mapping algorithm ends up classifying as ground points close to that plane. All other points are classified as non-ground. When the region under analysis is plenty of trees and/or building structures, few points are effectively detected at the ground level.

Figure 12 illustrates a segmented 3D cloud of points, where the ground is represented by gray points and the non-ground with colored points. The segmentation algorithm is able to distinguish individual trees or other elements in the non-ground points, but, for the proposal of drone application described in this chapter, the feature of separate different artifacts above the ground is not considered. The segmentation algorithm is detailed in [12].

6.4 Guidance and Navigation

For monitoring operations, paths must be evaluated to fully cover the target scenario. Based on the resulted 3D cloud of points already segmented, the space is classified into occupied, free and unknown areas, i.e., an occupancy 3D grid. Then, using the estimated location of the ground surface a 2D occupancy grid is created. It is a flat map used as reference for the flight height. Then, the Octomap [54] algorithm is used to compute a 3D occupancy grid with the points classified as non-ground.

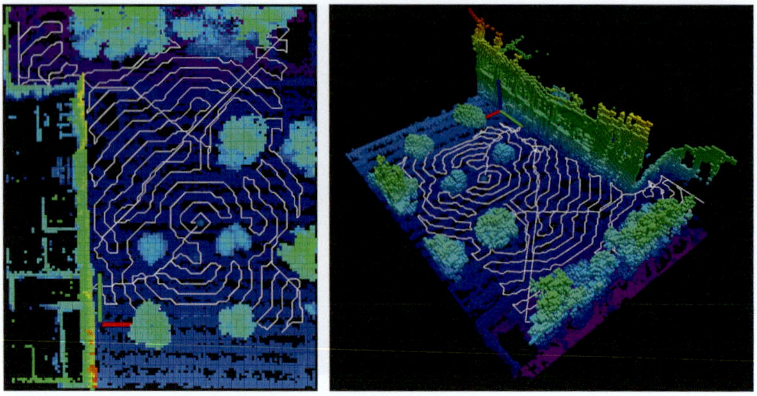

Fig. 13 Octomap and the optimized path in the garden at the IST campus

The 3D occupancy grid is used for guidance and navigation, with the path planned along the free area. During the project, one of the members of the team developed and presented in his Master Thesis, an adaptive informative path planner for surveying scalar fields. The algorithm, inspired on boustrophedon path for complete area coverage [53], is fundamentally based on maximizing mutual information to reduce the uncertainty of the estimated field, with an additional reward for exploring regions where the field magnitude is large. The path planner balances exploring hotspots versus increasing the total coverage area. When compared to boustrophedon paths, which are the standard for offline coverage path planning, the proposed solution results in better localization of radioactive sources on average (especially in the case of a single source), but worse estimation of source activity. However, the path planner can be used to locate sources with good accuracy and, subsequently, a more sensitive radiological sensor can be placed at the source locations to measure the activity with greater accuracy.

Figure 13 illustrates the resulted Octomap and the optimized path in the garden at the IST campus. The algorithm and more examples can be found in [14].

In [15] the problem of area coverage for multiple cooperative robots was addressed, dividing the workspace into convex cells, assigning each cell to a robot, and performing a boustrophedon path in each cell. A similar approach can be used here applying the adaptive informative path planner in each convex cell allocated for a different drone. Thus, the fleet of drones are able to operate autonomously and simultaneously in different regions of interest. However, this approach was not implemented and was also left for future work.

6.5 Setting up the Radiological Sensors

A calibration of the CZT and a verification of the GMC were conducted in order to assure the correct identification and activity estimation of the source to specify the most appropriate physical parameters of the drone flight (distance to the ground, time of exposition, drone speed, loitering time, etc.) for radiological data gathering, considering the characteristics of the sensors and location [17].

The verification of the GMC sensors was performed at Laboratório de Metrologia das Radiações Ionizantes (LMRI), a certified primary standard laboratory of the Portuguese National System of Metrology that performs activities such as quality control of devices that measure ionizing radiation. Each of the GMC was placed at three different distances (100, 150, and 200 cm) from three different Cs-137 sources with different activities (74, 740, and 7400 MBq) and tabulated equivalent dose rate (H) values. The drone are expected to fly at these distances from the ground during a field measurement, mainly because of the vegetation height. The linear regressions expressing the variation of $\dot{H}_T$ (μSv/h) as a function of activity in counts per minute (CPM) are $\dot{H}_T = 0.0134 \cdot \text{CPM}$ for the Sparkfun and $\dot{H}_T = 0.0024 \cdot \text{CPM}$ for the Mazur.

In order to analyse gamma radiation emission spectra, the system needs to be calibrated in three different ways: (1) energy calibration, (2) calibration of the shape of peaks, full width at half maximum (FWHM), and (3) calibration in efficiency. To obtain a sufficient number of peaks within the energy range, two certified calibration sources were used: (1) a Eu-152 source and (2) a mixed gamma aqueous solution. The CZT was placed with a Source Detector Distance (SDD) of 10 cm, on top of a holder with enough height to line the center of the detector with the center of both calibration sources, which were attached with duct tape. The activity of each source at the time of measurement was calculated using the radioactive decay law, considering the activity (A) and half-life ($t_{1/2}$) information present in the calibration certificates.

Figure 14 represents the acquired calibration spectrum previous to calibration. The spectra information was then later imported and analyzed using the open source software InterSpec v.1.0.1a, to calculate the peak mean energies, FWHM and areas, as well as to identify the peaks. In total, 12 peaks were considered, with energies between 59 keV and 1408 keV. The complete list of energy peaks emitted by the calibration sources and the setup used can be found in [17]. The energy calibration is represented by the linear fit in Eq. 1.1 (where c is the channel number):

$$E = 1.759\,c - 0.08 \quad [keV] \tag{1.1}$$

Figure 15 shows the efficiency value for each evaluated peak as well as the adjustment curve with a polynomial of degree three of the logarithmic dependency of Energy (E) and the Efficiency (ϵ). The number of counts measured by the detector was calculated as the peak area, according to InterSpec.

The output of this calibrations is that the generic monitoring should be performed in a field scenario at 1–2 m from the ground, as a compromise between the height of the vegetation and the GMC sensitivity. In this condition the confirmation of the existence of a hotspot

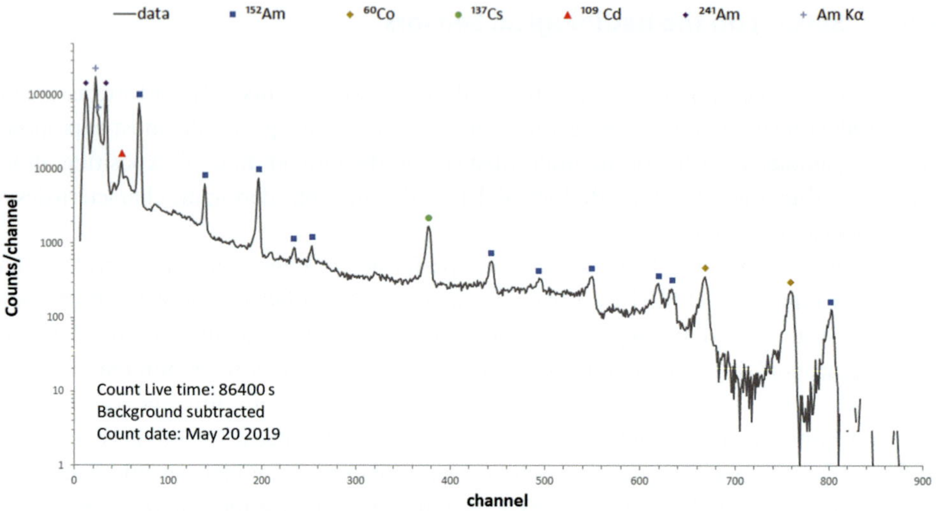

Fig. 14 Acquired calibration spectrum (background subtracted). Labeled peaks were considered for calibration. Time of acquisition was 86,400 s

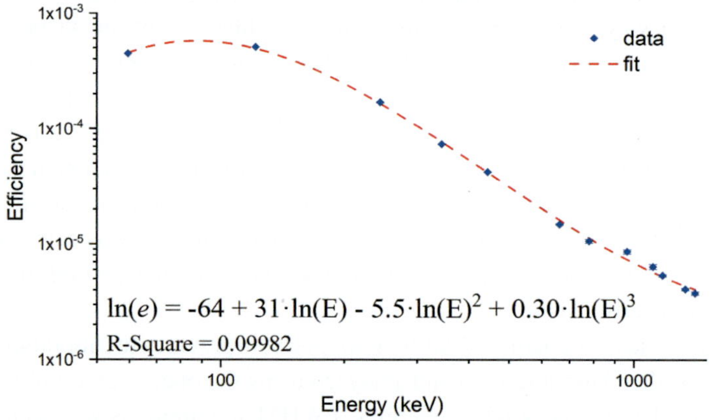

Fig. 15 Peak efficiency as a function of the energy for every considered peak. A logarithmic dependency was considered for both the Energy (E) and the Efficiency (ϵ)

found during generic monitoring can be performed by a drone with a GMC in a relatively short hover time (i.e., between 5 a 10 min). Considering a square with 1 m side, in order for a drone to detect a hotspot, it would have to hover the square during at least the response time, and thus the minimum speed for the drone would be 0.20 and 0.23 m/s for the Sparkfun and the Mazur, respectively.

The spectra acquisition with the CZT should be performed at a short SDD, between 10 and 100 cm. Given that the drone with the CZT will land on the ground where a hotspot is detected, a SDD between 10 and 20 cm is feasible. Preferably, the acquisition should be performed as close as possible, and in no less than 30 min, in order to acquire spectra with the maximum number of peaks and with a higher number of counts.

6.6 Heatmap Estimation

The heatmap estimation algorithm returns the scalar field of radiation across the AoI, given a number of locations where our GMC sensor collected radiation samples. Each measurement is assumed to aggregate the effect of all sources of radiation present on the ground within the AoI. In addition, it is assumed the radiation comes from the ground and decays with the distance (inversely squared). All the points of the 3D map, i.e., the cloud of points, are candidates to be sources of radiation and, hence, has a given radiation intensity. If, in fact, the point does not correspond to a source of radiation, the intensity must be zero. Thus, the problem can be modeled in a matrix form with the three set of parameters: (i) a vector with all the radiation measurements, (ii) a matrix with the inversely squared distances between all the points of the 3D map and the locations where radiation was measured, and (iii) a vector with the radiation intensities of all the points of the map. The matrix is large given the amount of points of the 3D map. Pseudo-inverse, sub-sample and voxel filter are used to improve the computation speed, as firstly detailed in [55] and then revisited in [12].

A 3D radiological heatmap is then composed using two elements: the non-ground (obstacles of the scenario such as trees, bushes, rocks, etc.) and the ground, colored based on the GMC heatmap. Figure 16 illustrates a 3D radiological heatmap achieved in a real olive grove.

Fig. 16 Example of an heatmap achieved in an olive grove

6.7 Hotspots Identification

The scenario may contain N radiological sources, and each source is placed in its (x, y, z) coordinates and with an activity representing the average number of particles emitted per second. The following four assumptions are made: (i) the radiological sources are isotropic point sources, (ii) radiation intensity measurements are independent random variables each following a Poisson distribution, (iii) all the measurements have an equal exposure time, and (iv) the activity of each source remains constant over time, i.e., the duration of any measurement session is negligible compared to the half-life of the sources.

A set of m observations are made in this scenario. Similarly, each observation has a position and a measured intensity corresponding to the number of particles detected within the exposure time interval, as a function of the detector geometry. Since the sources are isotropic point sources, the measured intensity only depends on (1) the distance between the position of observation and the position of each source, and (2) the background radiation intensity.

The problem is to estimate the number of sources N and the parameters (position and activity) of each source given a set of observations acquired in different locations, taking into account the sensors models, background radiation intensity values and noise. The authors have proposed and tested several approaches, namely:

- Maximum Likelihood approach, which is based on the estimation of the source parameters which have the maximum likelihood for explaining the observations. The likelihood is calculated through the joint probability distribution of N Poisson distributions (one for each potential source), where the samples correspond to the observed radiological intensity, and the distribution parameters are the positions and activities of the sources.
- Bayesian approach uses an initial guess for the source parameters which, through an iterative process, is updated until a convergence threshold is met or the limit of the number of iterations is reached.
- Robust Localization with Clustering approach is based on a Monte Carlo technique which uses a set of particles and is centered around the idea of updating the set of particles biased towards the best-classified particles.
- Least Square Error approach attempts to fit a model to the data while minimizing the sum of square errors.
- Neural Networks based model, called RHLnet inspired on Yolo state of the art [40], where a model based on classification and regression problems is applied for object detection tasks, with images used as input.

The first four approaches are detailed in [18], including the simulated and experimental results performed in the laboratory with controlled sources of radiation, as illustrated in Fig. 17. The last approach was performed more recently and is detailed in [10] and revisited

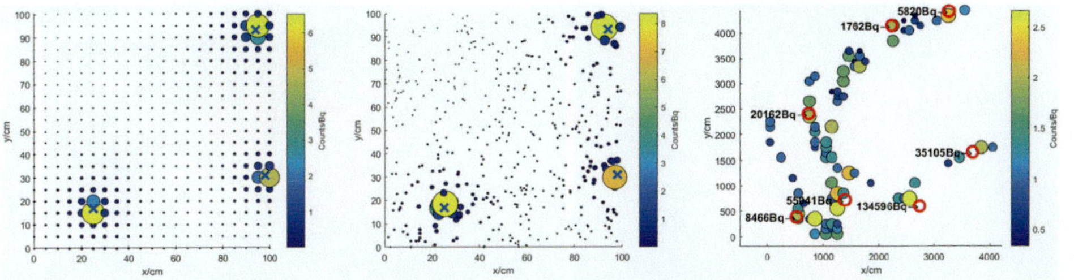

Fig. 17 Hotspots detection: simulated results (two first left images) and real scenario (right image)

in Sect. 6.10. The application of a collimator was also addressed and tested. However, a collimator can bring improvements to the estimation but may jeopardize the endurance of drone [18].

6.8 Safe Landing Site Selection

For the success of inspection operation, the CZT measurements must be taken as close as possible to the radiation hotspots. These hotspots are not guaranteed to be safe for landing drones, neither accessible or safe for human operators. An autonomous landing comes into play, where the drone is equipped with the needed sensor and an algorithm to identify if the location of the hotspot is safe for landing, otherwise, to find the best and closest place to land.

The task of landing a drone is complex, mainly in unstructured environments like the ones where radiological inspection is required. After the detection of hotspots and immediately before visiting each one, the terrain must be evaluated so that the chosen landing location is safe, whilst being as close as possible to the radiation hotspot.

Taking into account the available information, the proposed solution is divided into 3 steps. First, a global landing site selection algorithm receives as input a segmented and geo-referenced point cloud of the environment. Then, the best landing point in GPS coordinates is selected, based on terrain characteristics such as steepness and roughness, which allow for safety classification of each point, and distance to the radiation hotspot. This gives the drone an objective for the start of the landing procedure, to where it will move given the input of Proportional and Integral (PI) controllers [16]. Finally, the drone is equipped with a downwards facing depth camera for a re-assessment of the terrain when closer to the ground, with a local landing site selection algorithm, that provides the means for a thinner control of the landing, which will also be performed given the input of PI controllers.

In Fig. 18, the steps of the global algorithm are depicted. In this algorithm, the point cloud is divided into voxels, which are 3D volumetric shapes that can overlap each other. For each voxel, the equations of a fitting plane are obtained using the RANSAC algorithm [49] and, based on this plane, two terrain characteristics are computed: roughness and steepness. Those

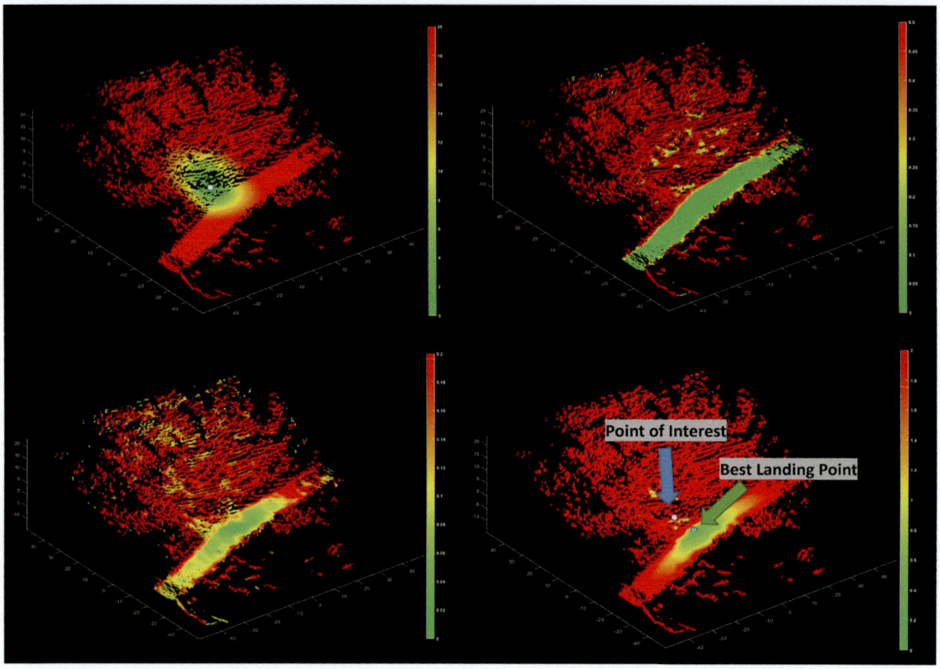

Fig. 18 Steps for the identification of the best place for landing. From left to right, top to bottom: proximity to a Point of Interest, Terrain Steepness, Terrain Roughness and Final Landing Cost, which is a combination of the last three

two characteristics are the inputs to a Support Vector Machine (SVM) [50] classifier that classifies each voxel as safe or unsafe for landing. Finally, to each safe voxel is attributed a cost, that weighs its average distance to danger within a certain radius, and distance to the initial point of interest. The chosen best landing spot is the center of the voxel with minimum cost, as depicted in the bottom right corner of Fig. 18, being a safe point that minimizes distance to the point of interest.

6.9 Control and Auto-Landing

In autonomous operation the drone follows its path computed in Sect. 6.4 for monitoring. A waypoint navigation is used for inspection, to visit the identified hotspots. The path following is implemented using the package "move_base" from ROS [51], with no further development. The auto-landing, developed by one of the co-authors of this work, is summarized in the following paragraphs of the current section and is detailed in his Master Thesis.

The safe landing points returned by the approach described in the previous section are used as inputs for auto-landing. The controller architecture of the auto-landing is composed

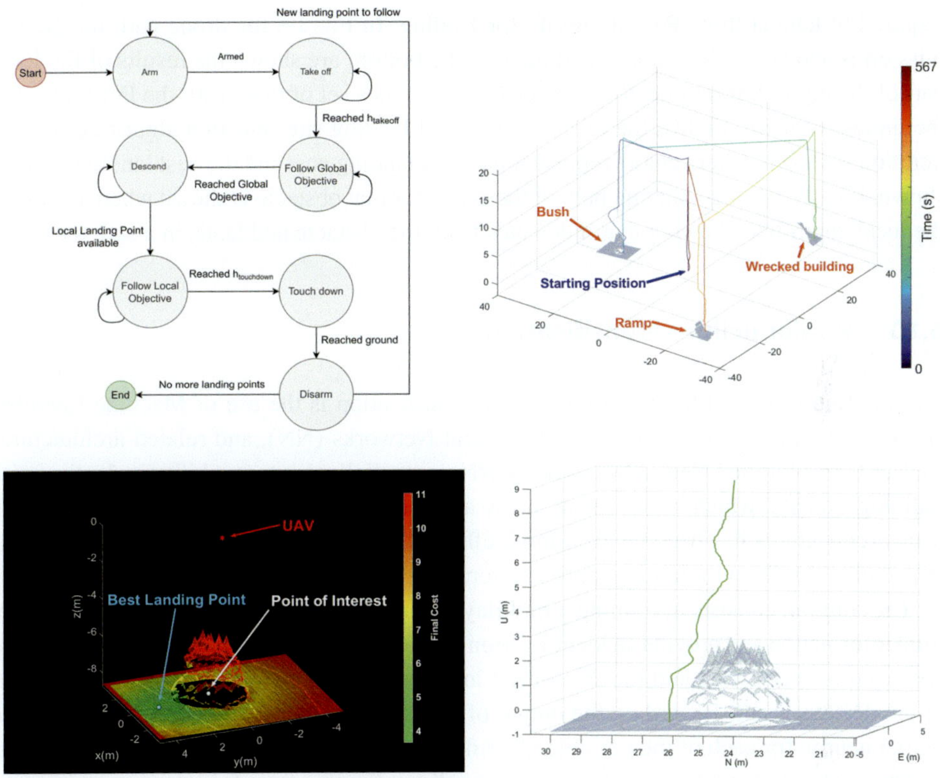

Fig. 19 Autonomous landing system (from left to right and top to down): high-level state machine, drone path in a full autonomous inspection mission with three points of interest, result of the local landing site selection for one iteration, and path of the drone when using the local controller

by a high-level state machine, a global controller, and a local controller. The implemented state machine shown in Fig. 19 ensures that the drone, for each point of interest: (i) takes off to a height high enough so that there are no obstacles; (ii) moves to the coordinates of the global landing point whilst maintaining a safe height, using a global controller; (iii) descends until receiving local objectives in real-time at a constant rate; (iv) follows the local landing point using a local controller; (v) performs the final touchdown and lands.

For each PoI, the global controller receives an objective from the global landing site selection algorithm. This controller uses a proportional velocity planner to guide the drone to the position of the global objective while maintaining a safe eight with the aid of a PI velocity controller. As shown in Fig. 19, upon reaching the global objective, the drone will descend and switch to the local controller, where it will receive objectives from the global landing site selection algorithm. The local controller used one PI controller for each axis of the drone's frame in order for it to follow the desired local objective. The landing system was tested in a simulated environment where a drone performed a mission where it was required to land at three PoI, all unsafe for landing. In Fig. 19, the drone path for the full mission is shown on the top left, and then on the bottom are shown the results of the local part of the algorithm for one of the obstacles, which is a set of trees with the PoI located in the middle of them. Firstly, the result of the local landing site selection algorithm for one iteration is shown, where colors represent the final landing cost and the best landing point is shown. Finally, the final landing path of the drone for that obstacle is shown when it is using the local controller, showing that it deviates from the obstacle and lands in a safe location.

6.10 Radionuclides Identification

The solution proposed to the radionuclides identification is the use of Machine Learning and Deep Learning techniques, namely Neural Networks (NN), and related architectures such as Convolutional Neural Networks (CNN), essentially due to its ability in dealing with radiological data, which is non-linear, noisy and inconsistent. Besides, the training of such deep architectures has become even more efficient with improvements and breakthroughs in techniques related to training, optimization and regularization [38].

The solution consists in training a multilayer perceptron, al in order to receive the spectrum information and provide an identification of the radionuclides present. The full acquired spectrum is normalized and the size of the input layer is equal to the number of channels of the detector itself. Regarding the output of the NN, the chosen approach was to have an output neuron for each of the possible identifiable radionuclides, i.e., a total of 15 neurons, assuming an equal number of potential radionuclides. For the case of single-isotope identification (multi-class) the activation of choice was Softmax [39], to predict the most probable radionuclide in the sample, whereas when trying to identify more than one radionuclide in the same sample (multi-label classification), the approach must be different. For this case the activation function chosen was Sigmoid. Then, the ReLu activation function was chosen for every neuron belonging to the hidden layers. Figure 20 illustrates a spectra with Cs137 represented in red and the peaks identified by the NN represented by blue bars. The full algorithm and resulted are detailed in [10].

6.11 Communication

In our architecture, two major components are considered: (i) the drone modules, which will run in the drone's companion computer, and (ii) the ground modules, which are launched in the ground station computer that will assign the missions to the drones. Multiple drones can be connected simultaneously and controlled through the ground station, each one running the drone module software locally. In the mission, accomplished in this article, the relay is a very important feature to assure the support of the communication to the drones, even if they get distanced from the ground station. If one drone needs to get further away from

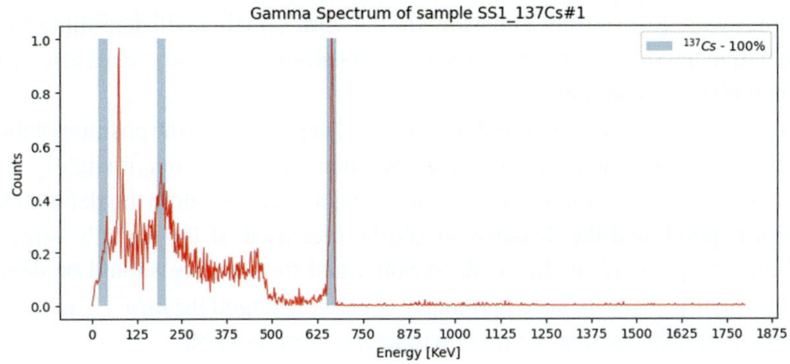

Fig. 20 Spectrum with identification peaks by neural networks

the ground station than the distance the network can reach, we need a way to extend the network's maximum range; otherwise, it will not be able to communicate. Therefore, a second drone will be dispatched for this purpose, maintaining a set distance between the first drone and the ground station. This second drone will act as a bridge between the first and the ground station. Any telemetry and data from the first drone will be forwarded to the ground station, and any instruction from the ground station will be forwarded to the first drone. The second drone might also get too far away from the ground station, requiring the usage of a third drone, and so on.

When a drone is on a mission and requires the assistance of a relay drone, it requests the ground station by publishing a message using ROS2. This message includes the sender's and receiver's identifiers, as well as the message's type, in this case, a relay request, and the Internet Protocol (IP) number and Media Access Control (MAC) addresses of the drone making the request. The ground station receives the request and searches for any drones that are available. If there are available drones, it responds affirmatively and sends a message to the relay drone to automatically be assigned to the mission. When a relay is no longer needed, the drone executing the mission will send a message to the ground station alerting it. Simultaneously, it delivers a message to the relay drone, which will return home. During the return journey, it sends a message to the ground station indicating that it is once again available to perform a relay.

A drone will use a network module to monitor its connection with the ground station while performing a mission. This module will calculate and publish various metrics, such as network quality and signal strength on a routine basis using ROS2 [11]. This data is then retrieved and analyzed by the drone. The drone will send a message to the ground station requesting a relay drone if the network's quality falls below a certain threshold. The drone continues its mission normally after sending this message. After experiencing a poor network and requiring a relay, the drone will continue to monitor the network's performance with the ground station, checking to see if the quality improves. If it does, and the value exceeds a certain threshold, the drone will no longer require a relay. The relay drone has more complex work to do during a mission: the network's quality will determine the distance this drone will maintain from the mission drone, meaning that these drones also monitor the connection with the mission drone.

The relay estimates the optimal spot to be located depending on the position of the mission drone after it takes off. Each time it calculates a new target position, it slightly adjusts the distance it is from the mission drone—if the signal's value is below the defined threshold, the network is good, and the distance should be increased; if the signal's value is above the threshold, the network quality is decreasing, and the distance should be decreased to improve the quality. This distance parameter is named "targetDistance". The user can also specify the initial distance that the relay will keep after taking off.

We used Batman Advanced routing protocol [56] to ensure that the relay drone properly forwards information between the necessary nodes, which is its primary role. When information does not reach every endpoint directly, this tool allows the nodes to act as a mesh network and automatically transmit information between nodes.

7 Experimental Results

Preliminary rehearsals with drones were performed in controlled scenario at IST and Universidade de Aveiro (UA) campi, and later in more realistic unstructured scenarios located in surrounds of Lisbon, but without sources of radiation. Finally, experimental results with drones were achieved in real scenarios of old mines of uranium ore, supervised by the Portuguese Environment Agency (APA) and the Portuguese Mining Exploration Company (EDM). Since some of the scenarios were inside Air Traffic Zones (ATZ) and Military Air Traffic Zones (MATZ), the flights were performed under the authorization of National Civil Aviation Agency (ANAC) and Portuguese Air Force (EMFA), respectively. All the experiments were performed in the centre of Portugal, but in different deposits of old mines. The locations of the selected deposits are intentionally not published here, as well as the absolute values of radiation measurements. Nonetheless, the highest values are always below the maximum allowed by safety standards for radiation protection.

Three scenarios were selected, the two first in November 2020 and the last during November 2021. The scenarios are depicted in Fig. 21, in the respective order. Each experiment was performed in a single day, including the time required for the briefing, explore the terrain, install the setup, perform the experiments, obtain results, debriefing, and uninstall the setup.

The first scenario is at an old uranium mine, where the ore extraction site is covered by water. The surroundings are hilly slopes with trees, bushes, and other vegetation. A small track is available and it is visible in the second image on the top of Fig. 21. The location of all the material removed in the past and considered wasteful was known. The area around the water is inaccessible, but the track was walkable and used to validate the results by expert operators from APA and EDM with their radiological and certified handheld devices.

A full equipped drone was used to perform the scouting operation and the 3D representation of the scenario was achieved, where the resulted map was segmented in to classes:

Fig. 21 The three real scenarios of old mines of uranium ore. The two first images correspond to the same scenario, but with different points of view

ground and not ground, as depicted in the first two images of Fig. 22. Then, a partially equipped drone, i.e., with a GMC, was used to perform the monitoring operation. As the first experimentation in a real scenario, both drones were teleoperated. The resulted heatmap is presented in the second image of Fig. 22, where an hotspot was identified. The inspection operation was not tested, since at the date of these experiments the inspection algorithms and auto-landing were not completely developed. Instead, and for validation, the CZT was manually placed at the coordinates of the identified hotspot and the following isotopes were identified in less than 15 min: ^{214}Pb and ^{214}Bi. A third partially equipped drone, only with the minimum sensors required to fly, was used to test the communication conditions in a real scenario to decide the strategies during next experiments. This drone has been used as the relay drone to assure the communication to the ground.

The second scenario is an olive groove, with sparse trees, ranging from 50 cm to 4 m in height. Trees are separated by 3 m or more. The ground is plowed, with only dirt and trees. The terrain is sloped but easy to walk to validate the results. The map of the ground radiation was not known, but public agencies knew that there is natural radiation above background levels.

A full drone and a partially equipped drone were used to perform the scouting and monitoring operations, respectively. The operations were performed in sequence. During the first operation the drone flied 20 m above the soil and took less than 5 min to cover the entire scenario, while in the second operation the flight was between 1 and 2 m of distance to the soil and with a longer time duration of approximately 20 min. The second scenario entailed a rough monitoring operation, not only because the proximity of the soil, but specially given the tree branches and leaves, not always detected by the proximity sensors. The resulted 3D segmented representation and heatmap were previously depicted in Fig. 16. A couple of hotspots were identified, but some of them presented a large blurred spot. Once again, the CZT was placed in the core of the hotspot and also in other hotspots with less intensity located in the opposite side of the scenario to validate the results. The identified isotopes were similar in all hotspots, namely ^{226}Ra, ^{214}Pb and ^{214}Bi, but with different intensities according to the hospot location. The second image of Fig. 22 illustrates a first person view close to some of the olive trees, with the main hotspot at the end.

The experiments in the first and second scenarios were performed during two consecutive days, one day for each scenario.

In the third scenario, the old uranium mine was located on the top of a hill, with a heap close to a local road. The experiments in this scenario were performed one year later. The map of the ground radiation around the mine was known, but an expected hotspot was preliminary identified by a manual survey using MARIA only along the road, in a recent past. There were suspicious additional hotspots located along the hill, but with harsh conditions for human access. The resulted 3D map and heatmap are presented in Fig. 23. The exact location of the hotspot was identified slightly above the road. Surprisingly no additional hotspots were detected along the hill. The results of the algorithms for auto-landing were

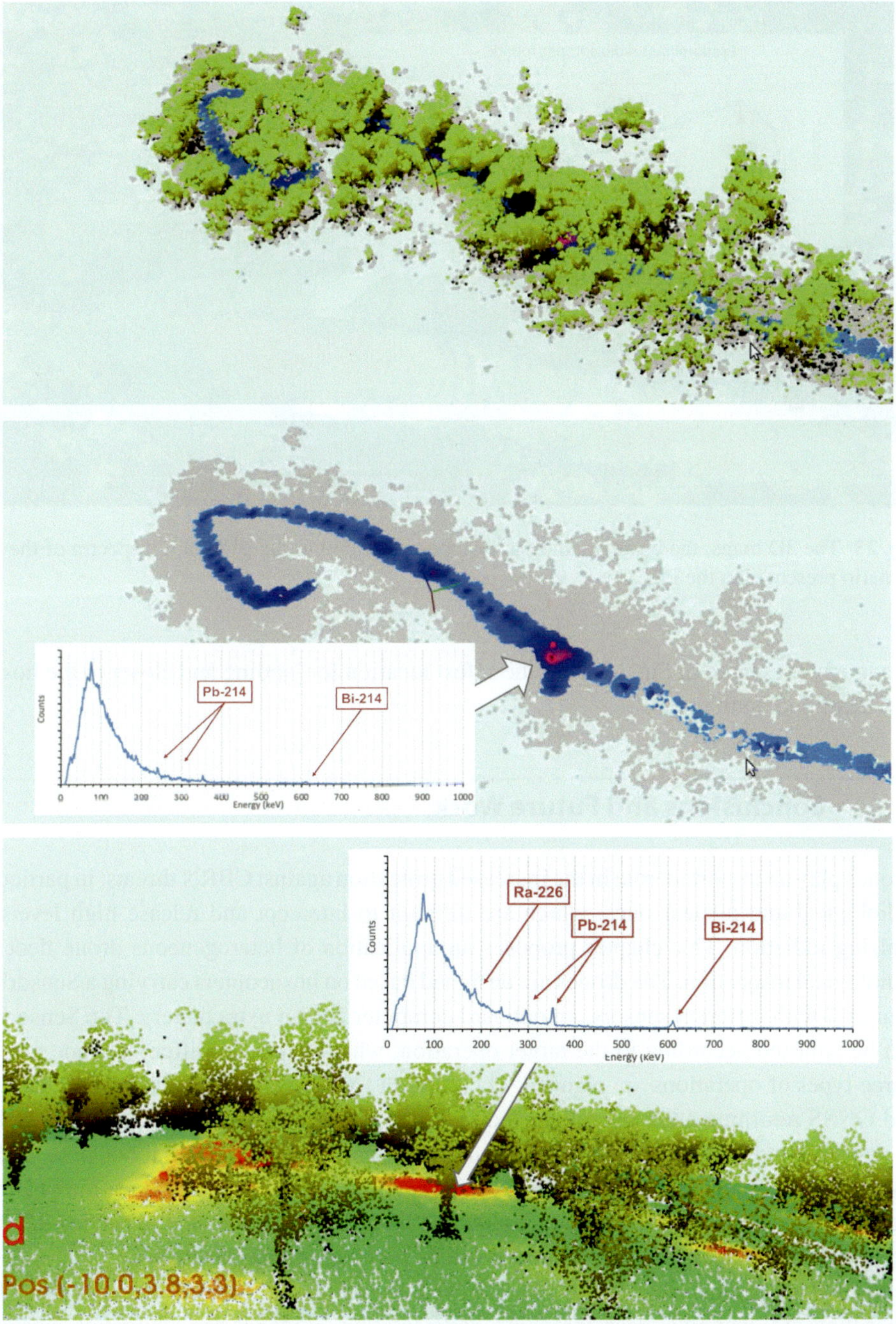

Fig. 22 The 3D maps, the heatmap and the isotopes identified in the gamma ray spectra of the first two scenarios presented in Fig. 21

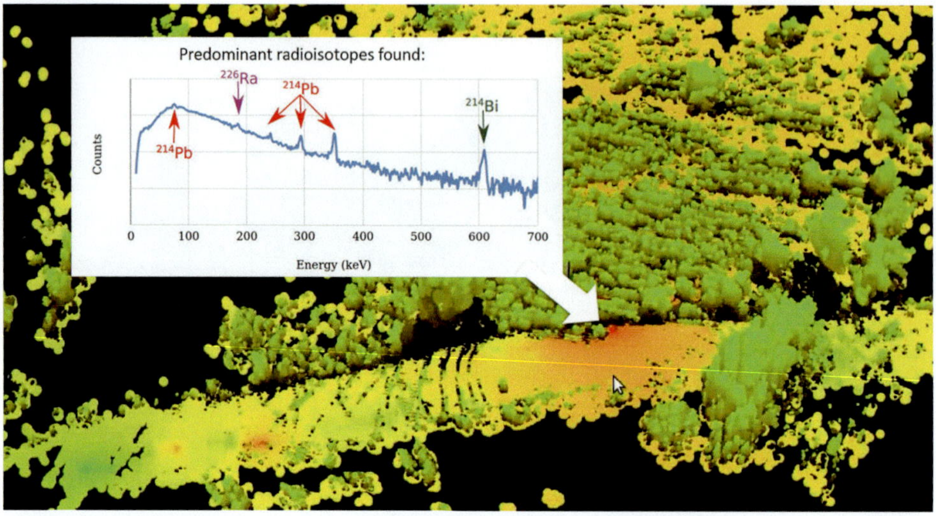

Fig. 23 The 3D maps, the heatmap and the isotopes identified in the gamma ray spectra of the last scenario presented in the Fig. 21

previously presented in Fig. 18, with the safest location for landing and closer to the hospot was identified on the road.

8 Conclusions and Future Work

Drones play an important role in the increased protection against CBRN threats, in particular radiological and nuclear ones, which are difficult to intercept and release high levels of ionizing radiation. This chapter proposes an application of heterogeneous drone fleet for radiological inspection. The drones are designed based on hexacopters carrying a Sensorbox equipped with different sensors, an onboard computer and an extra battery. The Sensorbox can be adjusted according to the target operation, which defines a different drone profile. Three types of operations are planned: (1) scouting for mapping the scenario with LIDAR and GNSS returning a 3D georeferenced map, (2) monitoring using the GMC and GNSS to provide a radiological intensity map and the estimation of the potential hotspots, and 3) inspection where the Gamma Ray Spectrometer (GRS) is used to characterize the potential radionuclides present in the scenario. The scouting operation is the first to be executed, while the monitoring and inspection operations can be performed simultaneously.

Several algorithms were developed and used according to the respective operation. The algorithms are: mapping, georeferencing, segmentation, guidance and navigation, heatmap estimation, hotspots identification, safe auto-landing, hotspots identification, radiological sensors calibration, radionuclides identifications and communication.

Experimental results were achieved in real scenarios of old mines of uranium ore. During the scouting operation the drones were teleoperated. The other two operations were performed in sequence, i.e., first the monitoring and then the inspection. All the algorithms are able to run in real time, but were tested offline, i.e., after the drones landing, except the guidance and navigation, safe auto-landing and communication. Different hotspots were identified and, when accessible for human being, validated by experts with radiological and certified handheld devices.

A full equipped drone performs any operation, but the endurance can be optimized when the Sensorbox is equipped with the required sensors only. The most expensive sensors are the LIDAR and the CZT: approximately 20x more than any other onboard sensor. One or few drones equipped with LIDAR can map a scenario. Multiple drones equipped with GMC, which requires a more affordable Sensorbox, i.e., without LIDAR neither CZT, can monitor the respective AoI. A single drone equipped with the CZT was used, but takes a very long time for inspection, specially when multiple hotspots are identified. Therefore, multiple drones equipped with CZT can speed up, but increases significantly the budget, since the CZT is one of the most expensive sensors.

The used LIDAR is suitable for the selected scenarios, returning a good 3D representation. The selected GMC has low sensibility for the reduced levels of radiation expected in the old mines, discontinued several years ago. However, heatmaps and hotspots were achieved with drones flying below 2 m to the soil and up to 30 cm/s of speed. Higher distances to the soil or higher speeds requires a more sensible sensor, which increases significantly the price.

Tracking of radiological activity along the time in the same terrain was not tested given the possible and reduced time duration allocated for the experiments of the project. However, the tracking of radiological activity is a potential feature available in the proposed drone application that should be explored in the future work. The supervision of radiological activity along the time is important, mainly for the identification of potential hazardous, such as leakage detection and other unexpected changes.

Drones confirmed to be an important tool to reach, monitor and inspect places with difficult accessibility. However, extra tools such as ground vehicles or human operators are still necessary and can be an important complement to improve the heatmap estimation, hotspot detection and its characterization, since drones are not able to land in all terrains.

Beyond radiological scenarios, a next challenge would be to extend the application of drone in the protection against chemical and biological threats.

Acknowledgements This chapter is a result of the project FRIENDS [35], supported by the Portuguese Foundation for Science and Technology (FCT), Compete 2020 and Lisboa 2020 under the PORTUGAL 2020 Partnership Agreement, through the European Regional Development Fund (ERDF). Institute for Plasmas and Nuclear Fusion (IPFN) and LARSyS Institute for Systems and Robotics (ISR) activities also received financial support from FCT though projects UIDB/50010/2020 and UIDP/50010/2020 and UIDB/50009/2020. The authors acknowledge to the APA and to the EDM for the authorization, consultancy and support to the experimental tests performed by the team of the project in the terrain.

References

1. R. Gladstone, Syria used chlorine in bombs against civilians, report says (24 Aug 2016). https://www.nytimes.com/2016/08/25/world/middleeast/syria-used-chlorine-in-bombs-against-civilians-report-says.html. Accessed 6 June 2022
2. N.D. Kristof, Poison gas fills Tokyo subway; six die and hundreds are hurt (20 Mar 1995). https://www.nytimes.com/1995/03/20/world/poison-gas-fills-tokyo-subway-six-die-and-hundreds-are-hurt.html. Accessed 6 June 2022
3. Legionella from Wikipedia. https://en.wikipedia.org/wiki/Legionella. Accessed 6 June 2022
4. Rapid Risk Assessment: Outbreak of Legionnairesdisease in the Lisbon area, Portugal, European Centre for Disease Prevention and Control, An Agency of the European Union (14 Nov 2014)
5. Radioactivity decay from Wikipedia. https://en.wikipedia.org/wiki/Radioactive_decay. Accessed 6 June 2022
6. Atomic bombings of Hiroshima and Nagasaki from Wikipedia. https://en.wikipedia.org/wiki/Atomic_bombings_of_Hiroshima_and_Nagasaki. Accessed 6 June 2022
7. Chernobyl and Fukushima: lessons not learned. https://www.euronews.com/2016/04/20/chernobyl-and-fukushima-lessons-not-learned. Accessed 6 June 2022
8. J. Chamot, R. Atkins, Leading engineers and scientists identify advances that could improve quality of life around the world, National Science Foundation (2008). https://www.nsf.gov/news/news_summ.jsp?cntn_id=111158. Accessed 6 June 2022
9. Grand Challenges for Engineering—Prevent Nuclear Terror, National Academy of Engineering. http://www.engineeringchallenges.org/challenges/nuclear.aspx. Accessed 6 June 2022
10. F. Mendes, M. Barros, A. Vale, B. Gonçalves, Radioactive hot-spot Localization and identification using deep learning. J. Radiol. Prot. **42**(1) (2022). https://doi.org/10.1088/1361-6498/ac1a5c
11. Robot Operating System (ROS). http://wiki.ros.org/. Accessed 16 Dec 2022
12. L. Pinto, A. Vale, Y. Brouwer, J. Borbinha, J. Corisco, R. Ventura, A. Silva, A. Mourato, G. Marques, Y. Romanets, S. Sargento, B. Gonçalves, Radiological scouting, monitoring and inspection using drones. Sensors **21**, 3143 (2021). https://doi.org/10.3390/s21093143
13. L. Marques, A. Vale, P. Vaz, State-of-the-art mobile radiation detection systems for different scenarios. Sensors **21**, 1051 (2021). https://doi.org/10.3390/s21041051
14. Y. Brouwer, A. Vale, R. Ventura, Informative path planner with exploration-exploitation trade-off for radiological surveys in non-convex scenarios. Robot. Auton. Syst. **136**, 103691 (2021). https://doi.org/10.1016/j.robot.2020.103691. (February)
15. I. Maza, A. Ollero, Multiple UAV cooperative searching operation using polygon area decomposition and efficient coverage algorithms, in *Distributed Autonomous Robotic Systems*, ed. by R. Alami, R. Chatila, H. Asama, vol. 6 (Springer, Tokyo, 2007). https://doi.org/10.1007/978-4-431-35873-2_22
16. P. Pereira, Vertical navigation for UAV that includes intelligent landing, M.Sc. Thesis, Instituto Superior Técnico, Portugal (2021)
17. J. Borbinha, Y. Romanets, P. Teles, J. Corisco, P. Vaz, D. Carvalho, Y. Brouwer, R. Luís, L. Pinto, A. Vale, R. Ventura, B. Areias, A. Reis, B. Gonçalves, Performance analysis of geiger-müller and cadmium zinc telluride sensors envisaging airborne radiological monitoring in NORM sites. Sensors **20**, 1538 (2020). https://doi.org/10.3390/s20051538
18. D. Macedo, A. Vale, Y. Brouwer, B. Gonçalves, H. Fernandes, Radioactive hot-spot detection using unmanned aerial vehicle surveillance, in *Proceedings of the International Conference on Advancements in Nuclear Instrumentation Measurement Methods and their Applications (ANIMMA2019), EPJ Web of Conferences*, vol. 225 (2020), p. 06005. https://doi.org/10.1051/epjconf/202022506005

19. J. Vilaça, A. Vale, F. Cunha, Design optimization of a ducted-drone to perform inspection operations, in *ROBOT2019: Fourth Iberian Robotics Conference. Advances in Intelligent Systems and Computing*, vol. 1092 (Springer, Cham, 2019). https://doi.org/10.1007/978-3-030-35990-4_1
20. Low/medium level waste from Koeberg dumped at Vaalputs, South Africa. https://koebergalert.org/nuclear-waste/. Accessed 5 June 2022
21. Retortillo mine exploitation of uranium in Spain-Portugal border, Spain. https://ejatlas.org/conflict/open-air-mine-exploitation-of-uranium-in-spain-portugal-border-spain. Accessed 5 June 2022
22. Increase radiation dose across Europe from Wikipedia. https://en.wikipedia.org/wiki/Effects_of_the_Chernobyl_disaster#cite_note-RFI_24-14. Accessed 6 June 2022
23. Radiation contour map of the Fukushima Daiichi Accident. This is arranged from the map made by Yukio Hayakawa with his permission 15 Mar 2012. https://www.mdpi.com/2227-9067/2/1/39/htm. Accessed 6 June 2022
24. IAEA Incident and Trafficking Database (ITDB), Incidents of Nuclear and Other Radioactive Material out of Regulatory Control. https://www.iaea.org/sites/default/files/20/02/itdb-factsheet-2020.pdf. Accessed 1 July 2020
25. Orano DS, Robots—Dismantling and Services. https://www.orano.group/en/press/press-kit/robots-dismantling-and-services. Accessed 8 June 2022
26. M. Lozano, A. Fernandez, Reward project. https://project-reward.eu/. Accessed 8 June 2022
27. D. Kanaan, S. Dogny, T. Varet, Robot for investigations and assessments of nuclear areas, in *4th International Conference on Advancements in Nuclear Instrumentation Measurement Methods and their Applications (ANIMMA)* (2015). https://doi.org/10.1109/ANIMMA.2015.7465612
28. I.S.Q. Gamma-Ex (2015). https://www.isq.pt/projeto/gamma-ex/. Accessed 8 June 2022
29. ENCIRCLE: European CBRN Innovation for the Market Cluster, ENCIRCLE Consortium—European Union. http://encircle-cbrn.eu/. Accessed 8 June 2022
30. C. Cai, B. Carter, M. Srivastava, J. Tsung, J. Vahedi-Faridi, C. Wiley, Designing a radiation sensing UAV system, in *2016 IEEE Systems and Information Engineering Design Symposium (SIEDS)* (IEEE, 2016). http://ieeexplore.ieee.org/document/7489292/
31. K. Boudergui, F. Carrel, T. Domenech, N. Guénard, J.-P. Poli, A. Ravet, V. Schoepff and R. Woo, "Development of a drone equipped with optimized sensors for nuclear and radiological risk characterization", in 2011 2nd International Conference on Advancements in Nuclear Instrumentation, Measurement Methods and their Applications, IEEE 2011. http://ieeexplore.ieee.org/document/6172936/
32. Mirion Technologies, SPIR-Explorer Sensor. https://www.mirion.com/products/spir-explorer-sensor-light-wide-range-radiological-detection-and-identification-sensor. Accessed 10 June 2022
33. Aerial Radiation Detection. https://usnuclearcorp.com/2019/02/aerial-radiation-detection/. Accessed 10 June 2022
34. G. Knoll, *Radiation Detection and Measurement*, 3rd edn. (Wiley, New York, NY, USA, 2010)
35. "Fleet of dRones for radIological inspEction, commuNication anD reScue", https://www.ipfn.tecnico.ulisboa.pt/FRIENDS/. Accessed 8 July 2022
36. Mobile Application for Radiation Intensity Assessment, IPFN-IST 2017. https://www.ipfn.tecnico.ulisboa.pt/MARIA/. Accessed 8 July 2022
37. A. Peeva, New drone technology for radiological monitoring in emergency situations, IAEA Department of Nuclear Sciences and Applications (2021), https://www.iaea.org/newscenter/news/now-available-new-drone-technology-for-radiological-monitoring-in-emergency-situations. Accessed 8 July 2022
38. M. Minar, J. Naher, Recent advances in deep learning: an overview (2018). https://doi.org/10.13140/RG.2.2.24831.10403

39. I. Goodfellow, Y. Bengio, A. Courville, *Deep Learning* (MIT, 2017)
40. J. Redmon, S. Divvala, R. Girshick, A. Farhadi, You only look once: united, real-time object detection. CoRR (2015). arXiv:1506.02640
41. K. Johnson, Humanitarian operations in Ukraine prompt drone design tweaks for draganfly 6 June 2022. https://www.flyingmag.com/humanitarian-operations-in-ukraine-prompt-drone-design-tweaks-for-draganfly/. Accessed 13 June 2022
42. 3D scanning, augmented reality, and more for mobile devices. https://structure.io/. Accessed 22 June 2022
43. Intel RealSense Depth Camera D415. https://www.intelrealsense.com/depth-camera-d415/. Accessed 16 June 2022
44. FLIR Duo R. http://www.flir.comsupportproductsduo-r/#Overview. Accessed 16 June 2022
45. NVIDIA Jetson Nano Developer Kit. https://developer.nvidia.com/embedded/jetson-nano-developer-kit. Accessed 16 June 2022
46. Mularie, World geodetic system 1984-its definition and relationships with local geodetic systems, in *Department Of Defense* (NIMA USA; National Imagery and Mapping Agency: St. Louis, MO, USA, 2000)
47. T. Qin, S. Cao, A-LOAM: advanced implementation of LOAM. https://github.com/HKUST-Aerial-Robotics/A-LOAM. Accessed 1 March 2021
48. J. Zhang, S. Singh, LOAM: lidar odometry and mapping in real-time, in *Proceedings of Robotics: Science and Systems* (MIT Press, Cambridge, MA, USA, 2014)
49. G. Konstantinos, Overview of the RANSAC Algorithm, Image Rochester NY (2010)
50. A. Mammone, M Turchi, N. Cristianini, Support vector machines, WIREs Computational Statistics (2009)
51. ROS move_base package documentation. http://wiki.ros.org/move_base. Accessed 16 Dec 2022
52. M. Fischler, R. Bolles, Random sample consensus: a paradigm for model fitting with applications to image analysis and automated cartography. Commun. ACM **24**, 381–395 (1981)
53. W. Huang, Optimal line-sweep-based decompositions for coverage algorithms, in *Proceedings 2001 ICRA. IEEE International Conference on Robotics and Automation (Cat. No.01CH37164)*, vol. 1 (Seoul, IEEE, 2001), pp. 27-32, https://doi.org/10.1109/ROBOT.2001.932525
54. B. Zhang, G.S. Sukhatme, Adaptive sampling for estimating a scalar field using a robotic boat and a sensor network, in *Proceedings 2007 IEEE International Conference on Robotics and Automation*. (Rome, IEEE, 2007), pp. 3673–3680. https://doi.org/10.1109/ROBOT.2007.364041
55. H. Carvalho, A. Vale, R. Marques, R. Ventura, Y. Brouwer, B. Gonçalves, Remote inspection with multicopters, radiological sensors and SLAM techniques, in *Proceedings of the 5th International Conference on Advancements in Nuclear Instrumentation Measurement Methods and their Applications (ANIMMA2017), EPJ Web of Conferences*, vol. 170 (2018), p. 07014. https://doi.org/10.1051/epjconf/201817007014
56. B, Sliwa, S. Falten, C. Wietfeld, Performance evaluation and optimization of B.A.T.M.A.N. V routing for aerial and ground-based mobile Ad-hoc networks, in *2019 IEEE 89th Vehicular Technology Conference (VTC2019-Spring)*. https://doi.org/10.1109/VTCSpring.2019.8746361

Alberto Vale Alberto Vale is Assistant Professor at Department of Electrical and Computer Engineering (DEEC) of Instituto Superior Técnico (IST) since 2021 and Senior Researcher at Institute of Plasmas and Nuclear Fusion (IPFN/IST) since 2008 working on the area of path planning and navigation of mobile robots applied to the Remote Handling systems of International Thermonuclear Experimental Reactor (ITER) and EUROfusion DEMO (DEMOnstration Power Station). He received the Licenciatura and Ph.D. degrees at IST in 1999 and 2005, respectively. From 1999 to 2005 he was a researcher at the Institute for Systems and Robotics (ISR), working on mobile robotics. From 2006 to 2008 he was Chief RD Engineer and co-founder of Albatroz Engineering S.A. From 2012 to 2017 he was responsible teacher of two classes in the graduation of Othoprosthesis course at Lisbon School of Health Technology (ESTeSL) of Polytechnic Institute of Lisbon (IPL).

Rodrigo Ventura Rodrigo Ventura (Assistant Professor) received the Licenciatura (1996), M.Sc. (2000), and Ph.D. degree (2008), in ECE from Instituto Superior Técnico (IST), Lisbon, Portugal. He is a (tenured) Assistant Professor at IST, and a member of Institute for Systems and Robotics (ISR). He has several publications in international journals and conferences, on various topics intersecting Robotics and Artificial Intelligence. He is founding member of the Biologically-Inspired Cognitive Architecture society. He participated in several international and national research projects. In particular, he was Principal Investigator of the national project AuReRo (Human-robot interaction with field robots using augmented reality and interactive mapping, 2011–2014), and coordinator of the the IST participation in the EU LdV project USORA (Unified Solution of Remote Access in Practical Vocational Engineering Education, 2013–2015). He is currently Principal Investigator of the national project HARODE (Human-aware service robots for domestic environments). His current research interests include human-robot interaction, human-aware navigation, decision-making under uncertainty, and cognitive robotics, targeting the space and service robot application domains. He is the co-author of two national patents on innovative solutions for robotic systems.

José Corisco José Corisco is a biologist graduated by the Faculty of Sciences in the University of Lisbon holding an auxiliary researcher position granted by the former Instituto Tecnológico e Nuclear. He has several years of experience in the fields of environmental radioactivity and experimental radioecology with participations in post-Chernobyl European funded projects related to the transfer of artificial radionuclides in the aquatic food chains. He has been recently involved in an academic and private enterprise partnership for the development of a robotic system for sediment monitoring in intertidal mudflats (http://www.introsys.eu/pt/#robotica), as a primary tool for the determination of radionuclides in estuarine environments. Other interests lay in the mobilization and bioaccumulation of uranium decay radionuclides with enhanced concentrations in stacks of industrial waste from phosphate industry and uranium mining.

Norberto Catarino Norberto Catarino received the B.Sc. in physics from the University of Porto, the M.Sc. in Nuclear and Particle Physics and Ph.D. in Physics Engineering from the University of Lisbon. In recent years, he has established himself in the field of ion beam techniques, focusing on plasma-facing component research. He started his research activity at CTN in 2009, working on magnetic nanolayers of wide bandgap semiconductors for spintronic applications. He specialized in the use of detector, electronic and nuclear acquisition systems. He worked in the European SPIRIT project, and gradually became involved in the research on materials for fusion reactors due to his participation in JET work program of the Ion Beam Laboratory. He is the Portuguese Chief Scientific Investigator of a Coordinated Research Projects (CRP-F11023) from the IAEA, to provide a platform for efficient collaboration and knowledge transfer between scientists from developing and developed Member States, with a focus on the use of accelerator-based techniques for fusion related technologies.

Nuno Veiga Nuno Veiga is a final year masters student in Electrotechnical and Computer Engineering at Instituto Superior Técnico, majoring in Control, Robotics and Artificial Intelligence. He is highly interested in autonomous systems, robotics and control and is eager to learn more and work hard in order to make things move on their own. His experiences include working in an autonomous Formula Student car and mapping with a drone in simulation.

Susana Sargento Susana Sargento is a Full Professor in the University of Aveiro and a senior researcher in the Institute of Telecommunications, where she is leading the Network Architectures and Protocols group. She was a visiting Ph.D. student in Rice University (2000–2001), and a Guest Faculty in Carnegie Mellon University (2008). Susana has been leading research projects with telecom operators and OEMs. She has organized several international conferences and workshops, and has also been a reviewer of conferences and journals, such as IEEE Networks, IEEE Communications. Susana has co-founded a vehicular networking company, Veniam (www.veniam.com), she is the winner of the 2016 European Prize for Women Innovators, and the winner of Femina 2020 prize in Science. She is also the co-coordinator of the national initiative of digital competences in the research axis INCoDe.2030, belongs to the evaluation committee of the Fundo200M (www.200m.pt), and she is one of the Scientific Directors of CMU-Portugal Programme. Her main research interests, with more than 400 scientific papers, are in the areas of self-organized networks, in ad-hoc and vehicular network mechanisms and protocols, Beyond 5G, ITS and content distribution networks. She regularly acts as an Expert for European Research Programmes.

Applications of UAVs in Search and Rescue

Reem Ashour, Sara Aldhaheri, and Yasmeen Abu-Kheil

Abstract

Unmanned aerial vehicles (UAVs) have grown in popularity over the last decade, with applications ranging from agriculture to commercial and widespread use in search and rescue (SAR) operations. The utilization of UAVs has been shown to enhance the effectiveness of these missions by providing a fast, agile, and cost-effective means of gathering information and performing tasks in challenging environments. This chapter aims to examine the current usage of UAVs in SAR missions, identify the challenges associated with their use, and provide a direction for future research on the topic. This chapter provides a detailed description of the various types of UAVs and sensors that are required to carry out successful SAR missions. Our research has found that UAVs are commonly utilized during the rescue and recovery phases in both indoor and outdoor settings for tasks such as mapping, detection, and intervention. While the number of studies on the deployment of UAVs in SAR operations has significantly increased in the past few years, further investigation is necessary to specifically focus on drone intervention in these missions.

R. Ashour (✉) · S. Aldhaheri
Autonomous Robotics Research Center, Technology Innovation Institute, Abu Dhabi, UAE
e-mail: reem.ashour@tii.ae

S. Aldhaheri
e-mail: sara.aldhaheri@tii.ae

Y. Abu-Kheil
Higher Colleges of Technology, Abu Dhabi, UAE
e-mail: yabukheil@hct.ac.ae

M. Abdelkader and A. Koubaa (eds.), *Unmanned Aerial Vehicles Applications: Challenges and Trends*, Synthesis Lectures on Intelligent Technologies,
https://doi.org/10.1007/978-3-031-32037-8_5

1 Overview

Improvements in the capabilities of unmanned aerial vehicles (UAVs) have enabled the creation of effective applications in missions when time is critical, such as search and rescue (SAR). A SAR mission must balance the need to act quickly enough to save as many potential survivors as possible with the desire to act cautiously enough to prevent putting anyone else in danger. UAVs can help achieve the primary goal of saving lives by engaging in activities such as detecting victims and risks, providing situational awareness for first responders, interacting with the situation, and automating support activities.

1.1 Motivation

The advances in the UAV's technological capabilities have allowed the development of efficient applications in different areas such as military, mapping, medical, and wireless communication, among other functions [1]. This additional support comes particularly useful when tackling time-sensitive missions. Historically, these missions would be performed by skilled experts and would cost money and effort to execute successfully. Still, such options have several drawbacks, including the time needed and amount of dispatch (i.e., trained dogs [2]) for locating missing people, which poses an issue as they are limited. Searches that are human-reliant can be time-consuming and would require many volunteers and professionals.

Over the past few years, UAVs have demonstrated their value in a variety of search and rescue scenarios, and they continue to do so. UAVs provide a variety of unique capabilities that are useful to different emergency response providers and have resulted in the prevention of the loss of numerous lives. For instance, in 2019, a hiker in his sixties became trapped in Utah's Snowy Canyon State Park, and a drone was dispatched as a light source to boost the visibility of the rescue crew [3]. In a separate incident, an 88-year-old man who had been reported missing in the state of Texas has been located thanks to a drone equipped with a thermal camera [4]. A rescue crew in Iceland in 2018 used a drone to locate two cousins who were stranded on the side of a high mountain [5]. With the help of the drone and its equipped lighting, the search and rescue team could detect the victims and communicate their location to rescuers on the ground [6].

Emerging technology has grown in recent years with increasing accuracy and durability. With applications similar to military operations, the progress from either area of knowledge is transferable.

The main reason for using UAVs in search and rescue operations is that these vehicles can significantly reduce the risk to humans, make search and rescue operations more effective, and expand the operation's capabilities beyond what humans can do.

1.1.1 Reduce Human Risk

Reducing human risk could be demonstrated by the fact that robots can reach regions that are either difficult for people to access or dangerous for humans to visit due to potentially hazardous materials. Although intervention by UAVs is a new research field, the use of UAVs has been essential for locating missing persons (victims) and providing life-saving resources. One of the early instances of using UAVs in search and rescue is the case of the missing man who was not found despite the efforts of the police and rescue team's search for three days but was found by a drone after only 20 min of searching [7]. In addition, rescue robot teleoperation while also looking for victims in such complicated situations may be highly stressful, resulting in cognitive and physical exhaustion [8, 9]. A minimum of two rescuers is suggested to manage a single robot due to the mental and physical overload one operator might feel when making the complete search effort utilizing teleoperated robots [9–11]. In search and rescue, dividing the task between humans and robots can reduce an operator's cognitive effort while simultaneously allowing rescuer robots to benefit from the operator's expertise and knowledge. Task sharing can minimize the number of activities that the rescuer must do. As a consequence, teleoperated robots minimizes the rescuers' mental stress.

1.1.2 Perform Search and Rescue More Efficiently

Rescue robots can complete tasks more quickly and accurately than humans. UAVs are ideally suited for capturing high-quality aerial images and videos and gathering massive quantities of imaging data. The benefits of using these high-resolution photos to construct 3D maps and interactive 3D models are numerous. Disaster relief organizations, for instance, might benefit from having access to 3D maps of affected areas before sending first responders into potentially dangerous circumstances. For example, the team led by Robin Murphy only needed seven hours to collect high-resolution data due to the use of UAVs. However, it would have taken them two to three times as long if they had used more traditional techniques [7, 12]. Also, the Global Positioning System (GPS) makes it possible to program and send unmanned aerial vehicles exactly where they need to go in outdoor environments so that they can do their jobs quickly and efficiently.

To extend the current rescuers' capabilities, sensors can generally work while being exposed to challenging situations such as a dusty atmosphere, physical action, toxic gases, harmful substances, or radiation [8, 13]. Second, the adaptability of robot mobility is critical for search and rescue tasks. For example, ground robots, humanoid robots, aerial robots, surface robots [14], underwater robots [15], and snake-like robots [16] are all specialized for specific environments [8]. Finally, robots can gather massive amounts of data in a short amount of time while maintaining a high level of quality [8]. This can be done with autonomous exploration and mapping, three-dimensional reconstruction, identifying victims, and evaluating the environment around them.

1.2 UAVs Capabilities

Multiple sensors are required to understand UAV flight information. The result of a survey sent to more than 300 people from different institutions with regards to the data to be displayed for the operator in the Graphical User Interface to optimize the performance of the SAR operations is as follows: GPS position, graph map/terrain map, flight trajectories (height, distance, speed, etc.), video transmission, thermal imaging, battery capacity, wind speed, starting position, 3D map, 3D speed projections of buildings, flight track on the map, number of people/signals found by drone, time of flight, air temperature [17].

To achieve suitable results depending on a reconnaissance assignment, appropriate sensors must be chosen. In addition to the aforementioned internal sensors that support the navigation of the UAVs (i.e. IMUs, GPS, tilt sensors, etc.), there is a range of sensors to choose from according to SAR type.

Acoustic sensors, optical cameras, and infrared (heat detecting) cameras are the traditional sensors that SAR operators utilize in their missions. Several of these instruments, apart from infrared cameras, require a line of sight or the detection equipment is situated quite close to the target. This prevents the UAV from helping rescuers find victims and instead forces it to analyze the damage caused by disasters. The acceptance and use of personal cell phones, on the other hand, have skyrocketed during the previous ten years [18]. When turned on and in use, cell phones release a variety of signals. In general, the phrase "wireless sensor" refers to a device that tries to detect the presence of wireless signals, much like a laptop can look for Wi-Fi signals and try to connect to them. Some writers have started to study the potential to passively or actively detect and interact with these signals. The possibility of adding wireless sensor technology to the SAR toolkit exists at the junction of technologies and needs created by the growth of cellular devices, their detection capabilities, and the rise in the popularity of UAVs. Responders may be able to swiftly scan a disaster area and find potential victims by equipping one or more UAVs with the wireless sniffer. The study in [19] shows an extensive review of the use of UAVs, their sensors, and flight operations optimization in the event of their need in SAR operations.

1.3 Rescue UAV Types

This section focuses on aerial robotic support in search and rescue using UAVs, and the different contributing factors of each UAV type. Rescue UAVs come in different types in order to satisfy the requirements of the application and fit the limitations of the environment. Some of the modern UAVs types include fixed wing, rotary wing, lighter-than-air, tethered, and vertical take off and landing (VTOL) platforms. These vehicles are similar to planes, helicopters, blimps, and kites, respectively. Each of these is further divided into more designs, which may include seaplanes and gliders for the fixed-wing UAVs or quadcopters for the rotary wings. The different types of UAVs are shown in Table 1.

Table 1 UAV types, characteristics, and limitations

Type	Examples	Characteristics	Limitations
Fixed-wing	Gliders	• Long range • Fast • Stable	• Few intervention options • Take-off and landing area required • Difficult to fly and maneuver
Rotary	Tricopters, quadcopters	• Hovering capabilities • Small - does not require large take-off • Easy to operate • Multiple agents capabilities	• Battery life • Can be unstable and complex • Small payload capacity • Slower than fixed-wing UAVs
Lighter-than-air	Blimps	• Hovering capabilities • Low-power consumption • Equips more sensors • Equips heavier sensors with higher resolution	• Relatively slower • Occlusions caused by own shadow
Tethered	Tethered quadcopters	• 24-hour of uninterrupted flight time • Automated flight profile • Hierarchy of multilevel fail-safe	• Short range
VTOL	rotor-craft and powered-lift	• Efficient manoeuvrability • Ability to hover steadily • Capability to takeoff/land in heavy wind	• Slower, and therefore less coverage

The five categories of UAVs offer particular characteristics, and, depending on these properties, their usage is determined as presented in Table 1. For long-range missions with little to no intervention required, a fixed-wing UAV may be utilized. Fixed wings are able to map points of interest. On the other hand, if the area of interest is determined and requires further more in-depth analysis, a hovering rotary-wing vehicle may be more useful [20].

Examples of autonomous quadrotors, fixed-wing and rotatory-wing vehicles equipped with GNSS sensors and RGB cameras for the search of people in emergency scenarios are available in [21–23]. A survey on UAVs for SAR operations reveals the predominance of RGB cameras as the main or only sensor in use [19], without taking inertial and GNSS units into account, and describes a description of the various types of aerial SAR robots used within the ICARIUS project [24].

The research in [25] provides a platform for quick UAV-based 3D mapping during SAR missions, and they show a typical quadrotor on that platform that has a full range of sensors. In [25], the UAV is a DJI S1000+ octocopter that communicates with a ground station mounted on a firefighter command vehicle using a proprietary radio link that can transmit data at speeds of up to 300 Mbps over a distance of 300 m. A color point cloud is generated by reflectance (from the laser measurements) and temperature (from the infrared camera). In [26], the Duckiefloat flying robot is presented as a collision-tolerant blimp-type UAV that can be easily maneuvered and carry several kilograms of payload for long-term autonomy in underground environments. The Duckiefloat has an RGBD camera (Intel D435) and an infrared sensor, which are used to avoid collisions in hard-to-reach subterranean areas of interest and to have Simultaneous Localization and Mapping (SLAM) capabilities with artifact search using a deep learning approach.

However, the majority of these and other investigations presume that aerial robots move across wide areas and are not overly confined by environmental factors (e.g. smoke, wind, or low light). The DARPA SubT challenge, discussed in detail in later sections, offers examples of robots made for conditions with greater difficulty [27].

2 Phase of Disasters and Search and Rescue Categories

The types of disasters that affect UAV platforms and payloads include natural catastrophes. These events may cover broad areas making a UAV's bird's-eye perspective important for creating situation awareness and identifying locations or people requiring rapid evacuation or aid. The most crucial parts of man-made disasters are hidden under rubble. Small UAVs with onboard sensors offer the most potential for man-made disasters. The primary goals of the rescue operation are to locate victims and collect information of a more general nature. A review of search and rescue operations is given, focusing on the disaster phase and the UAVs' role in these phases. After that, it shows the various search and rescue environments and their dangerous conditions. Then presents the conventional SAR scenario. Finally, it offers research on using UAV rescue robots in the real world.

2.1 Phases of Disaster Operations

The four phases of disaster operations are preparedness, prevention, rescue, and recovery, as shown in Fig. 1. Preparation and prevention are pre-incident activities, whereas rescue and recovery are post-incident activities. Rescue UAVs can assist with post-incident tasks. The term "rescue" encompasses a wide range of activities immediately following a catastrophic incident that directly addresses urgent dangers to the survival of individuals impacted by the disaster, such as finding and evaluating the survival health condition, expanding the scope of the risk, stabilizing the situation, and evacuating survivors. People use "rescue" and "search and rescue" interchangeably. "Search and rescue" refers to the critical disaster management tasks to help the rescue teams and victims. The term "search" refers to efforts to locate survivors, whereas "rescue" refers to activities aimed at extracting survivors. Finally, recovery works to reduce long-term hazards to life and property. The process in the rescue phase can be accurately classified into two tracks: strategic and tactical. Strategic operations

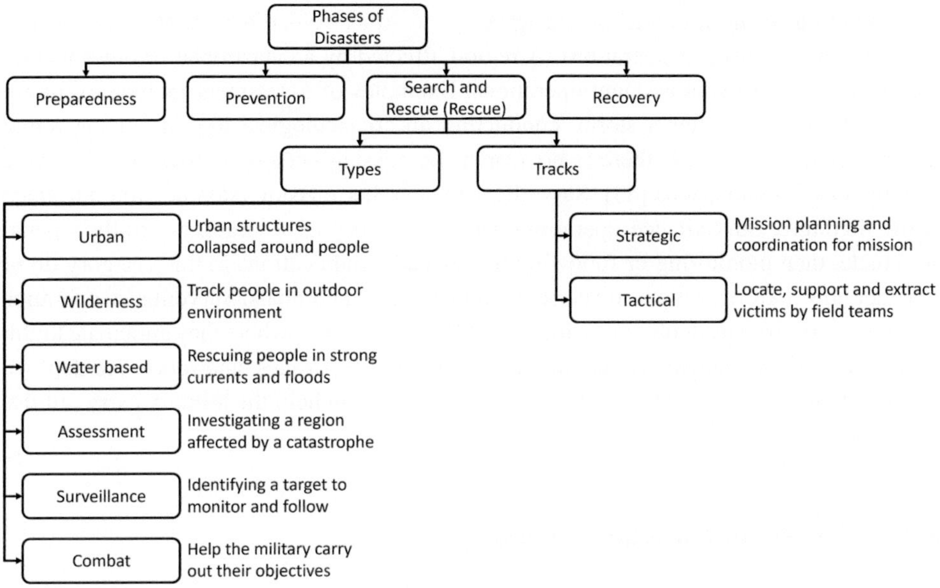

Fig. 1 Search and rescue phases

are concerned with mission planning and coordination for the entire mission and planning the entire rescue task, including the tools, logistics, and type of communication. In contrast, tactical operations are concerned with how field teams locate, support, and extract victims.

2.2 Operational Environments

SAR operations can be divided into various categories according to the environment under which they occur, as shown in Fig. 1 [28]. The types include those that take place in cities (Urban-USAR) [10, 29–40], those that are non-urban or distant (Wilderness-WISAR) [41], marine (Maritime-MSAR) [42], also referred to as air-sea (ASR), and those that take place underwater or on mountains, etc. As implied by its name, Urban is tied to the collapsing structure surrounding humans. In contrast, the wilderness, or non-urban, is primarily concerned with finding and following people who have been lost or wounded in deserts, mountains, rivers, lakes, or any other sparsely inhabited or isolated locations. Finally, the MSAR-ASR focuses on rescuing people caught in strong currents and floods. WiSAR and MSAR-ASR missions frequently involve finding a known number of mobile or semi-mobile victims in open-ended search areas. USAR operations are designed to find and help many stationary victims in a confined search environment. USAR varies from wilderness SAR (WiSAR) and air-sea rescue (ASR, the marine equivalent), as USAR is involved in finding numerous stationary victims who may be trapped in collapsed buildings or vehicles [19].

In addition to the aforementioned three types, other researchers contextualized their research with assessment operations. A region affected by a catastrophe or other incident is investigated during assessment operations. Examples of assessment include examining a hydroelectric dam after a storm, conducting an archaeological dig, and using remote sensing. Despite the SAR, there is no search for missing persons in these operations, as in numerous SAR missions [43]. Surveillance is another mission classification introduced in the literature. Surveillance operations include identifying a target, generally a person or vehicle, then monitoring or following it. Typically, publications in this category do not consider how SAR operations are carried out to achieve their goals. Surveillance operations are often carried out by deploying a drone or UAV to coordinate where they anticipate finding the target. The last category in the literature is that UAVs, drones, and robots are deployed in combat environments. The combat environment aims to help the military carry out their objectives [44].

2.3 Search and Rescue Scenario

In order to know where to place the rescue robots in the search and rescue phases, it is critical to understand the search and rescue scenario performed by the human team. The scenario is shown in Fig. 2 and it is detailed as follows:

1. When a disaster occurs, the first responders become aware of the victim's presence. This awareness can be triggered by the victim's friends, relatives, and coworkers based on the disaster's location. Also, the demographic patterns (for example, most people will be in apartment buildings at night and office buildings during the day) and a systematic search can help find victims.
2. After that, the responders' teams start to assess the damage by analyzing and investigating the disaster site. Hazardous materials, risks to rescuers, an extreme threat to trapped victims, and resource constraints are all factors that are assessed during this phase of the rescue operation. "Constrained resources" can mean a lot of different things. It could mean that it is hard to get resources to the location, that there isn't any nearby equipment or supplies that can be used, or that there are other problems that make it hard to get help quickly.
3. Following this, the rescue operation is planned.
4. After this, search and reconnaissance workers are sent to evaluate the environment and the situation. To best allocate medical needs in the field, the team need to be able to predict how much emergency medical care will be needed.
5. In the case of rubble presence, the rubble removal starts, taking into consideration the safety of the victims.
6. In the next phase, first responders will be able to get close to victims at the scene of the accident and give them the required medical care.

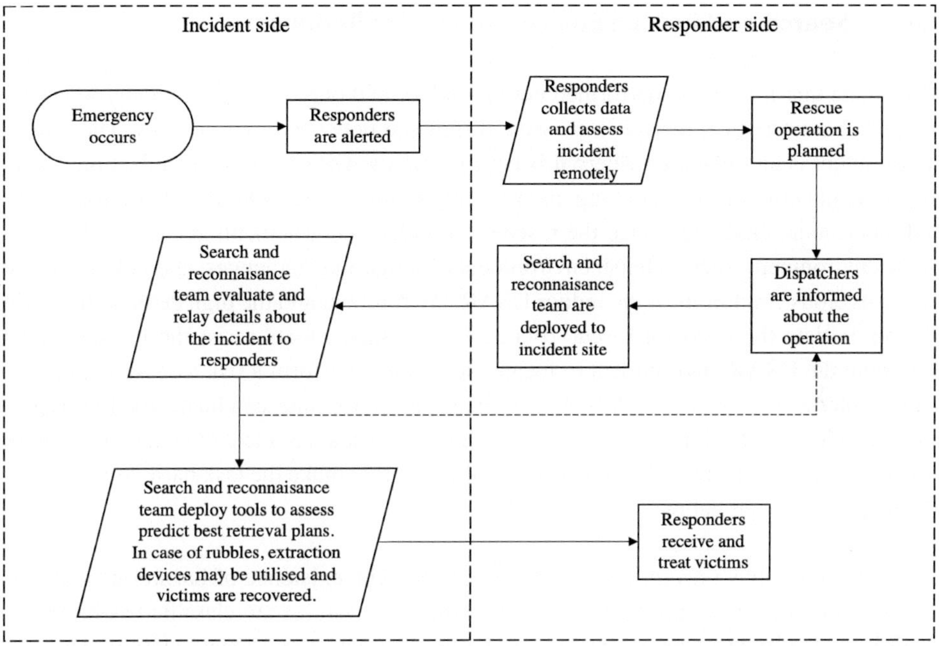

Fig. 2 Search and rescue scenario

7. Then the victims are sent to nearby medical facilities [28].
8. Field teams report operations frequently on their progress, and the command staff uses this information to adjust their plans.

UAV rescue robots are able to perform and contribute in the early phases of this scenario by assessing the damage, locating people, determining the risk, and allocating potential survivors. UAVs have the potential to be dispatched prior to the involvement of first responders and can be of assistance mainly in the second stage of this scenario. Furthermore, the fifth and sixth stages of this scenario demonstrate the tactical path that can include UAV duties. Before first responders make direct contact with the survivors, the UAV can identify whether the victims are in danger while the surrounding rubble is being cleared away and can also assess the level of the risk. In addition, UAVs are now being deployed in the first stages of the intervention operation, during which they are supplying potential survivors with medical assistance. And last, they may be put to use after a catastrophe has occurred to perform activities like surveying the affected regions and determining the extent of the damage.

2.4 Search and Rescue Environments Challenges

The rescue mission can be performed both in indoor and outdoor environments. According to their characteristics, there is a massive difference between environments. The outdoor environment is an open area where it is easy to use the GPS for UAV localization, but the rescue team suffers from searching and rescuing in huge areas. On the other hand, several and much more challenges face the rescuers in indoor environments related to the nature of the environment, such as harsh environments, navigation challenges, and unknown risks. The dangerous conditions of the indoor USAR environment and the high demand for saving victims initiate the need for rescue robots. UAV rescue robots assist the rescue team in searching the USAR environment to locate victims and perform a risk assessment process. The accelerated growth of UAVs makes them an appropriate candidate for this type of mission. Victim detection, exploration, mapping, and risk assessment of an unknown indoor environment are not trivial due to various challenges that make the task hard. Some of these challenges are listed below:

1. Harsh conditions prevent fast victim allocation: The existence of hazardous materials, radiation, dust, poisonous gases, and extreme temperatures are classified as harsh conditions. Different types of sensors can be used to detect and locate victims. However, using one sensor to detect victims' presence rapidly and accurately is not practical, as relying on one sensor can slow the overall mission. For example, monocameras to detect victims visually may not be capable of locating victims in dark environments. Also, the detection by thermal cameras can be deceived by other heat sources, like fire. Hence, utilizing multi-model sensing to explore the USAR environment is required for victim positioning to facilitate the first responders' intervention.
2. Navigation challenge: The navigation challenge is embodied in the existence of collapsed structures, narrow openings, water leakage, and obstructed pathways. Exploring and mapping this environment is challenging because of the lack of awareness. Therefore, an efficient exploration and mapping system is required to efficiently explore the environment and create an informative and rich map that labels the objects of interest, utilizing the contextual information captured by the onboard sensors and using the high-level abstraction of the environment objects.
3. Unknown risks and dynamic environments: USAR environments are continuously changeable in their structure and risk. The rescue team used to intervene without any risk assessment or scene analysis. Without a risk assessment, the risk to the rescue team is undefined. Therefore, an efficient risk assessment is required to increase situational awareness of the rescue environment.

Consequently, UAV rescue robots are considered a promising solution as they can reduce human risk, extend current human capabilities, and perform complex search and rescue missions safely and efficiently. In this context, UAV rescue robots are devoted to exploring the environment, locating victims and sources of danger, neutralizing hazards and sources of risk, building situational awareness of the environment, and providing a danger-free route for the first responders to intervene and save the victim's life.

2.5 Real World Deployment

Small UAVs offer a promising framework for SAR missions and can be crucial to their success. UAVs can rapidly and easily enter inaccessible locations, proving to be an essential additional tool to have. They are quick, portable, and exhibit autonomous behavior, providing accurate and dependable flying support. UAVs also lend their modularity to offering a range of adaptations to different types of missions [28]. The following list shows some of the varying real-world applications of UAVs over the past two decades:

1. The first use of UAVs in disaster relief was in August 2005, two days after Hurricane Katrina hit the Gulf Coast. Equipped with cameras, microphones, and other sensors, they searched for victims stranded in flooded homes and rooftops, assessed the damage, and transmitted images from inaccessible places to rescuers, thus providing a crucial link between the emergency responders and survivors [20]. The drone used to inspect structural damage is shown in Fig. 3a
2. 2014—Balkans flooding Serbia: UAV was employed to assess the disaster's scope and actively look into why drainage mitigation efforts failed. A pump that a farmer had installed that would have been invisible from the ground was discovered by the UAS [45]. UAV deployed is shown in Fig. 3b
3. 2017—Hurricane Harvey used UAVs shown in Fig. 3c. UAS were heavily utilized by the Office of Emergency Management in Fort Bend, Texas, to assist with all emergency assistance tasks before, during, and after the flood crest. There were 112 flights [20].
4. 2018—A mother and her daughter were saved from drowning when a drone delivered them life jackets [46] as shown in Fig. 3d.
5. In Snowy Canyon State Park in Utah in January 2019, a search and rescue team assisted a hiker caught at night on a ledge by using a drone [47].
6. 2019—Drone assisted Firefighters at Notre Dame fire [48] using drone shown in Fig. 3e
7. 2022—Kayakers found and saved using a thermal sensing drone in the heat-detection images [4] shown in Fig. 3f.

(a) iSENSYS IP3 rotary-wing UAV for Hurricane Katrina response to search rural areas cut off by the flooding [50]

(b) UAS used in the Balkan flooding incident. It was equipped with 3D mapping capabilities [51]

(c) UAS used by Fort Bend, Texas, Office of Emergency Management for Hurricane Harvey flooding [52]

(d) UAV delivering life jacket to save victims from drowning [47]

(e) DJI Mavic 2 Enterprise Dual - used in the surveying the 2019 Notre Dame fire [49]

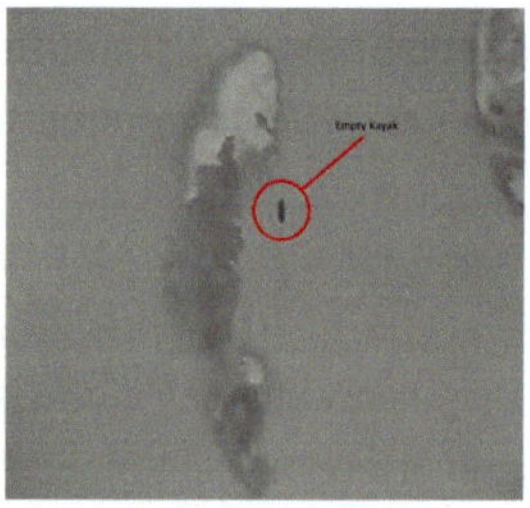

(f) Footage from thermal drone assisting rescue detecting an empty kayak [4]

Fig. 3 Real world deployment flying vehicles

3 UAVs Application in SAR Areas

The UAV's SAR application consists of several pipelines that aid the mission, from object detection to manipulation. In the event of a natural disaster, UAVs can assist with victim retrieval by performing a variety of tasks. In search and rescue scenarios, UAVs aim to save lives. Research has shown that the first 72 h after a disaster are critical for finding and saving victims. Therefore, equipping first responders with high-technology facilities can improve survival probabilities. However, the motivation behind specific UAV designs and their hardware and software capabilities depends on the potential tasks they may perform and the requirements of the environment. This section provides the UAVs' tasks in search and rescue environments. Tasks currently performed by rescue robots can be classified into sensing, reconnaissance, and intervention. In sensing and reconnaissance, the robot can search for victims or potential hazards, which will result in speeding up and completing the mission without increasing the risk to victims or rescuers. Additionally, the rescue robot can perform reconnaissance and mapping to provide situational awareness. This section highlights four areas of UAV applications in search and rescue scenarios: victim detection mapping, risk assessment, and intervention.

3.1 Searching—Victim Detection and Identification

In a SAR operation, the data gathered from initial assessments define the decisions of the following steps and how to approach the surveillance stage. In search-and-rescue missions, identifying victims as fast as possible is of utmost importance so that first responders can intervene and save them. Hence, examining the operational environment, which may involve object detection (victim detection) tools, is one of the primary assessment components. Object detection (victim detection) has been extensively researched, and state-of-the-art objection detection has been extended to victim detection and localization applications. The human detection problem has been extensively studied from thermal and visual perspectives. Several color and infrared cameras, among other sensors, made it possible to employ distinct detection strategies. In the beginning, machine learning techniques were used, but now deep learning techniques are heavily utilized.

From the perspective of thermal detection using machine learning, the study in [52] used background subtraction and a trained Support Vector Machine (SVM) classifier to find people in thermal images. Similarly, in [53], the SVM classifier for thermal detection was trained with a combination of Histogram of Oriented Gradients (HOG) and CENsus TRansform Histogram (CENTRIST) features. In deep learning approaches, various types of research were performed; for example, in [54], the YOLOV3 algorithm was selected as the neural network (NN) for training and classification. The system was created for SAR operations, where the pipeline performs object detection and classifies people in outdoor environments. The system is tested on thermal video data from ground-based and test flight footage. Also, in [55], the work analyzed the performance evaluation of faster R-CNN and single-shot multi-box detector (SSD) algorithms with different backbone networks to detect human targets in aerial view thermal images. It showed how the performance of these algorithms changed when the backbone network and anchor parameters were altered.

Conversely, in terms of visual detection, in the context of machine learning approaches, the SVM classifier, and HOG features are used to locate victims in various poses and lighting conditions in [56]. Furthermore, variant machine learning algorithms are used to perform human visual detection, as seen in [57, 58]. Machine learning-based detection methods are limited in different ways, making them less accurate and precise. First, machine learning methods don't work well in real-world disaster areas because they are trained on uniform backgrounds. Human objects are also modeled by computer programs using machine learning and already-set characteristics. This assumption might be incapable of generalizing human features, which could lead to mistakes in detection. So, instead of using the process of "feature extraction," deep learning models are used to make detection more accurate [59].

On the other hand, deep learning allows for more accurate approaches to human detection by locating the most valuable patterns of features previously learned from visual input, in contrast to traditional techniques. Object detection by recognizing predefined categories of objects is a primary task for a SAR mission, which increases the dataset size for human detection. Human detection using images taken by UAVs and trained to produce deep-learning models has seen impressive results in recent years. Most publicly available online object detection models are trained to find ordinary inanimate objects in distinctive scenarios. Unfortunately, such information is difficult for humans to detect during emergencies. Models for human detection in novel and difficult-to-observe environments have been trained using innovative approaches incorporating visual compositions. Image compositions and other creative methods have been used to train models to recognize humans in complex, difficult-to-observe environments.

Accurate approaches to human detection achieved using deep learning can be found in [60–63]. In [60], the work reports the results of an experimental assessment utilizing the most recent lightweight version of the YOLO detection algorithm, YOLOv5, to identify humans in danger using two of the newest benchmark datasets developed for SAR with UAVs. The acquired results are encouraging since they are competitive with the state-of-the-art in terms of detection accuracy and speed. In [64], a deep-learning model called Fast R-CNN succeeds in running at 5 fps with reasonable detection accuracy. One fast real-time deep-learning network for object detection was introduced in [61] and called a Single-Shot Multi-Box Detector (SSD). The model comprises several convolutional layers, each performing the same detection but at a different aspect ratio, resulting in higher overall detection accuracy. However, one of the most significant drawbacks of deep-learning classifiers is the trade-off between accuracy and speed. To put it another way, an accurate convolutional neural network (CNN) model tends to have more layers with a larger size to learn deep features. As a result, evaluating the input image streams requires more time according to real-time requirements. Recent work has been done to address and improve the trade-off between accuracy and speed.

In [62], the work aimed to design more robust face detectors that produced synthetic 3D faces in different poses, scenes, and occlusions that addressed real-world face detection challenges. Another work that takes inspiration from the cut-paste-learn method found in [63] approaches the expansion of lifelike data by creating manipulated images of what might be found in post-disaster situations. The background and foreground of the composite image, typical of varying color and illumination, are harmonized through a novel framework to generate reliable data. Using simulated composite data for training, a victim detection model with an average precision of 44.9% was presented. A disaster victim detection network was developed by image harmonization. This process was accomplished by composing images of victims on pictures of disaster-torn areas. The current dataset for human detection does not include people in poses typically found in debris [65]. A successful detection may not have a high IoU in a real emergency. Still, as long as it can identify (part of) a victim, it is considered successful. This also raises the question of whether there is a more suitable evaluation statistic for this particular victim detection task than the average precision.

Utilizing UAVs in SAR operations greatly minimizes time, money, and effort expenditures while also having the potential to save lives. To meet this difficulty, UAVs can serve as flying base stations and provide emergency wireless services in disaster-affected areas [66]. As shown in [67], a group of UAVs can be used to identify situations on city highways that require emergency vehicle assistance. UAVs expedite the rescue effort by giving the rescue teams useful information. While transmitting the distress signal, a reliable routing system guarantees communication stability [68].

Works to assist search and rescue operations for a large natural disaster area have been proposed in [69]. A UAV scans the disaster area for wireless signals (WiFi) coming from any mobile phone. The issues of maximizing the search area and minimizing location errors with various search pathways are also addressed in this work. Coverage, location errors, average speed, average searching time, and power consumption have all been examined for various search paths. Our preliminary findings demonstrated that by combining several search paths, the suggested UAV system could cover a big disaster area and decrease location mistakes. Future research will look more closely at dynamic search path adjustment to improve search results and coverage. More multi-agent systems studies can further aid such systems [69].

The fusion of multiple sensors is another approach that can be utilized to improve object and human detection accuracy. Sensor fusion makes it possible to benefit from the advantages of numerous sensors while mitigating the drawbacks of each sensor [70, 71]. For example, in [70], the sensor fusion was performed at the feature level, where an initial RGB-D image of the environment is captured. Following this, a depth region segmentation is carried out on the image to obtain sub-region segmentation. The image is then processed to extract skin features, body part associations, and geometric features. In the last step, the extracted features are fed into a support vector machine (SVM) classifier to find the victim in an RGB-D image. In [71], the thermal image is utilized to create the primary area of interest (ROI) coordinates of the human position. The estimated ROI is then fused with the depth image to get the person's actual location and handle occlusions. In [59], a multi-sensor system is used to find victims in an urban SAR simulation. The suggested multi-sensor design fuses data from many sensors (vision, thermal, and wireless) into a single, more accurate map by utilizing a probabilistic Bayesian framework to locate victims.

After the victim has been detected, victim locating can be a valuable tool to find more specific information. Victim detection finds the victim, while victim localization identifies the exact location of the victim. This could be performed by mapping the search and rescue environment.

3.2 Mapping

Maps are depicted in a variety of formats, including topological maps, which show the connections that exist between various points; metric maps, which present the environment in a variety of dimensions and indicate which areas are occupied and which are empty; and semantic maps, which encapsulate characteristics that are compatible with human conception. When it comes to representing the environment and conducting research, metric maps are widely regarded as the most useful type of map [72–76]. On the other hand, these maps do not contain sufficient information to direct the robot toward the items that have been spotted (i.e., the location of victims or sources of danger). Therefore, a more advanced ability than metric mapping is required, and the robot should understand its operating environment. Semantic maps offer a solution such as this, incorporating semantically labeled objects into metric maps. As a result, the user will have a deeper understanding of the environment that is all around them [59, 77–80].

In [81], the study proposed a method for topological mapping that was inspired by the way that bees construct their colonies. As only one bee builds and lives in each place of a honeycomb, a topological map had to be created so that the UAVs used in the mapping process would behave like bees. A metric is used to determine which honeycomb a UAV should map. In the simulations carried out, FIFO and Euclidean distance were both considered as potential metrics for determining which honeycomb should be mapped. Altering the total number of simulated unmanned aerial vehicles was another method of investigation. This revealed that setting the exploration order directly impacts the number of offsets and a UAV in the environment, taking into consideration its position on the map. This may result in saving both energy and time during the investigation process. The ability to generate RGB-D and thermal-reading information not only helps rescue workers prepare for barriers and dropped objects but also enables them to be prepared for elements that pose a hazard to life, such as high temperatures. In [59], three different sensors (RGB camera, a thermal camera, and a wireless sensor) are used to make three distinct 2D victim-location maps. The Bayesian probabilistic framework determines the probability that the victim is present on each map. After that, the Bayesian-probabilistic framework is applied to these maps. It results in producing a merged 2D map that reflects the likelihood of the existence of humans.

The work in [79] presents a methodology for 3D semantic mapping and exploration of unknown indoor environments. The study focuses on creating a three-dimensional map of the surrounding area and carrying out semantic labeling at the grid level to classify all accessible objects. The approach reported in this paper considers semantic information, such as the class of items, to direct the search toward segmentation and object identification. This is in contrast to typical exploration algorithms, which use geometric heuristics and information gain theory on an occupancy grid map. Deep learning is utilized in the proposed method to map 2D semantically segmented images into 3D semantic point clouds that encapsulate occupancy and semantic annotations. Similarly, the work presented in [80] examines the creation of a 3D semantic map for the purpose of risk assessment in USAR contexts since it reflects the location of the risk in the environment as well as its corresponding level. The focus of the work was on creating a three-dimensional map of the surrounding area and doing grid-level semantic labeling to identify all of the different kinds of hazards present in the scene and to differentiate the levels of danger they provide. The strategy that has been presented makes use of a deep learning model to generate semantic segments for hazardous objects in 2D images. Additionally, the method generates semantically annotated point clouds that encapsulate semantic annotations such as hazard type and risk severity level. The annotated point cloud is then used to construct a three-dimensional semantic map that gives situational awareness of the hazards presented in the surrounding environment. The proposed method is tested in a realistic and accurate simulation of an indoor setting. The results show that the system can successfully make a risk assessment map. Another approach that creates a grid map, however, for the outdoor environment is presented in [82]. In this work, an intelligent characterization of the searching area is performed to estimate the potential risk and occupancy degree of the gridding map. This estimation is mainly based on fuzzy logic, considering different factors. Each of these cells will have an assigned value of potential risk and occupancy that will be used to calculate the waypoints that the UAV will follow.

3.3 Risk Assessment

Risk assessment is a crucial step in search and rescue operations. In addition to being useful for post-disaster mapping and surveying, UAVs can also be used for disaster risk management and mitigation. This can be accomplished by gathering pre-disaster surface information and capturing real-time imagery during disasters. This information could be used to direct relief organizations and assist government decision-making. The risk assessment aims to increase knowledge of environmental hazards and risks to help prioritize the first responders' actions and interventions. Also, the risk assessment could determine who is in danger, whether or not the existing control models are sufficient, and the amount of damage or harm that is possible to avoid. The risk assessment technique is performed during urban search and rescue (USAR) operations.

The process consists of detecting hazards, analyzing, evaluating, and appropriately controlling risks [83]. The ranking or prioritizing of the dangers helps to determine the degree of the risk. As a result, appropriate management has to be applied. Prioritizing is helpful, especially when several threats are present in the environment. No single technique can be used to characterize the level of risk definitively. One of the most common ways to prioritize is to use a risk matrix as shown in Fig. 4, which shows how the likelihood and severity of risk are related [83]. Using UAVs to conduct risk assessments indoors and outdoors during search and rescue operations is becoming common. The significant technical breakthroughs that have led to the production of extremely sophisticated UAVs that can be purchased at affordable rates and equipped with high-resolution cameras and other sensing devices are

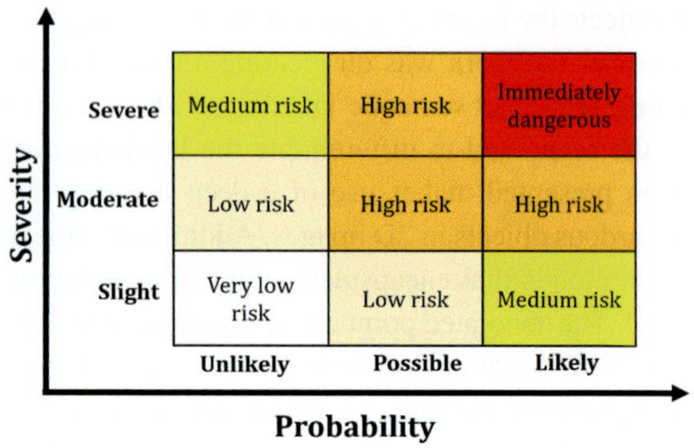

Fig. 4 Risk matrix and risk quantification [80]

principally responsible for their involvement. Due to the absence of human influences, UAVs are also safer than conventional risk assessment techniques.

UAVs have been employed in real-world disaster scenarios for risk assessment. In March of 2019 [84], Terra Drone Indonesia conducted surveys of around 750 hectares of land utilizing UAVs equipped with LiDAR and photogrammetry techniques. The data and images acquired from these surveys have been used to build maps that accurately indicate the land's current state and highlight damaged regions subjected to the most significant devastation. With the use of these maps, government officials will be able to devise strategies that will ensure the reconstruction of homes and other infrastructure in a manner that is both secure and adequate. In [85], the days following the earthquake that occurred in April 2016 in Ecuador, numerous drone pilots from the commercial sector offered to support the response. For the risk management process in some of the most severely impacted areas, manually operated UAVs took aerial photographs, contributed to creating maps, and helped with risk assessment. Studies on the potential for landslides have been carried out using UAVs, as described in [86]. The rock cuts, cliffs, and rough mountains were all planned to be inspected by the UAV. The UAV was capable of acquiring information effectively and in a flexible manner.

For the purpose of creating a map that shows the risks in different areas, the risks need to be calculated or estimated. In [82], the goal of the potential risk/occupancy map is to put a number on the chance that people will be in each cell and, if they are, the risk to their lives if they are there. The types of emergencies can help you figure out the possible danger. This information can be found by looking at how experts have looked at the area or by historical data. Two primary aspects define the map: the terrain and the emergency considerations presented by fuzzy inference systems (FIS). All risk assessment methods provided in the literature were performed on large outdoor areas or the inspection of buildings from the outside. Additionally, risk assessment using UAVs requires trained pilots. In [80], a risk assessment method based on a deep neural network is used to detect numerous sources of danger and categorize these sources of risk into various levels of severity to raise situational awareness of the interior environment. Despite the traditional risk assessment and estimating the risk using the risk matrix, a supervised machine learning approach is used to overcome this problem. This problem is challenging to tackle due to the nature of the USAR environment. The USAR environment suffers from a large variety of structures in varying configurations.

3.4 Rescue and Intervention

A UAV's SAR rescue and intervention portion come in different forms. The rescue may be for medical delivery, such as in the case of blood delivery in Rwanda [87] or chemotherapy by the NHS in the UK [88]. Other recent initiatives attempt to create a lightweight, high-endurance UAV that can carry a floatation device to a drowning offshore worker and return to the oil rig in a variety of weather situations, as well as assess the structural integrity and flight endurance of UAVs [89]. This section explores two types of intervention performed by UAVs in SAR: firefighting and cave exploration.

3.4.1 Firefighting

In 2020, Mohamed Bin Zayed International Robotics Challenge (MBZIRC) conducted a competition for designing urban firefighting technology that focused on perception, navigation, and mobile manipulation [90]. The motivation behind MBZIRC recognizes the significance of the implementation of robotic systems in real-life search and rescue scenarios, such as the use of UGVs when firefighters had to retreat from the Notre Dame de Paris cathedral fire in 2019. Additionally, the reliance on UAVs equipped with thermal sensors and cameras for monitoring tasks and navigation without a direct line of sight during the 2012 earthquake in Mirandola, Italy [91].

Due to their limited payload capacity, micro-scaled UAVs can only transport a water bag that is partially loaded with water. Even if this total amount were discharged flawlessly into an actual fire, it is doubtful that the fire would be put out. The MAV's size must be altered to improve its payload capacity. However, because of the MAV's larger size, it will be highly challenging, if not impossible, to fly through windows and inside structures. So, in most situations, using water to put out a fire is not the best course of action.

The suggested autonomous system was chosen to be the center of an industrial firefighting MAV system using fire-extinguishing capsules based on our performance in the competition [92].

As a result of this system's requirement that the fire is hit directly, precise systems for identifying, approaching, and aiming at fires are needed. Combined with a throwing mechanism that can swiftly and accurately deliver the active firefighting capsules, these methods are being modified from work reported here. Figure 5 depicts a model of a complicated industrial platform [93].

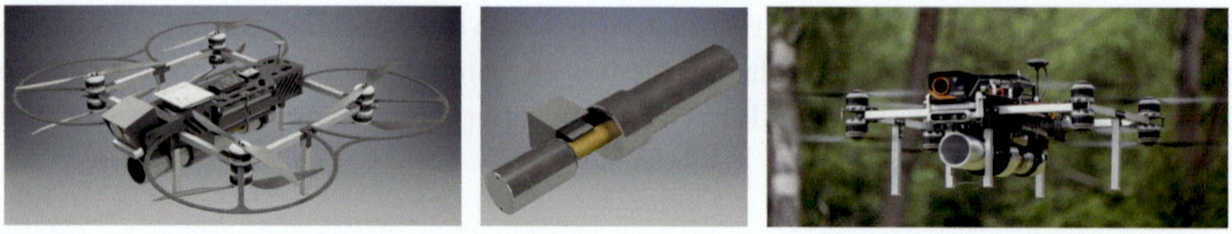

Fig. 5 Concept for firefighting MAV [90]

3.4.2 Cave Explorations—DARPA

The literature on wilderness SAR (WiSAR) operations, SAR in the mountains, underground mines, caves [94–96], woods, and other rural or remote locations is frequently discussed. The DARPA SubT challenge brought to light the requirement for heterogeneous MRS in WiSAR situations [25, 97].

In post-avalanche situations, SAR missions are most frequent in mountainous areas. Avalanche transmitters (AT) are frequently carried by mountaineers in regions at risk of avalanches. ATs have been searched for using UAVs designed for extreme weather (strong winds, high altitudes, and low temperatures) [97]. A multi-UAV autonomous system for locating avalanche victims was created in [98].

SAR for mining applications is another particular situation that has drawn scientific interest. The restrictions of a wireless connection and potentially poisonous gases present two difficulties for SAR operations in underground settings. An overview of wireless robotic communication networks for underground mines has been provided by [96]. The DARPA SubT challenge has given researchers a chance to create unique inter-robot communication methods that use breadcrumb nodes.

A self-contained system for effectively using flying robots in the autonomous investigation of cave environments is presented in another research. This system will aid human explorers, first responders, and speleologists. The suggested system interconnects essential robotic subsystems to enable complete autonomy of the robots and is generally suitable for an arbitrary exploration mission within an uncharted and unstructured subterranean environment.

The proposed system's perception is based on a multi-channel LiDAR sensor that is employed in both the mapping and Lidar Odometry and Lidar Mapping (LOAM) modules for motion estimates and for creating a spatial representation of the immediate environment [99].

By outfitting the robot with vertically facing RGBD cameras that can fill in the blind spots in the limited vertical field of view of the LiDAR [100], the system's vertical navigation skills can be considerably enhanced. These optional sensors can also be used for search and rescue operations or for the visual documentation of recently discovered cave systems in addition to navigation.

In order to achieve complete autonomy in challenging, uncharted subterranean areas without GNSS, the suggested self-sustaining system connects solutions for all essential robotic operations. Among other things, this offers robust path-planning for unknowable dynamic situations to enable flights in restricted cavities and open caverns of natural caves, as well as laser-data processing that manages high humidity and dustiness within underground habitats.

4 Fundamental Challenges and Future Directions

This section discusses the extent of cutting-edge SAR technology implemented in UAVs. It also explores the difficulties experienced in their implementation and the future directions of UAVs in SAR missions using multiple agents.

4.1 Fundamental Challenges

As with most devices, UAVs have limitations that pose hardware and software challenges. This section expands on the limitations and some solutions for UAVs in a SAR assignment.

- **Portability Issues**
  Although light and small, UAVs can pose a portability issue if equipped with specific payloads. For example, cutting-edge object recognition or detection methods typically necessitate system loading and incorporating complex models. They are typically not portable as a result. Power and portability are thus crucial shortcomings in use cases involving UAVs and mobile platforms for smartphones. In order to detect things in close to real-time [101], establishes a baseline for the performance obtained by a smartphone attached to a UAV.
  Common UAVs' payload constraints prevent the speedy transmission of material (e.g. first aid supplies) and data. As a result, data quality suffers from increased noise and deformations as it travels from the aircraft to a ground station [20].
- **Power Consumption/Long Distance Missions**
  High battery consumption restricts the ability to stream videos while flying. Deploying functions outside of the UAV may prolong the battery's lifetime so the vehicle can fly and stream missions longer. This is due to the limited battery life on a drone [101]. UAVs are vulnerable to environmental and meteorological risks because of their low weight and propulsion. Extreme cold negatively impacts battery capacity, which can reduce the UAV's flying duration by up to 50%. A typical issue for which technical solutions are being researched includes maintaining an adequate pre-flight battery temperature (using, for example, a heated and thermally insulated box) [28]. The capability of UAVs to fly in challenging conditions has been demonstrated in several experiments in situations of varying degrees of rainfall. Although various changes are being made to prevent these concerns, it is acknowledged that moisture in electrical circuits and components may

result in short circuits or other complications [102]. The typical UAV employed in SAR missions is also susceptible to specific meteorological circumstances, such as winds and turbulence brought on by airflow encountering obstacles like terrain and vegetation at the ground's surface, which can severely impact their ability to fly [103, 104]. In addition, these disruptions can significantly lower the likelihood of detection (jitter impact of computational vision) [105]. UAVs may be delayed because many SARs can be conducted in low-light conditions, making navigation and situational awareness difficult. The use of external lighting on the ground or the installation of a payload within the UAV itself (depending on its payload) are two potential solutions to these operational problems.
For UAVs to cover a longer distance, other methods of charging the UAVs should be applied, such as Wireless Power Transfer or Laser Power Transfer (LPT)—primarily employed in space and military applications. Laser beams of specific wavelengths and frequencies charge PV cells found on the UAVs on either rotary or fixed-wing UAVs [106].

- **Area Mapping**
The geological characteristics of an area may have considerably changed during SAR missions due to landslides or floods, for example. To ensure the security of the SAR crew and accelerate the rescue process, the area's local map has to be revised [23]. Existing mapping methods can be improved, which is a matter of discussion among scholars. The limited UAV energy capabilities in terms of flying time, which sometimes need to be improved to cover the complete search area, are essential issues that must be addressed. Additionally, due to their short battery lives, sensors (like cameras) have a limited functioning period. However, it is preferable to extend the time when a UAV is utilized in search and rescue mode because retrieving, getting ready for another flight, and resetting might take much time. The researchers' main challenge is to solve a flight path problem that enables the UAV's built-in sensors to cover as many crucial locations as feasible within the allotted flight time. That is, the programming algorithms use the probability distribution map to automatically build the flight plan, ensuring that areas with a high possibility of discovering a missing person are searched first and maximizing the efficiency of the search. Concurrently, the operator's choices in tactics must allow the UAV to generate freshly accurate probability distribution maps in a short amount of time [21, 107–109]. Research-oriented topics include using tools (e.g. Geographic Information System (GIS)) in SAR operations to create probability distribution maps or as a platform for integrating data on the missing person's behavioral profile [110, 111]. The flight plan is adjusted for avalanche SAR missions based on the strength of the signal [112]. This is done following the typical methods and procedures for rescue operations, and it considers how important it is to act quickly [97].
Two critical objectives for the success of SAR operations are quick coverage (rapid ground mapping) and detection (victim tracking). However, these objectives have a trade-off, especially on WiSAR missions. Fast mapping of a region will not provide as much detail regarding detecting people. On the other hand, the slower mapping will help locate the

person, but it will take too long to be practical for search and rescue activities. A significant challenge for academics is determining the optimal balance between "coverage" and "detection." The likelihood of finding a person during a search is a critical component that must be converted into a real limitation to maximizing its effectiveness. In other words, the UAV can fly as high as the optical sensor can recognize a human shape. UAV height may increase if the requirement for human form detection is dropped in favor of other distinguishing characteristics such as odd clothing colors and other man-made items [21, 113]. As a result, the flight plan over a region should be configured to function at altitudes acceptable for optical sensing of the sensors [28].

- **Optimization of Sensing Technology**
  Most of the articles in the literature either talked about sensor technology in a vague way, like in [41], which stated that UAVs have 100% accurate and uniform sensors but did not say what they were. Other papers, like [114], consider an undefined set of sensors, and their model formulation tries to match the suitable sensors with the right tasks. As mentioned in the UAV Capabilities, optical sensors (cameras) and heat-sensing cameras (infrared) are the most common sensors used in USAR operations.
  The images captured from the UAV sensors present considerable challenges, making it extremely challenging to recognize specific characteristics or objects of interest in the scene [105]. Environmental and meteorological conditions also play a role in the sensing limitations, such as the density of vegetation, lighting conditions, and weather. There is a possibility of missing crucial data and receiving false alerts [21, 115–117]. Field testing has demonstrated a need for an effective procedure for human detection in UAV images, making it a challenging problem to solve. Although significant progress has been made in specialized fields such as pedestrian detection, most systems function more effectively when humans are fully visible and appear in a narrow range of poses, such as walking or standing.
  Numerous models have been developed using and developing multiple visible-spectrum computational vision algorithms that incorporate motion and color features, body sensing, barriers in moving surroundings, etc. However, in the complicated settings of search and rescue operations, where people's positions are constantly changing, it is difficult to generalize these models [118]. Other sorts of sensor-based knowledge are encountered in the literature. Significant efforts have been made to employ UAVs' laser scanners for human detection because so many have them. While thermal cameras are most commonly associated with nighttime surveillance, their benefits extend to daylight monitoring as well, as proven by their widespread usage by the military and law enforcement. They can be found wherever a person's body temperature is different from their surroundings. It has become common practice to employ thermal imaging for people detection through indirect approaches developed for visible-light pictures or by a novel use of these techniques [118]. Additionally, the conventional e-vision issues caused by shadows are reduced. Trials in the field have shown that the standard thermal sensors have their specific limitations to the findings acquired, such as low signal-to-noise ratio (SNR) and

white/black/cold polar variations happening around very hot or cold objects, especially when capturing outdoor thermal surveillance photographs of the same area on various days (morning and afternoon). This may disprove the first hypothesis that the body area is significantly warmer and brighter than the backdrop. If this is the case, then the method will only help determine the precise positions of human shapes and bodies.
Researchers are required to address the problems accumulated due to the collection of issues associated with optical sensors. This can be done through the collaboration of different types of sensors, the design of appropriate algorithms, or the optimization of algorithms already used for identification and search [23, 105, 119–122].

4.2 Future Directions

In future projects that use UAV technology, there may be a need to overcome the limitations mentioned above and combine multiple systems into the mission, which can introduce new challenges.

In civil applications, for example, different architectures are suggested based on the use, quality-of-service (QoS), and type of UAVs, which makes it possible for multiple UAVs to meet the needs of civil applications and be easily deployed. From an autonomy perspective, teleoperation navigation is a known obstacle, frequently diverting the operator's attention from more significant search and assessment problems. Hence, the increased autonomy through reduced human intervention will allow for easier missions, which is likely to happen if each UAV can do its detection while sending metadata to each other and the base. Communication between the multi-UAVs can be achieved through wireless communication. A multi-UAV system must be very reliable to be considered robust so that even if one part breaks, the mission does not fail. In search and rescue situations, time is of the essence, so strict QoS standards for robustness and visual quality must be met.

Another future direction could be to solve the problems of limited payload and power consumption. The small size of UAVs restricts the amount of processing power, communication technology, and sensors that can be used, which is considered one of the most crucial areas of future research in the field of UAV application [123].

On the other hand, search and rescue is quite demanding regarding robotic capabilities and operator working conditions, regardless of the type of disaster [10]. By definition, the robot functions in a hostile environment with limited mobility. The cliché "simpler is better" is especially true in rescue robotics because the presence of abrasive dust and water, the corrosive effects of wet cement, and the large variety of barriers in the environment conspire to accelerate the wear and tear on a robot [20, 124, 125].

Man-made disasters (like terrorist explosions or major accidents) occur in a smaller region than natural disasters. The difficulty often lies in determining what cannot be seen, such as the interior of the rubble, its whereabouts, the health of the survivors, and the condition of potentially dangerous utilities (e.g., electricity, gas lines) [124].

5 Conclusions

Rescue robots are evolving from a novel concept to a crucial disaster relief component. Much focus has been on aerial robots due to their potential usefulness in disaster response. In addition to human-robot interaction, UAVs in SAR face issues in all critical subsystems (mobility, communications, control, sensors, and power). This chapter illustrated the different SAR environments and SAR scenarios. It discussed the role of UAVs in this scenario. In addition, it highlighted some of the real-world UAV deployments around the world. Furthermore, it comprehensively discussed UAV applications in search and rescue operations. Additionally, it examined the UAV's capabilities and limitations in general.

Even though UAV intelligence is relatively high, many problems still need to be resolved before they can be relied on to give complex SAR operations full credibility. In this area, novel and forward-thinking approaches to problem-solving are being developed, and traditional research procedures reveal future opportunities. However, none of the risks linked with the utilization of UAVs are considered particularly severe, mainly when the anticipated benefits to rescue workers are considered.

References

1. A. Khan, S. Gupta, S.K. Gupta, Emerging uav technology for disaster detection, mitigation, response, and preparedness. J. Field Robot. (2022)
2. M.B. Bejiga, A. Zeggada, A. Nouffidj, F. Melgani, A convolutional neural network approach for assisting avalanche search and rescue operations with UAV imagery. Remote Sens. **9**(2), 100 (2017)
3. H. Kesteloo, DJI mavic 2 enterprise dual helps hurt hiker rappel 400 ft during rescue at snow canyon state park, UT (2020). https://dronedj.com/2020/01/21/drone-helps-hurt-hikerrappel-400-ft-snow-canyon-state-park/
4. H. Kesteloo, Thermal camera drone finds lost kayaker within minutes (2022). https://dronexl.co/2022/09/02/thermal-camera-drone-lost-kayaker-minutes/
5. K.H. Frith, A. Amiri, Emerging technologies center. Nurs. Educ. Perspect. **43**(3), 203 (2022)
6. H. Mohapatra, Offline drone instrumentalized ambulance for emergency situations. IAES Int. J. Robot. Autom. **9**(4), 251 (2020)
7. V. Bowman, I. McTaggart, J. Warrington, C. Hymas, M. McGrath, Drone finds missing man in 20 min (2014). Accessed 23 Aug 2022
8. B. Siciliano, O. Khatib, *Springer Handbook of Robotics*, in *IEEE Robotics & Automation Magazine, Robotics & Automation Magazine, IEEE, IEEE Robotics & Automation Magazine*, vol. 3, (2008), p. 110. issn: 1070-9932. https://doi.org/10.1109/mra.2008.928399
9. J.L. Casper, R.R. Murphy, Workflow study on human-robot interaction in usar, in *IEEE International Conference on Robotics and Automation, 2002. Proceedings. ICRA'02*, vol. 2 (IEEE, 2002), pp. 1997–2003

10. J. Casper, R.R. Murphy, Human-robot interactions during the robotassisted Urban search and rescue response at the world trade center. IEEE Trans. Syst., Man, Cybern., Part B (Cybern.) **33**(3), 367–385 (2003)
11. A hierarchical reinforcement learning based control architecture for semiautonomous rescue robots in cluttered environments, in *2010 IEEE Conference on 2010 IEEE International Conference on Automation Science and Engineering, Automation Science and Engineering (CASE)* (2010), p. 948. ISSN: 978-1-4244-5447-1. http://doi.org/10.1109/coase.2010.5584599
12. J.-i. Meguro, K. Ishikawa, Disaster information collection into geographic information system using rescue robots (2006), pp. 3514–3520
13. Y. Liu, G. Nejat, Robotic urban search and rescue: a survey from the control perspective. J. Intell. Robot. Syst. **72**(2), 147 (2013)
14. B. Robert, Search and rescue and disaster relief robots: has their time finally come? Ind. Robot.: Int. J. **2**, 138 (2016). ISSN: 0143-991X. http://doi.org/10.1108/ir-12-2015-0228
15. D. Lang, How i learned to make underwater robots (2012). http://edition.cnn.com/2012/10/10/tech/how-ilearned-to-make-underwater-robots/index.html
16. R. Maruyama, Semi-autonomous snake-like robot for search and rescue (1995)
17. M. Półka, S. Ptak, Ł Kuziora, The use of UAV's for search and rescue operations. Procedia Eng. **192**, 748–752 (2017)
18. W. Stecz, K. Gromada, UAV mission planning with SAR application. Sensors **20**(4), 1080 (2020)
19. S. Grogan, R. Pellerin, M. Gamache, The use of unmanned aerial vehicles and drones in search and rescue operations–a survey. Proc. PROLOG (2018)
20. R.R. Murphy, et al., Search and rescue robotics, in *Springer Handbook of Robotics*, ed. by B. Siciliano, O. Khatib (Springer Berlin Heidelberg, Berlin, Heidelberg 2008), pp. 1151–1173, ISBN: 978-3-540-30301-5. https://doi.org/10.1007/978-3-540-30301-5_51
21. M.A. Goodrich et al., Supporting wilderness search and rescue using a camera-equipped mini UAV. J. Field Robot. **25**(1–2), 89–110 (2008)
22. J. Qi, et al., Search and rescue rotary-wing UAV and its application to the Lushan Ms 7.0 earthquake. J. Field Robot. **33**(3), 290–321 (2016)
23. J. Sun, B. Li, Y. Jiang, C.-Y. Wen, A camera-based target detection and positioning UAV system for search and rescue (SAR) purposes. Sensors **16**(11), 1778 (2016)
24. R. Konrad, D. Serrano, P. Strupler, Unmanned aerial systems. *Search and Rescue Robotics—From Theory to Practice* (2017), pp. 37–52
25. J.P. Queralta, et al., Collaborative multi-robot search and rescue: planning, coordination, perception, and active vision. IEEE Access **8**(191), 617–643 (2020). https://doi.org/10.1109/ACCESS.2020.3030190
26. Y.-W. Huang, et al., Duckiefloat: a collision-tolerant resource-constrained blimp for long-term autonomy in subterranean environments (2019). arXiv:1910.14275
27. T. Rouček, et al., Darpa subterranean challenge: multi-robotic exploration of underground environments, in *International Conference on Modelling and Simulation for Autonomous Systems* (Springer, Berlin, 2019), pp. 274–290
28. S. Waharte, N. Trigoni, Supporting search and rescue operations with UAVs, in *2010 International Conference on Emerging Security Technologies* (IEEE, 2010), pp. 142–147
29. K. Hatazaki, M. Konyo, K. Isaki, S. Tadokoro, F. Takemura, Active scope camera for urban search and rescue, in *2007 IEEE/RSJ International Conference on Intelligent Robots and Systems* (IEEE, 2007), pp. 2596–2602
30. G. Loukas, S. Timotheou, Connecting trapped civilians to a wireless ad hoc network of emergency response robots, in *2008 11th IEEE Singapore International Conference on Communication Systems* (IEEE, 2008), pp. 599–603

31. X. Zhang, M. Wu, J. Peng, F. Jiang, A rescue robot path planning based on ant colony optimization algorithm, in *2009 International Conference on Information Technology and Computer Science*, vol. 2 (IEEE, 2009), pp. 180–183
32. F.S.N. Fard, H. Parvar, M.E. Shiri, E. Soleimani, Using self configurable particle swarm optimization for allocation position of rescue robots, in *2010 Second International Conference on Computer and Network Technology* (IEEE, 2010), pp. 362–366
33. H.N. Pishkenari, S. Mahboobi, A. Alasty, Optimum synthesis of fuzzy logic controller for trajectory tracking by differential evolution. Sci. Iran. **18**(2), 261–267 (2011)
34. Y.F. Ding, Q. Pan, Path planning for mobile robot search and rescue based on improved ant colony optimization algorithm, in *Applied Mechanics and Materials*, vol. 66 (Trans Tech Publications, 2011), pp. 1039–1044
35. E. Olson et al., Progress toward multi-robot reconnaissance and the magic 2010 competition. J. Field Robot. **29**(5), 762–792 (2012)
36. P. Mirowski, T.K. Ho, S. Yi, M. MacDonald, Signalslam: simultaneous localization and mapping with mixed wifi, bluetooth, lte and magnetic signals, in *International Conference on Indoor Positioning and Indoor Navigation* (IEEE, 2013), pp. 1–10
37. A. Özgelen, E.I. Sklar, An approach to supervisory control of multirobot teams in dynamic domains, in *Conference Towards Autonomous Robotic Systems* (Springer, Berlin, 2015), pp. 198–203
38. L. Pineda, T. Takahashi, H.-T. Jung, S. Zilberstein, R. Grupen, Continual planning for search and rescue robots, in *2015 IEEE-RAS 15th International Conference on Humanoid Robots (Humanoids)* (IEEE, 2015), pp. 243–248
39. M. Elbanhawi, A. Mohamed, R. Clothier, J.L. Palmer, M. Simic, S. Watkins, Enabling technologies for autonomous MAV operations. Prog. Aerosp. Sci. **91**, 27–52 (2017)
40. C. Mouradian, J. Sahoo, R.H. Glitho, M.J. Morrow, P.A. Polakos, A coalition formation algorithm for multi-robot task allocation in large-scale natural disasters, in *2017 13th International Wireless Communications and Mobile Computing Conference (IWCMC)* (IEEE, 2017), pp. 1909–1914
41. S. Hayat, E. Yanmaz, T.X. Brown, C. Bettstetter, Multi-objective UAV path planning for search and rescue, in *2017 IEEE International Conference on Robotics and Automation (ICRA)* (IEEE, 2017), pp. 5569–5574
42. S. Aldhaheri, G. De Masi, È. Pairet, P. Ardón, Underwater robot manipulation: advances, challenges and prospective ventures, in *OCEANS 2022-Chennai* (IEEE, 2022), pp. 1–7
43. G. Wang, et al., Coastal dam inundation assessment for the yellow river delta: measurements, analysis and scenario. Remote Sens. **12**(21) (2020). ISSN: 2072-4292. https://doi.org/10.3390/rs12213658. https://www.mdpi.com/2072-4292/12/21/3658
44. I. Moir, *Military Avionics Systems* (Wiley, 2019)
45. R. Murphy, Summary of 47 known disaster robot deployments 2001–2015 (2015). http://crasar.org/2015/
46. T. Stickings, Lifeguard drone saves a drowning woman and six other swimmers, in *Mail Online* (2018). https://www.dailymail.co.uk/news/article-6070655/Incredible-moment-lifeguard-DRONE-saves-drowningwoman-six-swimmers.html
47. The Salt Lake Tribune (2019). https://www.sltrib.com/news/2019/01/11/crews-use-dronerescue/
48. R. Bogue, The role of robots in firefighting. Ind. Robot.: Int. J. Robot. Res. Appl. (2021)
49. R.R. Murphy, et al., Use of small unmanned aerial systems for emergency management of flooding [techbrief], United States. Federal Highway Administration, Technical Report (2019)

50. G. De Cubber, H. Balta, D. Doroftei, Y. Baudoin, UAS deployment and data processing during the balkans flooding, in *2014 IEEE International Symposium on Safety, Security, and Rescue Robotics* (IEEE, 2014), pp. 1–4
51. O. Fernandes, R. Murphy, J. Adams, D. Merrick, Quantitative data analysis: crasar small unmanned aerial systems at hurricane harvey, in *2018 IEEE International Symposium on Safety, Security, and Rescue Robotics (SSRR)* (IEEE, 2018), pp. 1–6
52. J. Portmann, S. Lynen, M. Chli, R. Siegwart, People detection and tracking from aerial thermal views, in *2014 IEEE International Conference on Robotics and Automation (ICRA)* (2014), pp. 1794–1800. https://doi.org/10.1109/ICRA.2014.6907094
53. I. Riaz, J. Piao, H. Shin, Human detection by using centrist features for thermal images, in *International Conference Computer Graphics, Visualization, Computer Vision and Image Processing* (Citeseer, 2013)
54. J. McGee, S.J. Mathew, F. Gonzalez, Unmanned aerial vehicle and artificial intelligence for thermal target detection in search and rescue applications, in *2020 International Conference on Unmanned Aircraft Systems (ICUAS)* (IEEE, 2020), pp. 883–891
55. K. Akshatha, A.K. Karunakar, S.B. Shenoy, A.K. Pai, N.H. Nagaraj, S.S. Rohatgi, Human detection in aerial thermal images using faster R-CNN and SSD algorithms. Electronics **11**(7), 1151 (2022)
56. Y. Uzun, M. Balcılar, K. Mahmoodi, F. Davletov, M.F. Amasyalı, S. Yavuz, Usage of hog (histograms of oriented gradients) features for victim detection at disaster areas, in *2013 8th International Conference on Electrical and Electronics Engineering (ELECO)* (IEEE, 2013), pp. 535–538
57. D.-X. Xia, S.-Z. Su, S.-Z. Li, P.-M. Jodoin, Lying-pose detection with training dataset expansion, in *2014 IEEE International Conference on Image Processing (ICIP)* (IEEE, 2014), pp. 3377–3381
58. D.-X. Xia, S.-Z. Li, Rotation angle recovery for rotation invariant detector in lying pose human body detection. J. Eng. (2015). ISSN: 2051-3305. https://doaj.org (Visited on 27 Aug 2017)
59. A. Goian, R. Ashour, U. Ahmad, T. Taha, N. Almoosa, L. Seneviratne, Victim localization in USAR scenario exploiting multi-layer mapping structure. Remote Sens. **11**(22), 2704 (2019)
60. S. Caputo, G. Castellano, F. Greco, C. Mencar, N. Petti, G. Vessio, Human detection in drone images using yolo for search-and-rescue operations, in *International Conference of the Italian Association for Artificial Intelligence* (Springer, Berlin, 2022), pp. 326–337
61. W. Liu, et al., SSD: Single shot multibox detector, in *European Conference on Computer Vision* (Springer, Berlin, 2016), pp. 21–37
62. J. Han, S. Karaoglu, H.-A. Le, T. Gevers, Object features and face detection performance: analyses with 3d-rendered synthetic data, in *2020 25th International Conference on Pattern Recognition (ICPR)* (2021), pp. 9959–9966. https://doi.org/10.1109/ICPR48806.2021.9412915
63. A. Borji, M.-M. Cheng, Q. Hou, H. Jiang, J. Li, Salient object detection: a survey. Comput. Vis. Media **5**(2), 117–150 (2019)
64. L. Zhang, L. Lin, X. Liang, K. He, Is faster R-CNN doing well for pedestrian detection? (2016)
65. N. Zhang, F. Nex, G. Vosselman, N. Kerle, Training a disaster victim detection network for UAV search and rescue using harmonious composite images. Remote Sens. **14**(13) (2022). ISSN: 2072-4292. https://doi.org/10.3390/rs14132977. https://www.mdpi.com/2072-4292/14/13/2977
66. S. Hayat, E. Yanmaz, R. Muzaffar, Survey on unmanned aerial vehicle networks for civil applications: a communications viewpoint. IEEE Commun. Surv. Tutor. **18**(4), 2624–2661 (2016)

67. O.S. Oubbati, A. Lakas, P. Lorenz, M. Atiquzzaman, A. Jamalipour, Leveraging communicating UAVs for emergency vehicle guidance in urban areas. IEEE Trans. Emerg. Top. Comput. **9**(2), 1070–1082 (2019)
68. S.H. Alsamhi et al., UAV computing-assisted search and rescue mission framework for disaster and harsh environment mitigation. Drones **6**(7), 154 (2022)
69. Y.-H. Ho, Y.-R. Chen, L.-J. Chen, Krypto: assisting search and rescue operations using Wi-Fi signal with UAV, in *Proceedings of the First Workshop on Micro Aerial Vehicle Networks, Systems, and Applications for Civilian Use* (2015), pp. 3–8
70. W.-Y.G. Louie, G. Nejat, A victim identification methodology for rescue robots operating in cluttered USAR environments. Adv. Robot. **27**(5), 373–384 (2013)
71. H.S. Hadi, M. Rosbi, U.U. Sheikh, S.H.M. Amin, Fusion of thermal and depth images for occlusion handling for human detection from mobile robot, *in 2015 10th Asian Control Conference (ASCC)* (2015), pp. 1–5. https://doi.org/10.1109/ASCC.2015.7244722
72. A. Rosinol, M. Abate, Y. Chang, L. Carlone, Kimera: an opensource library for real-time metric-semantic localization and mapping (2019). arXiv:1910.02490
73. I. Kostavelis, A. Gasteratos, Semantic mapping for mobile robotics tasks: a survey. Robot. Auton. Syst. **66**, 86–103 (2015)
74. C. Cadena et al., Past, present, and future of simultaneous localization and mapping: toward the robust-perception age. IEEE Trans. Rob. **32**(6), 1309–1332 (2016)
75. D. Lang, D. Paulus, Semantic maps for robotics, in *Proceedings of the Work-shop "Workshop on AI Robotics" at ICRA* (2014)
76. I. Kostavelis, K. Charalampous, A. Gasteratos, J.K. Tsotsos, Robot navigation via spatial and temporal coherent semantic maps. Eng. Appl. Artif. Intell. **48**, 173–187 (2016)
77. T. Dang, C. Papachristos, K. Alexis, Visual saliency-aware receding horizon autonomous exploration with application to aerial robotics, in *2018 IEEE International Conference on Robotics and Automation (ICRA)* (2018), pp. 2526–2533. https://doi.org/10.1109/ICRA.2018.8460992
78. T. Dang, C. Papachristos, K. Alexis, Autonomous exploration and simultaneous object search using aerial robots, in *2018 IEEE Aerospace Conference* (IEEE, 2018)
79. R. Ashour, T. Taha, J.M.M. Dias, L. Seneviratne, N. Almoosa, Exploration for object mapping guided by environmental semantics using UAVs. Remote Sens. **12**(5), 891 (2020)
80. R. Ashour, M. Abdelkader, J. Dias, N.I. Almoosa, T. Taha, Semantic hazard labelling and risk assessment mapping during robot exploration. IEEE Access **10**(16) 337–349 (2022). https://doi.org/10.1109/ACCESS.2022.3148544
81. R. da Rosa, M. Aurelio Wehrmeister, T. Brito, J.L. Lima, A.I.P.N. Pereira, Honeycomb map: a bioinspired topological map for indoor search and rescue unmanned aerial vehicles. Sensors **20**(3) (2020). ISSN: 1424-8220. https://doi.org/10.3390/s20030907. https://www.mdpi.com/1424-8220/20/3/907
82. V. San Juan, M. Santos, J.M. Andújar, Intelligent UAV map generation and discrete path planning for search and rescue operations. Complexity (2018)
83. Risk Assessment: Osh Answers (2022). Accessed 25 Aug 2022
84. In a first, indonesia is using lidar drones for disaster recovery efforts 2020 (2020). https://www.terra-drone.net/global/2019/05/15/terra-drone-indonesia-lidar-drones-for-disaster-recovery-palu/
85. T. Luege, Case study no. 14: using drones to create maps and assess building damage in ecuador capacity 4 dev (2020). https://europa.eu/capacity4dev/innov-aid/blog/case-study-no-14-using-drones-create-maps-and-assess-building-damageecuador
86. I. Jaukovic, A. Hunter, Unmanned aerial vehicles: a new tool for landslide risk assessment (2017)

87. W. Max, G. Levy, Drones have transformed blood delivery in Rwanda (2022). https://arstechnica.com/science/2022/04/drones-have-transformed-blood-delivery-in-rwanda/2/
88. G. NHS, Greener NHS » world's first chemo drone delivery announced on NHS birthday (2022). https://www.england.nhs.uk/greenernhs/2022/07/worlds-first-chemo-drone-deliveryannounced-on-nhs-birthday/
89. S.E. Eid, S.S. Dol, Design and development of lightweight-high endurance unmanned aerial vehicle for offshore search and rescue operation, in *Advances in Science and Engineering Technology International Conferences (ASET)* (IEEE, 2019), pp. 1–5
90. V. Spurny, et al., Autonomous firefighting inside buildings by an unmanned aerial vehicle. IEEE Access **9**(15), 872–890
91. J. Quenzel, et al., Autonomous fire fighting with a UAV-UGV team at MBZIRC 2020, in *2021 International Conference on Unmanned Aircraft Systems (ICUAS)* (IEEE, 2021), pp. 934–941
92. C. Corrado, K. Panetta, Data fusion and unmanned aerial vehicles (UAVs) for first responders, in *IEEE International Symposium on Technologies for Homeland Security (HST)* (IEEE, 2017), pp. 1–6
93. L. Merino, F. Caballero, J.R. Martı-nez-de-Dios, I. Maza, A. Ollero, An unmanned aircraft system for automatic forest fire monitoring and measurement. J. Intell. Robot. Syst. **65**(1), 533–548 (2012)
94. A. Chikwanha, S. Motepe, R. Stopforth, Survey and requirements for search and rescue ground and air vehicles for mining applications, in *2012 19th International Conference on Mechatronics and Machine Vision in Practice (M2VIP)* (IEEE, 2012), pp. 105–109
95. J. Zhao, J. Gao, F. Zhao, Y. Liu, A search-and-rescue robot system for remotely sensing the underground coal mine environment. Sensors **17**(10), 2426 (2017)
96. A. Ranjan, H. Sahu, P. Misra, Wireless robotics networks for search and rescue in underground mines: taxonomy and open issues, in *Exploring Critical Approaches of Evolutionary Computation* (IGI Global, 2019), pp. 286–309
97. M. Silvagni, A. Tonoli, E. Zenerino, M. Chiaberge, Multipurpose UAV for search and rescue operations in mountain avalanche events. Geomat. Nat. Haz. Risk **8**(1), 18–33 (2017)
98. G. Bryant, An autonomous multi-UAV system for avalanche search, M.S. Thesis, NTNU (2019)
99. A. Koval, C. Kanellakis, E. Vidmark, J. Haluska, G. Nikolakopoulos, A subterranean virtual cave world for gazebo based on the DARPA SubT challenge (2020). arXiv:2004.08452
100. F. Pece, J. Kautz, T. Weyrich, Three depth-camera technologies compared, in *First BEAMING Workshop, Barcelona* (Citeseer, 2011), p. 9
101. I. Martinez-Alpiste, P. Casaseca-de-la-Higuera, J. Alcaraz-Calero, C. Grecos, Q. Wang, Benchmarking machine-learning-based object detection on a UAV and mobile platform, in *2019 IEEE Wireless Communications and Networking Conference (WCNC)* (2019), pp. 1–6. https://doi.org/10.1109/WCNC.2019.8885504
102. A. Albrigtsen, The application of unmanned aerial vehicles for snow avalanche search and rescue, M.S. Thesis (UiT The Arctic University of Norway, 2016)
103. M.T. DeGarmo, Issues concerning integration of unmanned aerial vehicles in civil airspace, Center for Advanced Aviation System Development, vol. 4 (2004)
104. H.B. Abrahamsen, A remotely piloted aircraft system in major incident management: concept and pilot, feasibility study. BMC Emerg. Med. **15**(1), 1–12 (2015)
105. D. Gerhardt, Feature-based mini unmanned air vehicle video euclidean stabilization with local mosaics (Brigham Young University, 2007)
106. S.A.H. Mohsan, N.Q.H. Othman, M.A. Khan, H. Amjad, J. Żywiołek, A comprehensive review of micro UAV charging techniques. Micromachines **13**(6), 977 (2022)

107. L. Lin, M.A. Goodrich, UAV intelligent path planning for wilderness search and rescue, in *2009 IEEE/RSJ International Conference on Intelligent Robots and Systems* (IEEE, 2009), pp. 709–714
108. L. Lin, M.A. Goodrich, A Bayesian approach to modeling lost person behaviors based on terrain features in wilderness search and rescue. Comput. Math. Organ. Theory **16**(3), 300–323 (2010)
109. S. Clark, M.A. Goodrich, A hierarchical flight planner for sensor-driven UAV missions, in *2013 IEEE RO-MAN* (IEEE, 2013), pp. 509–514
110. E. Soylemez, N. Usul, Utility of GIS in search and rescue operations, in *ESRI Users Group Conference* (2006)
111. D. Ferguson, GIS for wilderness search and rescue, in *ESRI Federal User Conference*, vol. 2012 (2008), p. 10
112. F. Serre, G. Pollin, F. Blanc-Paques, D. Drone, Device and method for seeking targets (2012)
113. T. Niedzielski et al., A real-time field experiment on search and rescue operations assisted by unmanned aerial vehicles. J. Field Robot. **35**(6), 906–920 (2018)
114. F. Mufalli, R. Batta, R. Nagi, Simultaneous sensor selection and routing of unmanned aerial vehicles for complex mission plans. Comput. Oper. Res. **39**(11), 2787–2799 (2012). ISSN: 0305-0548. https://doi.org/10.1016/j.cor.2012.02.010. https://www.sciencedirect.com/science/article/pii/S0305054812000366
115. J. Wu, G. Zhou, Real-time UAV video processing for quick-response to natural disaster, in *2006 IEEE International Symposium on Geoscience and Remote Sensing* (IEEE 2006), pp. 976–979
116. C.-c. Li, G.-s. Zhang, T.-j. Lei, A.-D. Gong, Quick image-processing method of UAV without control points data in earthquake disaster area. Trans. Nonferrous Metals Soc. China **21**, s523–s528 (2011)
117. L. Lin, M.A. Goodrich, Hierarchical heuristic search using a gaussian mixture model for UAV coverage planning. IEEE Trans. Cybern. **44**(12), 2532–2544 (2014)
118. M. Andriluka, et al., Vision based victim detection from unmanned aerial vehicles, in *2010 IEEE/RSJ International Conference on Intelligent Robots and Systems* (IEEE, 2010), pp. 1740–1747
119. D.-J. Lee, P. Zhan, A. Thomas, R.B. Schoenberger, Shape-based human detection for threat assessment, in *Visual Information Processing XIII*, vol. 5438 (SPIE 2004), pp. 81–91
120. K. Mikolajczyk, C. Schmid, A. Zisserman, Human detection based on a probabilistic assembly of robust part detectors, in *European Conference on Computer Vision* (Springer, Berlin, 2004), pp. 69–82
121. J.W. Davis, M.A. Keck, A two-stage template approach to person detection in thermal imagery, in *2005 Seventh IEEE Workshops on Applications of Computer Vision (WACV/MOTION'05)*, vol. 1 (IEEE 2005), pp. 364–369
122. P. Rudol, P. Doherty, Human body detection and geolocalization for UAV search and rescue missions using color and thermal imagery, in *2008 IEEE Aerospace Conference* (IEEE, 2008), pp. 1–8
123. J. Scherer, et al., An autonomous multi-UAV system for search and rescue, in *Proceedings of the First Workshop on Micro Aerial Vehicle Networks, Systems, and Applications for Civilian Use* (2015), pp. 33–38
124. R. Murphy, Gaps analysis for rescue robots, ANS 2006: Sharing Solutions for Emergencies and Hazardous Environments (2006)
125. R.R. Murphy, Trial by fire [rescue robots]. IEEE Robot. Autom. Mag. **11**(3), 50–61 (2004)

Dr. Reem Ashour received her MSc. in haptic teleoperation of Remotely Operated Vehicles (ROVs) in unknown environments at Khalifa University of Science and Technology, UAE, in 2013, and her Ph.D. degree in Engineering concentration in Robotics at Khalifa University of Science and Technology in 2020. Her doctoral research investigates semantic exploration and mapping in urban search and rescue environments. Since 2020, she has been a senior researcher at Technology Innovation Institute at the Autonomous Robotics Research Centre, UAE. Also, she worked as a post-doctoral at Khalifa University from 2020 to 2021. She was a research associate in the Electrical and Electronics Department at Khalifa University from 2015 to 2020. Her research interests include autonomous exploration, semantic mapping, deep learning, and artificial intelligence.

Sara Aldhaheri is a Robotics Master of Science student at King's College London and Assistant Researcher at Technology Innovation Institute (TII). She joined TII in January 2021 and has been contributing to projects with the Autonomous Robotics Research Center. Sara graduated from New York University (Abu Dhabi) in 2020 with a BSc in Mechanical Engineering and a minor in Computer Science. Her research interests include assistive robotics, manipulation, and deep learning.

Dr. Yasmeen Abu-Kheil received her B.Sc. Degree in electronics engineering in 2005 and an M. Sc. degree in Mechatronics Engineering in 2008, both from the American University of Sharjah, UAE. Dr. Yasmeen had a Ph.D. in engineering degree with a Robotics concentration from Khalifa University (2017). Her research is focused on developing a robotic capsule for non-invasive endoscopy. She was awarded the best poster award in the Intelligent Robot Interactions with the Anatomy Workshop, held at the IEEE/RSJ International Conference on Intelligent Robots and Systems (IROS 2019). She was awarded several internal and external funds such as HCT Seed Fund, HCT SURF Fund, ADEK Award for Research Excellence, and Sandooq al Watan Fellowship program. Currently, she is working as a lecturer at the Higher Colleges of Technology (Abu-Dhabi Women's Branch). Before joining HCT, she worked as a postdoctoral fellow at the Visual Signal Analysis and Processing Lab (VSAP) at Khalifa University and as a department head for the Electrical and Mechanical Departments at the Institute of Applied Technology (IAT), UAE. She played a full and active role in various extra-curriculum activities. She was the project lead in the "Site Inspection Drone", a finalist in the Drones for Good Competition, 2015, and "ColCAM: Spherical capsule for endoscopic applications", a semi-finalist in the 2016 Robotics for Good competitions. She was also the UAE Electronics expert in World Skills (2013-2018).

Autonomous Inspection of Flare Stacks Using an Unmanned Aerial System

Muaz Al Radi, Hamad Karki, Naoufel Werghi, Sajid Javed, and Jorge Dias

Abstract

Flare stack systems are a crucial component in the operation of oil refineries and petrochemical plants as they safely release the excess gas generated during the plant's operation. Performance and structure inspection of these systems is an essential and challenging task due to the flare stacks' harsh operation environment. Flare stacks go through various types of faults both in their mechanical structure and combustion operation, including cracks in the structure, abnormal pilot flame operation, and incomplete combustion of the released gas. The occurrence of such faults could lead to leakage of dangerous emissions to the environment and deadly fires and explosions in the petrochemical plant. The application of autonomous robotic systems in the inspection of such systems is a promising solution for minimizing the involved hazards and costs. In this work, an autonomous

M. Al Radi (✉) · N. Werghi · S. Javed · J. Dias
Department of Electrical Engineering and Computer Science, Khalifa University, Abu Dhabi, UAE
e-mail: 100059660@ku.ac.ae

N. Werghi
e-mail: Naoufel.Werghi@ku.ac.ae

S. Javed
e-mail: Sajid.Javed@ku.ac.ae

J. Dias
e-mail: Jorge.Dias@ku.ac.ae

Center for Autonomous Robotic Systems, Khalifa University, Abu Dhabi, UAE

H. Karki
Department of Mechanical Engineering, Khalifa University, Abu Dhabi, UAE
e-mail: hamad.karki@ku.ac.ae

M. Abdelkader and A. Koubaa (eds.), *Unmanned Aerial Vehicles Applications: Challenges and Trends*, Synthesis Lectures on Intelligent Technologies,
https://doi.org/10.1007/978-3-031-32037-8_6

low-cost unmanned aerial system (UAS) is used for inspecting the flare stack. The proposed UAS uses the video stream obtained from an on-board camera for analyzing the observed scene, controlling the UAS's movement, and obtaining conclusions on the flare stack's operation. Visual servoing control was employed for autonomously collecting visual inspection databases for structure inspection and multi-view operation inspection. For inspecting the flare stack's operation, the novel YOLOv7 model is implemented for analyzing the flare stack's scene and obtaining useful conclusions regarding the operation.

1 Introduction

Energy obtained from oil and natural gas has became one of the most important resources of energy in the world. Huge amounts of oil and natural gas are produced everyday to cover the world's increasing demand for energy in all types of applications, including manufacturing, industrial processes, transportation, and domestic uses. With this huge production and consumption of fossil fuels comes the need for technologies that guarantee the safety and efficiency of these processes. An important component in all oil refineries and petrochemical plants that secures the safety of their operation is the flare stack. The flare stack converts the harmful excess gas that is released from the plant to a safer form before discharging it to the atmosphere. In the cases of abnormal plant operation, such as water and power shortage of equipment malfunctioning, the plant stops its gas production, and due to the high pressure of gas, storage or reuse of this gas becomes inaccessible. Thus, this excess gas must be released to the atmosphere until the plant returns to its functioning state. The produced gas consists mostly of methane and other large hydrocarbons that are hazardous and are of extreme danger of the environment. To reduce the negative effect on the environment, the flare stacks combusts these released gases in open air thus converting them to safer water vapor and carbon dioxide.

Even though flare stacks cover a crucial spot in securing the safe operation of petrochemical plants, they are commonly less monitored compared to other components. This is due to many reasons, including the high temperatures at which the flaring process takes place and the flaring in open air at high elevations. The monitoring system of a flare stack should satisfy the requirements of collecting accurate and comprehensive data on the system's operation while not interfering with the system's operation and while consuming the minimal resources. Considering these technological obstacles and requirements, vision-based inspection of flare stack using robotic systems emerge as a promising option for obtaining accurate inspection data during the plant's operation. Rotary-wing Unmanned Aerial Systems (UAS) have attracted huge attention in the last years and was applied in a wide variety of applications, including photography, surveillance, delivery systems, and remote inspection. UASs can be implemented for collecting comprehensive visual data of the flare stack system using its on-board sensors and Artificial Intelligence (AI) techniques can be used for analyzing the collected data and obtaining useful conclusions on the overall system's state.

Several studies in literature discussed the applications of UASs and Artificial Intelligence (AI)-based data processing methods for inspection and monitoring of combustion-involving systems in general and operational flare stacks in specific [1, 2]. A study by Esfahlani [3] discussed the design of an autonomous smoke and fire detection system employing an UAS. An image processing technique was implemented for analyzing the visual data obtained from the UAS's on-board cameras for detecting the regions of flame and smoke. Moreover, simultaneous localization and mapping (SLAM) was used for constructing a 3D map of the observed environment while monitoring for possible fires. Augmented reality tests were conducted for validating the proposed technique in which the system showed high detection efficiency and excellent failures resistance. In a similar study, Wang et al. [4] proposed a UAS-based forest flame detection system employing a deep learning detector. The flame detection system uses imagery collected using the UAS and an improved YOLOv3 detection network analyzes the captured data and detects the occurrence of flames. The improved YOLOv3 network utilizes more shallow features in the obtained imagery which helps in enhancing the flame detection performance. The improved model was able to achieve an excellent midpoint between more accurate detection and faster data processing. In another work, Gu et al. [5] proposed the application of an image processing-based detection system for detecting soot release from flare stacks. The steps of the illustrated image processing technique localizes the flare, segments the flare's pixels, and draws conclusions on the release of soot during the flaring process. The obtained results indicated the accuracy of the proposed system in soot detection and its superior performance compared to eight state-of-the-art deep learning models. Another similar work by Frizzi et al. [6] discussed the application of Convolutional Neural Network (CNN) models for detecting and segmenting smoke and flame from captured video. They implemented two stages of CNNs to firstly extract local features of smoke and flame from the scene and secondly concatenate the features for obtaining deeper information about smoke and flame in the scene. Comparison of the proposed method with two segmentation models in literature, namely the models of Ronneberger et al. and Yuan et al. [7], showed its improved performance in segmentation tasks.

In this work, a UAS is designed for automating the process of flare stacks inspection by employing vision-based control in the data collection stage and deep learning detection models for analyzing the obtained data and generating useful conclusions on the system's operation. The main contributions of this work can be summarized as follows:

- This work proposes a novel complete system architecture of a UAS-based flare stack inspection system implementing vision-based control and video-based operation analysis for automating the inspection of flare stack systems.
- This work implements Image-Based Visual Servoing (IBVS) visual control technique which was not applied previously in literature for automating the control of UASs for the inspection of vertical flare stack systems.

- This is the first work that applies the novel YOLOv7 deep learning model [8] in the problem of flame and smoke detection both in general and in the context of flare stacks operation inspection.

2 Methodology

2.1 Inspection Data Collection Using a UAS

In this work, visual inspection data are collected using an autonomous Unmanned Aerial System (UAS). The UAS autonomously collects images of the flare stack's structure and flaring process. The UAS's movement is controlled using the visual data obtained from an on-board camera. In this section, the details of the autonomous inspection data collection using an UAS are included.

2.1.1 UAS Architecture

The proposed UAS-based inspection systems employs an on-board visual data analysis architecture. This architecture helps in securing faster control commands generation for the vision-based control. An on-board computer is utilized for controlling the UAS's movement and analyzing the obtained data. During the operation, the on-board sensors return the estimated UAS's state and the mounted cameras return a continuous video stream to the on-board computer. Processing of the obtained data is carried out by the on-board computer for generating control commands and conclusions on the flare stack system's operation. The resulting control commands are sent to the UAS's autopilot for continuing the inspection mission and the conclusions are both recorded and transmitted to the mission monitoring station through a Wi-Fi-based data link interface. The monitoring station performs several tasks including real-time monitoring of the UAS's movement and operation throughout the inspection mission, sending high-level commands to the UAS such as mission starting and emergency landing, and providing a comprehensive view on the inspected flare stack system's state and operation. Figure 1 shows the overall architecture of UAS-based inspection data collection system.

2.1.2 Autonomous UAS Control Using Visual Sensors

The proposed autonomous UAS-based inspection system uses a vision-based control technique for controlling the UAS's movement throughout the inspection mission based on the observed scene from the UAS's on-board camera. In this subsection, the implemented vision-based control technique is presented and its application for autonomous inspection data collection is highlighted.

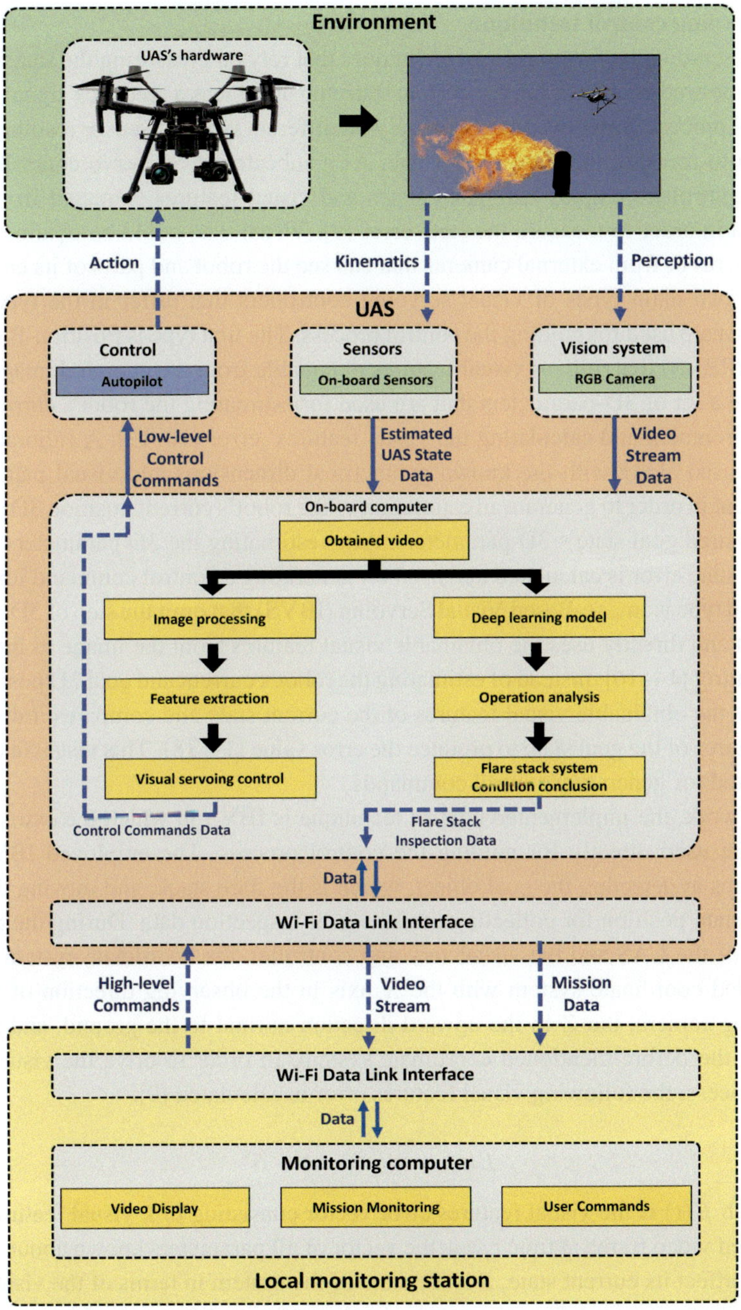

Fig. 1 The overall system architecture of the designed UAS-based flare stacks inspection system consisting of three main blocks (UAS, Monitoring station, Environment). The UAS uses its onboard sensors and vision system to obtain information about the environment, analyzes the data onboard, and sends control commands to the UAS's hardware while keeping communication with a monitoring station for mission monitoring

Visual servoing control technique:
The term "servoing" refers to control techniques that rely on measuring the state error value between a desired goal state and the current state and using it as a feedback signal for driving the control process. Servoing controllers rely on different types of sensor readings and state estimators to measure the state error value. As a subcategory of servo controllers, visual servoing controllers employ estimated states and visual features extracted from obtained visual data in order to measure the state error [9]. Visual data could be obtained from on-board cameras or from external cameras that can see the robot and parts of its environment. There are two main types of visual servoing controllers that differ in the type of visual features that are used for guiding the control process. The first type is Position-Based Visual Servoing (PBVS) that utilizes visual features obtainable from the captured images in order to generate a set of 3D-parameters that are used for estimating the robot's current position in the environment and calculating the visual features' error [10–13]. A calibrated camera system is used along with the known geometrical dimensions and visual patterns of the environment in order to generate an estimation of the robot's current position 3D parameters and the desired goal state's 3D parameters. After estimating the 3D parameters of the two states, the state error is calculated and used for generating a control command for the robot. The second type is Image-Based Visual Servoing (IBVS) that omit the step of 3D parameters estimation and directly uses the obtainable visual features from the image to influence the robot's control [14–16]. Instead of estimating the robot's current and goal 3D position states, IBVS uses the obtainable visual features of the current state and compares it directly with visual features of the goal state to produce the error value [17, 18]. This visual features error value is used for generating control commands.

In this work, the implemented control technique is IBVS in which the extracted visual features are used directly for guiding the control process. The employed IBVS control method aims at detecting the goal object, which is the flare stack, and moving the UAS to an appropriate position for collecting useful visual inspection data. During the design and operation of the UAS and the visual servoing controller, the coordinate system is set as a right-handed coordinate system with the X axis in the observing direction of the UAS's front-facing camera, the Z in the upward direction normal to the ground, and the Y axis completes the before-mentioned coordinate system. In order to drive the visual servoing control process, the following visual features error is calculated [9]:

$$E(t) = S(m(t), a) - S^* \tag{1}$$

In which $E(t)$ is the visual features error vector consisting of k visual features, $m(t)$ is the obtained video frame at time t, a is the vector of all parameters known about the system and could affect its current state, S is the state of the system in terms of the visual features at the current time t, and S^* is the required values of visual features which correspond to the desired state. The employed detector returns three visual features from the input image which are the area of the detected object's bounding box extracted from the video frame $m(t)$ (A) and the coordinates of the detected object's bounding box center relative to the center of the image plane (x, y). Thus, the number of visual features k is equal to three and visual features vector can be defined as follows:

$$S = [x \; y \; A]^T \tag{2}$$

In order to generate control commands for the UAS, the following velocity vector $V(t)$ is defined:

$$V(t) = [v_{yaw} \; v_z \; v_x]^T \tag{3}$$

In which v_{yaw} is the UAS's rotational yaw velocity, v_z is the UAS's linear velocity in the Z axis direction, and v_x is the UAS's linear velocity in the x axis direction. For controlling this velocity vector, a Proportional-Integral-Derivative (PID) controller that uses the features error vector $E(t)$ is introduced as follows:

$$V(t) = K_p E(t) + K_I \int E(t)dt + K_d \frac{d}{dt} E(t) \tag{4}$$

In which the three coefficients K_p, K_I, and K_d correspond to the weights of the proportional, integral, and derivative control components, respectively. As the obtained video stream from the on-board camera system return a discrete number of frames per second, the feature error vector and PID control law are discretized as follows:

$$E[n] = S(m[n], a) - S^* \tag{5}$$

$$V[n] = K_p E[n] + K_I \sum_{i=1}^{n} E[i] + K_d(E[n] - E[n-1]) \tag{6}$$

In which $m[n]$ is the obtained video frame at the sample time n, $E[n]$ is the visual features error vector computed at the sample time n, and $V[n]$ is the velocity commands vector computed by the PID controller at sample time n. Convergence of the control process is achieved when the values in the features error vector $E[n]$ are minimal and the generated velocity commands vector values $V[n]$ are approaching zero as the current state's visual features S are close to the visual features of the goal state S^*. When the required position is reached, the UAS starts collecting visual inspection data of the flare stack.

Data collection cases:
In this subsection, the data collection cases that were designed for application using the proposed UAS-based flare stacks inspection system are overviewed.

Multi-view flaring performance evaluation:
In this data collection case, the UAS is programmed to implement the proposed visual servoing control technique to detect the flare stack's body, approach it, and autonomously collect the required visual inspection data for evaluating the flaring process. Figure 2 presents a state diagram of this data collection case. After receiving the take off signal and completing

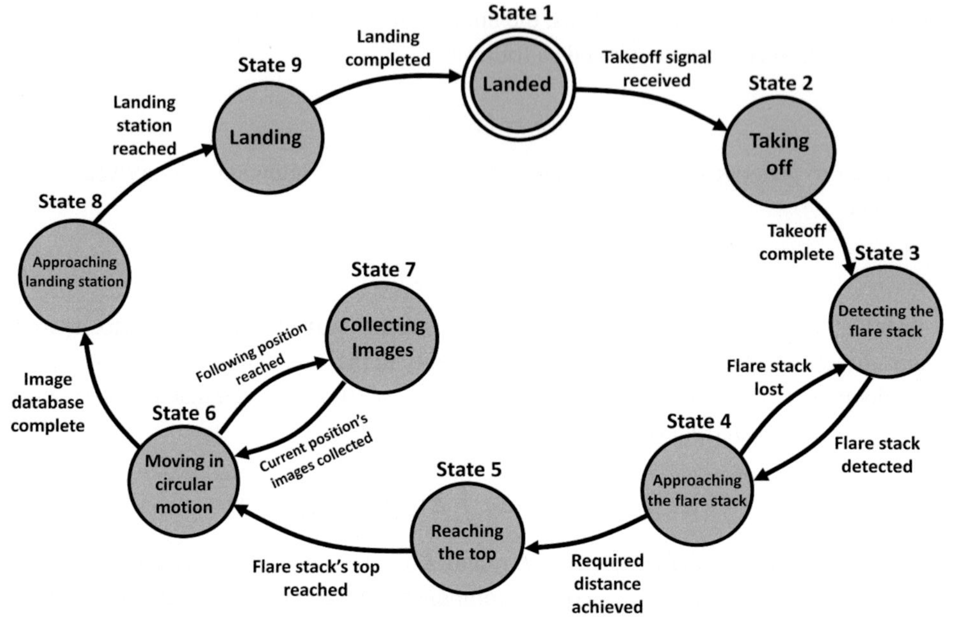

Fig. 2 State diagram of the operation inspection data collection case

the taking off (State 1 and 2), the UAS uses its on-board camera to detect the flare stack and approach it autonomously (State 3 and 4). When the UAS successfully reaches the required distance from the flare stack, the UAS increases its elevation until the flaring operation can be seen in the on-board camera's stream (State 5). The UAS then moves in a circular path while keeping the flare stack in the center of the on-board camera's field of view and collects images of the flaring process from all sides (State 6 and 7). When a complete image database of the flaring process is obtained, the UAS moves toward a landing station and lands (States 8 and 9).

Flare stack's structure inspection:
In this case, The UAS was programmed for utilizing visual servoing control to autonomously approach the flare stack and collect a comprehensive image database for inspecting the flare stack's structure. Figure 3 shows a state diagram of this data collection case. Similar to the previous data collection case, the UAS first takes off when the taking off signal is received (States 1 and 2). Next, the UAS approaches the flare stack using the visual features obtained when the flare stack's body is detected (States 3 and 4). When the UAS reaches the required position for starting the inspection process, the UAS starts moving in a circular movement around the flare stack at the current height while collecting visual inspection data from all sides (States 5 and 6). After the UAS completes a full circle around the structure and collects images from all sides of the flare stack, the UAS slightly increases its elevation to collect more inspection data of higher parts of the flare stack (State 7). The UAS keeps collecting

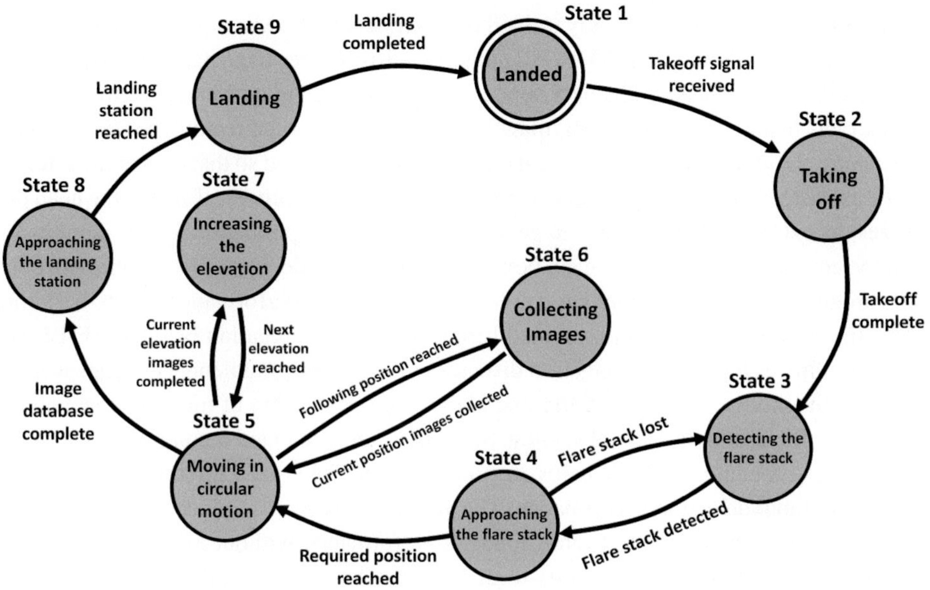

Fig. 3 State diagram of the structure inspection data collection case

inspection data and increasing the elevation until a comprehensive visual structure inspection database is collected. Finally, the UAS approaches the landing station and lands (States 8 and 9).

2.2 Inspection Data Analysis Using YOLOv7 Detector

After collecting the visual inspection data using the UAS, a deep learning-based visual data analysis technique employing the novel YOLOv7 detection model was used. In this section, the details of the visual data analysis technique are included.

2.2.1 YOLOv7 Model

In order to analyze the collected visual inspection data, the novel YOLOv7 object detection model is implemented [8]. The You Only Look Once (YOLO) deep learning models have attracted huge research attention in all types of problems involving object detection due to the excellent detection performance that they achieved throughout the years [19, 20]. The recent YOLOv7 model was able to surpass the previous YOLO models and all state-of-the-art object detection models in terms of inference speed and detection accuracy. According to the authors [8], the YOLOv7 model outperformed several object detection models, including the Detection Transformer (DETR) [21], deformable DETR [22], YOLOR [23], YOLOv4 [24], and YOLOX [25], while being only trained from scratch on the Microsoft COCO (common objects in context) dataset. The YOLOv7 model consists of three main parts: the "backbone", the "neck", and the "head". The backbone is a large Convolutional Neural Network (CNN) that extracts features from the input image and feeds the extracted features to the neck part. In the neck, the extracted features are concatenated and combined so that deeper information can be extracted from the input image to enhance the detection process. Finally, the head part utilizes the extracted deep features to generate bounding box prediction for the input image with the corresponding class for each box. The YOLOv7 model contained several novelties in its architecture and operation that resulted in its improved performance. In the backbone CNN, the authors proposed an Extended Efficient Layer Aggregation Network (E-ELAN) which minimizes the back-propagation distance of the learning gradient. This improvement enhances the model's learning ability and allows for the learning of more diverse features in the data. Regarding the model scaling, the authors implemented a compound width and depth scaling technique which helps in providing multiple sizes for the network while ensuring optimal detection performance. The authors employed a trainable bag-of-freebies which helps in enhancing the overall model's performance without extra operation cost. The authors first employed a module-level re-parameterization technique which involves averaging the learned weights of multiple models to achieve higher generalization capability. The authors also utilized an auxiliary head part that is located shallower in the model before the actual head which helps in monitoring the model's performance with more details during the training process. By supervising this auxiliary head along with the actual head using coarse-to-fine auxiliary head supervision, the overall model's performance was improved. The YOLOv7 employs the leaky-Rectified Linear Unit (leaky-ReLU) activation function which is given by the following equation:

$$leakyReLU(x) = max(0.01x, x) \tag{7}$$

Also, the model uses the Sigmoid Linear Unit (SiLU) activation function which is give by the following equation:

$$SiLU(x) = \frac{x}{1 + e^{-x}} \tag{8}$$

In this work, the YOLOv7 base model with 36.5 million parameters is used for ensuring excellent detection accuracy with fast inference.

2.2.2 Visual Analysis Using the Detector Network

In order to analyze the flare stack system's operation, the following performance quality measures were utilized by the proposed UAS-based operation inspection technique.

The existence of visible flame:

One of the important quality measure to consider in the operation flare stacks is the existence of visible flames. The flare pilot is one of the important components of the flare stack which guarantees the ignition of the released gases and the start of the combustion process. However, faults in the operation of the flare pilot results in the release of the excess gases without combustion which is of extreme danger. The released gases could result in fires and explosions while being extremely hazardous [26]. The inspection of the pilot flare's operation is of important and must be analyzed accurately during the inspection of the flare stack system. The operation of the pilot flare can be observed through the existence of visible flames released from the flare stack. In this work, the measure is estimated via detection from the obtained visual data.

The existence of visible smoke/soot:

Another important quality measure is the existence of visible smoke released from the flare stack. As the flaring process converts the excess gas to water vapor and carbon dioxide in the ideal case, the released gases are supposed to be transparent with no visible color. However, incomplete combustion results in the release of larger carbon compounds and carbon monoxide which are hazardous on the environment. Another carbon compound that is released in abnormal operation is soot, which is a combination of large carbonaceous compounds that result from incomplete gas combustion. The release of soot can be observed via the existence of dark smoke released from the flare stack during the combustion. In this work, the release of smoke or soot is assessed via visual detection from the obtained image data.

2.2.3 Model Training and Evaluation Protocol

In order to train the YOLOv7 object detection model for flame and smoke detection, a dataset of flame and smoke images from different scenes was collected from various inline resources and labelled. As building a dataset of actual flare stack images was inaccessible due to the very limited number of publicly available operational flare stacks images, the detection dataset was constructed of general images of flame and smoke objects to simulate the appearance of operational flare stack systems. The dataset consists of 1742 images of flame and smoke with an average resolution of 1024×768 that was labelled for multi-class object detection of "Flame" and "Smoke" classes. The Computer Vision Annotation Tool (CVAT) was used for labeling the data. The novel YOLOv7 model was trained on a PC workstation equipped with an Nvidia Geforce RTX 3050 graphics card, an Intel i7-10700 processor, and 16GB of RAM. The PyTorch deep learning library and framework was used for training and implementing the model. Training of the model was carried out for 300 epochs on our flame and smoke dataset and the model with the best performance throughout the training was used. A split of 80% of the data was used for training and the remaining 20% were used for testing the model.

For evaluating the performance of the implemented object detection models, the following evaluation measures were implemented:

Precision:
The value of precision can be found using the following equation:

$$P = \frac{TP}{TP + FP} \tag{9}$$

In which FP is the number of false positives and TP is the number of true positives.

Recall:
The value of recall can be found using the following equation:

$$R = \frac{TP}{TP + FN} \tag{10}$$

In which FN is the number of false negatives and TP is the number of true positives.

Mean Average Precision (mAP):
The Mean Average Precision (mAP) is the average value of Average Precision (AP) obtained for all classes in the considered detection problem. It is found using the following equation:

$$mAP = \frac{1}{|\text{classes}|} \sum_{c \in \text{classes}} \frac{|TP_c|}{|TP_c| + |FP_c|} \tag{11}$$

In which FP_c is the number of false positives and TP_c is the number of true positives. Two values of the mAP were calculated for evaluating the performance of the employed models. The first value is mAP@0.5 which is the mAP value calculated through setting the Intersection over Union (IoU) threshold at 0.5. The second value is mAP@0.5:0.95 in which the mAP value is calculated starting from an IoU threshold of 0.5 and increasing in steps of 0.05 up to a value of 0.95 and then computing the average of these values.

Inference time and Frames Per Second (FPS):
The inference time is the time that is required by the object detection model to generate the predictions for bounding boxes. The value of the inference time indicates the speed of the implemented model and its ability to run in real-time. On the other hand, a large value of the inference time results in a slow system operation. As the inference time values are mostly very small, an equivalent evaluation measure is the Frames Per Second (FPS) which is found using the following equation:

$$FPS = \frac{1}{\text{Inference Time}} \tag{12}$$

The inference time and FPS values were calculated on a PC with an Nvidia Geforce RTX 3050 graphics card for all the implemented models.

2.3 Proposed System Evaluation in a Simulated Environment

In order to validate the proposed method and evaluate the implemented techniques, tests using the AirSim simulator in a simulated petrochemical plants environment were conducted. This section illustrates the employed simulation software and simulated environment.

2.3.1 AirSim Simulator

In order to simulate the operation of the system and assess its overall performance, the AirSim simulator was implemented [27]. AirSim accelerates the processes of performance evaluation of robotic systems as it allows for building a realistic robot testing environment and performing the required testing. It is as a simulator for Unmanned Aerial Vehicles (UAVs) and cars which was developed by Microsoft Research and built on the Unreal Engine 4. The simulator can be used for collecting data for various AI applications, such as autonomous mobile robots, reinforcement learning, and computer vision. It provides access to a real-time stream of various image data types, including RGB images and depth images. It also provides a wide variety of instantaneous vehicle's parameters, such as orientation, position, and speed. In this work, the AirSim's built-in object detector was utilized for extracting the needed visual features for controlling the UAS through visual servoing; however, for actual application of such system, training of a deep learning-based object detector for detecting actual flare stacks' bodies is required and will be investigate in future works.

2.3.2 Petrochemical Plant Model with Operational Flare Stacks

In the AirSim simulator, a petrochemical plant model with operational flare stacks was designed using the Unreal Engine 4. Figure 4a shows the built simulated environment. The simulated petrochemical plant environment contains two flare stacks with simulated flame and smoke to represent the flaring process, as shown in Fig. 4b. This model was employed

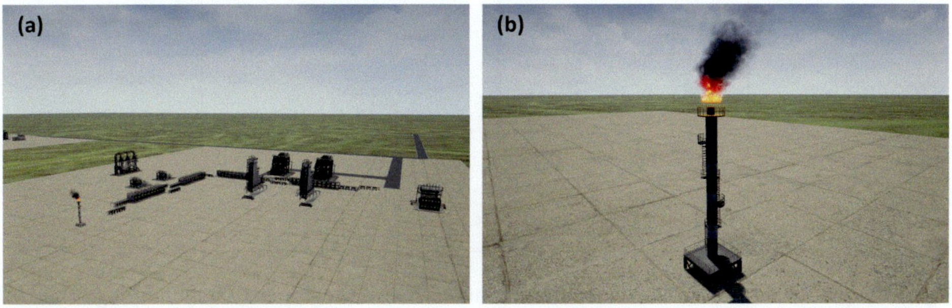

Fig. 4 **a** The built petrochemical plant environment in the simulator and **b** the flare stack model with simulated flame and smoke

for evaluating the proposed UAS's visual servoing control technique and data collection performance. In future works, the implementation of the developed inspection technique on actual flare stacks system will be discussed.

3 Results

In this section, the obtained results regarding the UAS-based inspection data collection and deep learning-based flare stack operation inspection are included and discussed.

3.1 Visual Data Collection Using an UAS

In this subsection, the results obtained regarding the visual data collection using the UAS are presented.

3.1.1 Multi-view Flaring Operation Monitoring

This data collection case was implemented on the AirSim simulator using Python programming language. For guiding the vision-based control, AirSim's built-in detector function was used for detecting the flare stack's body. Figure 5 illustrates the 3D movement of the UAS during the multi-view flaring operation monitoring and Fig. 6 shows sample frames obtained from the UAS's front camera during the data collection case. First, the UAS could

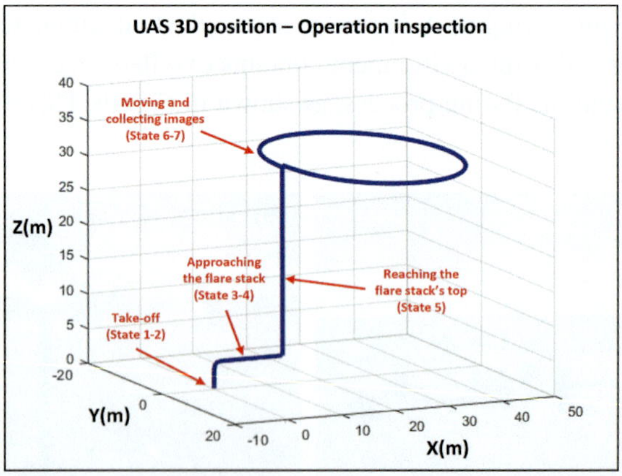

Fig. 5 3D plot of the UAS's movement during multi-view flaring operation monitoring. The UAS increases its elevation, starts visual servoing to approach the flare stack autonomously, starts increasing its elevation again when the required distance from the flare stack's body is achieved, and starts collecting multi-view inspection data when the flaring process is visible by the onboard cameras

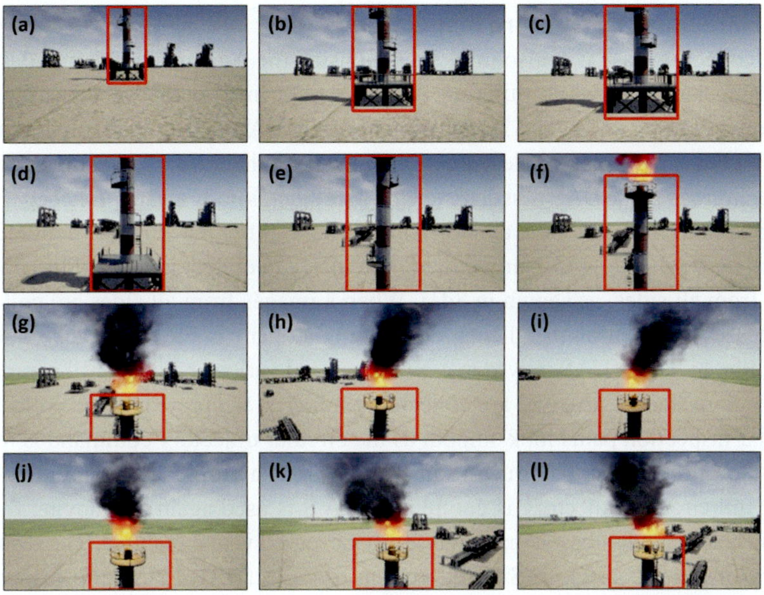

Fig. 6 Sample frames from the UAS's movement during the flaring operation data collection case, the UAS **a** detects the flare stack's body, **b** moves in the direction of the flare stack, **c** reaches the required distance from the flare stack, **d–g** increases the elevation until the flare stack's top is seen, **h–l** moves in a circular path to collect images of the flare for multi-view data collection

successfully detect the flare stack's body, move in its direction, and reach a suitable distance from it, as illustrated in Fig. 6a–c. Next, the UAS increased its elevation until the flare stack's top and the flaring process could be observed, as illustrated in Fig. 6d–g. After reaching the required position, the UAS started moving in a circular path while focusing its front camera on the flare stack's top, as shown in Fig. 6h–l. Images of the flaring process are collected from all sides during the circular movement which can be used for multi-view flaring performance inspection. It can be noticed that the circular motion of the UAS appears smooth. The reason of the smooth path is that the UAS controller sends a circular trajectory command when the required position relative to the flare stack's body is achieved while using IBVS for centering the UAS's front facing camera on the flare stack's body. The circular path appears smooth as the UAS receives a predefined full circle trajectory command at the positions of initiating the multi-view data collection. The results showed that the UAS could successfully utilize visual servoing control to collect the required visual inspection data for analyzing the flare stack's operation.

3.1.2 Flare Stack's Structure Inspection

This data collection case was also implemented using the AirSim simulator and the Python programming language while utilizing AirSim's built-in detector for detecting the flare stack's body. Figure 7 shows the 3D movement of the UAS during the structure inspection process and Fig. 8 presents sample frames obtained from the UAS's front camera during this data collection case. The UAS first detected the flare stack's body and approached it similar to the previous data collection case, as shown in Fig. 8a–c. After reaching the required distance from the flare stack, the UAS starts moving circularly to collect images of the flare stack's structure from all sides, as shown in 8d–i. When a images from all sides are collected, the UAS increases its elevation to collect images of higher parts of the flare stack's structure, as shown in Fig. 8j–l. The UAS moves in circular paths at various heights until comprehensive data of the flare stack's structure are collected. Similar to the previous data collection case, the circular inspection paths look smooth as the circular trajectory command is received at different elevations when the structure inspection from multiple sides should be initiated. After the full circular trajectory is completed, IBVS is used again for moving the UAS to the next inspection point based on the currently observed visual features. The results shows the ability of the proposed UAS-based system to collect visual structure inspection data autonomously and efficiently.

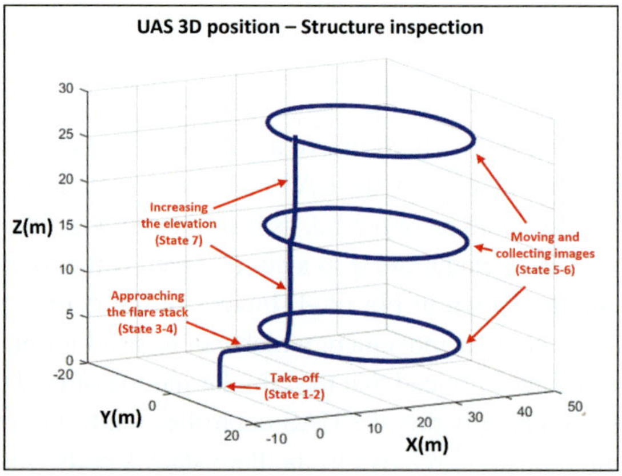

Fig. 7 3D plot of the UAS's movement during the structure inspection. The UAS increases its elevation, starts visual servoing to approach the flare stack autonomously, starts collecting visual inspection data from all sides of the flare stack's structure when the required distance from the flare stack's body is achieved, starts increasing its elevation again when the visual data collection process is completed at the current elevation, and continues inspecting the flare stack's body at multiple elevations until a visual database of the entire flare stack's structure is collected

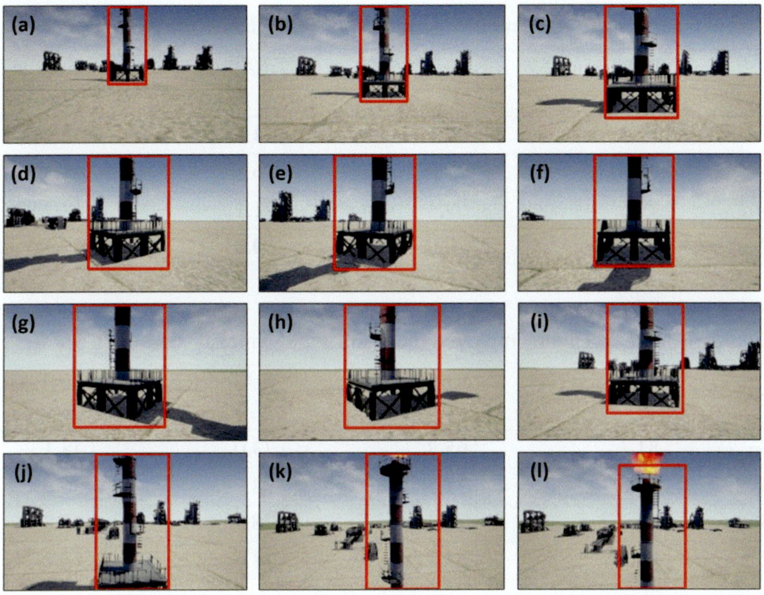

Fig. 8 Sample frames from the UAS's movement during the structure inspection data collection case, the UAS **a** detects the flare stack's body, **b** moves in the direction of the flare stack, **c** reaches the required distance from the flare stack, **d–i** moves in a circular path to collect images of the flare stack's body from all sides for structure inspection, **j–l** increases the elevation to collect inspection data at a higher elevation

3.2 Obtained Data Analysis Using YOLOv7

The YOLOv7 model was trained on the collected flame and smoke dataset for the analysis of flare stacks operation. In order to compare the performance of the proposed model with other models, three YOLOv5 object detection models with similar network sizes were trained

Table 1 Comparison between the employed models for all classes (*P*: Precision, *R*: Recall, *mAP*: Mean Average Precision)

Model	Parameters (M)	P (%)	R (%)	mAP @0.5 (%)	mAP @0.5:0.95 (%)	Inf. time (s)	FPS
YOLOv5n	1.9	80.2	68.4	73.8	43.4	**0.0040**	**250.0**
YOLOv5s	7.2	84.6	**71.1**	77.6	**52.4**	0.0067	149.3
YOLOv5m	21.2	82.9	70.1	75.0	49.0	0.0145	69.0
YOLOv7	36.5	**86.3**	67.6	**79.3**	51.7	0.0328	30.5

Table 2 Comparison between the employed models for the "Flame" class

Model	Precision (%)	Recall (%)	mAP @0.5 (%)	mAP @0.5:0.95 (%)
YOLOv5n	79.2	**71.1**	74.6	38.7
YOLOv5s	81.6	71.0	76.5	48.8
YOLOv5m	77.5	69.4	71.1	42.2
YOLOv7	**81.8**	70.0	**80.0**	**49.5**

Table 3 Comparison between the employed models for the "Smoke" class

Model	Precision (%)	Recall (%)	mAP @0.5 (%)	mAP @0.5:0.95 (%)
YOLOv5n	81.1	65.7	73.0	48.1
YOLOv5s	87.5	**71.3**	78.6	**56.0**
YOLOv5m	88.4	70.7	**78.9**	55.8
YOLOv7	**90.8**	65.1	78.5	53.8

on the same dataset following the same training protocol. Table 1 shows the values of the different evaluation measures for all models on the testing set for all classes together. The YOLOv7 model reached a precision value of 86.3%, a recall value of 67.6%, an mAP @ 0.5 value of 79.3%, and an mAP @ 0.5:0.95 value of 51.7% for all classes, thus achieving the best overall performance compared to the other considered models as indicated by the highest value of the mAP @ 0.5 and the high value for the mAP @ 0.5:0.95 which is very close to that of the YOLOv5s model. Regarding the inference speed, the model was able to perform predictions in real-time as indicated by its FPS value of 30.5. Regarding the performance on the "Flame" class, the trained YOLOv7 model achieved the best performance among the models as indicated by the its increased values for mAP @ 0.5 and mAP @ 0.5:0.95 for the flame class which are equal to 80.0 and 49.5%, respectively, as shown in Table 2. Similarly, the model achieved comparable performance to the other models for the "Smoke" class as it achieved the highest precision value of 90.8% among the other models and comparable values for mAP @ 0.5 and mAP @ 0.5:0.95 of 78.5 and 53.8%, respectively, as compared to 78.6 and 56.0% for the YOLOv5s model as shown in Table 3. In order to illustrate the performance of the trained YOLOv7 model for flame and smoke detection, Fig. 9 shows some detection predictions of the model both on simulated flare stacks scenes and real flare stacks scenes. The model was able to accurately detect the bounding boxes both for flame class and smoke class and thus it could be used for analyzing the overall system's operational performance as indicated by the existence of visible flame and smoke.

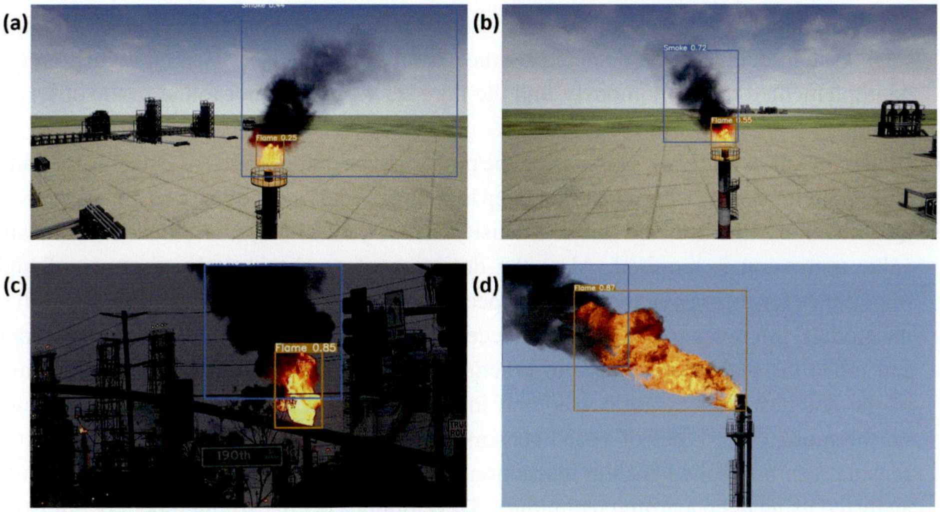

Fig. 9 Obtained predictions of the trained YOLOv7 model in **a**, **b** simulated flare stacks scenes and **c**, **d** actual flare stacks scenes (Brown: "Flame" class, Blue: "Smoke" class)

4 Conclusions and Future Research Directions

In this work, an autonomous UAS-based inspection technique for inspecting flare stack systems using vision-based control was developed. An IBVS technique was implemented for controlling the UAS throughout the inspection mission using the UAS's on-board camera stream. A vision-based inspection technique for flare stacks operation inspection using the novel YOLOv7 deep learning object detection model was proposed. The performance of the system was evaluated using the AirSim simulation software. The developed UAS system could successfully collect the required images autonomously using visual servoing control both for structure and operation inspection and the vision-based inspection technique could successfully recognize abnormal flare stack operation scenarios. In our future works, the following directions will be considered for improving the proposed UAS-based flare stack inspection system:

- **Inspection data collection using a UAS**: The proposed UAS-based data collection technique could be tested on an actual flare stack system using a real UAS. Performing detailed comparisons between the proposed IBVS-based autonomous data collection process and other visual servoing techniques such as PBVS [13] and Hybrid Visual Servoing (HVS) [28] is required for validating the proposed methods. Designing a visual servoing control method while taking into consideration the environment's effects both on the UAS's movement and the quality of the obtained images is of crucial importance. Moreover, a Simultaneous Localization and Mapping (SLAM) technique could be applied to track the

UAS and flare stack's locations throughout the mission to avoid failures when the flare stack system could not be detected by the visual system. Also, a search algorithm can be implemented to autonomously find the flare stack if it was not found in the observed frame.

- **Inspection data analysis**: Comparing the performance of the YOLOv7 detector for flame and smoke detection tasks with other deep learning models must be carried out. Moreover, Quantization of the detected smoke's density either by multi-class classification or density value regression is required. Also, the detection of smoke in early stages is of high importance for increasing the systems speed in detecting abnormal operation. Regarding the models' inference speed, hardware acceleration of the employed deep learning models can move the techniques closer to practical implementation. Analyzing the dynamics of the combustion was not considered in this work and must be discussed in future. Furthermore, other types of visual sensors, such as thermal cameras and hyperspectral sensors, can be applied for the inspection of flare stack system's operation and can be analyzed using deep learning techniques. Developing a deep learning model or adjusting the YOLO object detection model to analyze such types of visual data is recommended.

Acknowledgements This work was supported by Khalifa University through the funding to the project RC1-2018-KUCARS (Center of Autonomous Robotic Systems) and Abu Dhabi National Oil Company (ADNOC) through the funded project named "Intelligent Image and Video Analytics for Corrosion and Flare Monitoring".

References

1. M. Al Radi, H. Karki, N. Werghi, S. Javed, J. Dias, Video analysis of flare stacks with an autonomous low-cost aerial system, in *ADIPEC*, OnePetro (2022)
2. M. Al Radi, H. Karki, N. Werghi, S. Javed, J. Dias, Vision-based inspection of flare stacks operation using a visual servoing controlled autonomous unmanned aerial vehicle (UAV), in *IECON 2022–48th Annual Conference of the IEEE Industrial Electronics Society* (IEEE, 2022), pp. 1–6
3. S.S. Esfahlani, Mixed reality and remote sensing application of unmanned aerial vehicle in fire and smoke detection. J. Ind. Inf. Integr. **15**, 42–49 (2019)
4. Z. Wang, H. Zhang, M. Hou, X. Shu, J. Wu, X. Zhang, A study on forest flame recognition of UAV based on yolo-v3 improved algorithm, in *Recent Advances in Sustainable Energy and Intelligent Systems* (Springer, Berlin, 2021), pp. 497–503
5. K. Gu, Y. Zhang, J. Qiao, Vision-based monitoring of flare soot. IEEE Trans. Instrum. Meas. **69**(9), 7136–7145 (2020)
6. S. Frizzi, M. Bouchouicha, J.-M. Ginoux, E. Moreau, M. Sayadi, Convolutional neural network for smoke and fire semantic segmentation. IET Image Proc. **15**(3), 634–647 (2021)
7. F. Yuan, L. Zhang, X. Xia, B. Wan, Q. Huang, X. Li, Deep smoke segmentation. Neurocomputing **357**, 248–260 (2019)
8. C.-Y. Wang, A. Bochkovskiy, H.-Y.M. Liao, Yolov7: trainable bag-of-freebies sets new state-of-the-art for real-time object detectors (2022). arXiv:2207.02696

9. F. Chaumette, S. Hutchinson, P. Corke, Visual servoing, in *Springer Handbook of Robotics* (Springer, Berlin, 2016), pp. 841–866
10. C. Lee, H. Seo, H.J. Kim, Position-based monocular visual servoing of an unknown target using online self-supervised learning, in *2019 IEEE/RSJ International Conference on Intelligent Robots and Systems (IROS)* (IEEE, 2019), pp. 4467–4473
11. R.S. Sharma, S. Shukla, L. Behera, V.K. Subramanian, Position-based visual servoing of a mobile robot with an automatic extrinsic calibration scheme. Robotica **38**(5), 831–844 (2020)
12. S. Heshmati-alamdari, A. Eqtami, G.C. Karras, D.V. Dimarogonas, K.J. Kyriakopoulos, A self-triggered position based visual servoing model predictive control scheme for underwater robotic vehicles. Machines **8**(2), 33 (2020)
13. X. Xu, R. Tang, L. Gong, B. Chen, S. Zuo, Two dimensional position-based visual servoing for soft tissue endomicroscopy. IEEE Robot. Autom. Lett. **6**(3), 5728–5735 (2021)
14. T. Xu, Y. Guan, J. Liu, X. Wu, Image-based visual servoing of helical microswimmers for planar path following. IEEE Trans. Autom. Sci. Eng. **17**(1), 325–333 (2019)
15. C.P. Bechlioulis, S. Heshmati-Alamdari, G.C. Karras, K.J. Kyriakopoulos, Robust image-based visual servoing with prescribed performance under field of view constraints. IEEE Trans. Rob. **35**(4), 1063–1070 (2019)
16. Y. Zhang, S. Li, A neural controller for image-based visual servoing of manipulators with physical constraints. IEEE Trans. Neural Netw. Learn. Syst. **29**(11), 5419–5429 (2018)
17. A. Ghasemi, P. Li, W.-F. Xie, Adaptive switch image-based visual servoing for industrial robots. Int. J. Control Autom. Syst. **18**(5), 1324–1334 (2020)
18. H. Xie, A.F. Lynch, K.H. Low, S. Mao, Adaptive output-feedback image-based visual servoing for quadrotor unmanned aerial vehicles. IEEE Trans. Control Syst. Technol. **28**(3), 1034–1041 (2019)
19. J. Redmon, A. Farhadi, Yolo9000: better, faster, stronger, in *Proceedings of the IEEE Conference on Computer Vision and Pattern Recognition* (2017), pp. 7263–7271
20. J. Redmon, A. Farhadi, Yolov3: an incremental improvement (2018). arXiv:1804.02767
21. N. Carion, F. Massa, G. Synnaeve, N. Usunier, A. Kirillov, S. Zagoruyko, End-to-end object detection with transformers, in *European Conference on Computer Vision* (Springer, Berlin, 2020), pp. 213–229
22. X. Zhu, W. Su, L. Lu, B. Li, X. Wang, J. Dai, Deformable detr: deformable transformers for end-to-end object detection (2020). arXiv:2010.04159
23. C.-Y. Wang, I.-H. Yeh, H.-Y.M. Liao, You only learn one representation: unified network for multiple tasks (2021). arXiv:2105.04206
24. A. Bochkovskiy, C.-Y. Wang, H.-Y.M. Liao, Yolov4: optimal speed and accuracy of object detection (2020). arXiv:2004.10934
25. Z. Ge, S. Liu, F. Wang, Z. Li, J. Sun, Yolox: exceeding yolo series in 2021 (2021). arXiv:2107.08430
26. X. Xie, K. Gu, J. Qiao, A new anomaly detector of flare pilot based on siamese network, in *2020 Chinese Automation Congress (CAC)* (IEEE, 2020), pp. 2278–2238
27. S. Shah, D. Dey, C. Lovett, A. Kapoor, Airsim: high-fidelity visual and physical simulation for autonomous vehicles, in *Field and Service Robotics* (Springer, Berlin, 2018), pp. 621–635
28. A.A. Hafez, E. Cervera, C. Jawahar, Hybrid visual servoing by boosting IBVS and PBVS, in *2008 3rd International Conference on Information and Communication Technologies: From Theory to Applications* (IEEE, 2008), pp. 1–6

Muaz Al Radi is a Graduate Research and Teaching Assistant (GRTA) and a Ph.D. in Electrical and Computer Engineering student at Khalifa University. He received his B.Sc. degree in Sustainable and Renewable Energy Engineering from the University of Sharjah, Sharjah, UAE, in 2020 and his M.Sc. in Electrical and Computer Engineering from Khalifa University, Abu Dhabi, UAE, in 2022. His research is focused on Computer Vision, Vision-based Control, Artificial Intelligence (AI), and Robotics.

Hamad Karki received the Ph.D. degree from the Tokyo University of Technology, Tokyo, Japan, in 2008 with robotics group. His Ph.D. work was primarily focused on field robots which are basically tracked vehicles that have various off-road running features applicable in rescue missions. His research interests include robotics, mechatronics, and control systems.

Naoufel Werghi received the Ph.D. degree in computer vision from the University of Strasbourg, Strasbourg, France, in 1996.He has been a Research Fellow with the Division of Informatics, The University of Edinburgh, Edinburgh, U.K., and a Lecturer with the Department of Computer Sciences, University of Glasgow, Glasgow, U.K. He has also been a Visiting Professor with the Department of Electrical and Computer Engineering, University of Louisville, Louisville, KY, USA. He is currently an Associate Professor with the Department of Electrical and Computer Engineering, Khalifa University of Science and Technology, Abu Dhabi, United Arab Emirates. His main area of research is image analysis and interpretation, where he has been leading several funded projects in the areas of biometrics, medical imaging, and intelligent systems.

Sajid Javed received the B.Sc. degree in computer science from the University of Hertfordshire, U.K., in 2010, and the master's and Ph.D. degrees in computer science from Kyungpook National University, Republic of Korea, in 2017. He is currently an Assistant Professor of computer vision with the Department of Electrical and Computer Engineering (ECE), Khalifa University of Science and Technology, United Arab Emirates. Prior to that, he was a Research Scientist at Khalifa University Center for Autonomous Robotics System (KUCARS), from 2019 to 2021. Before joining Khalifa University, he was a Research Fellow at the University of Warwick, U.K., from 2017 to 2018, where he worked on histopathological landscapes for better cancer grading and prognostication. His research interests include visual object tracking in the wild, multi-object tracking, background-foreground modeling from video sequences, moving object detection from complex scenes, cancer image analytics including tissue phenotyping, nucleus detection, and nucleus classification problems. His research themes involve developing deep neural networks, subspace learning models, and graph neural networks.

Jorge Dias received the Ph.D. degree in electrical engineering from the University of Coimbra, Coimbra, Portugal, in 1994. He coordinated the Artificial Perception Group, Institute of Systems and Robotics, University of Coimbra. He is currently a Full Professor with the Khalifa University of Science and Technology, Abu Dhabi, United Arab Emirates, where he is also the Deputy Director of the Center of Autonomous Robotic Systems. His expertise is in the area of artificial perception (computer vision and robotic vision) and has contributions on the field since 1984. He has been a principal investigator and a consortia coordinator from several research international projects and coordinates the research group on computer vision and artificial perception from the Khalifa University Center for Autonomous Robotic Systems (KUCARS). He has published several articles in the area of computer vision and robotics that include more than 300 publications in international journals and conference proceedings and recently published book on probabilistic robot perception that addresses the use of statistical modeling and artificial intelligence for perception, planning, and decision in robots. He was the Project Coordinator of two European Consortium for the Projects "Social Robot" and "GrowMeUP" that were developed to support the inclusivity and wellbeing for of the elderly generation.

Improving Distributed Parameter System State Estimation Using UAV-Based Mobile Sensors: Application in Air Pollution Monitoring

Hassan Iqbal, Kun Qian, and Christian Claudel

Abstract

Monitoring distributed parameter systems is a complex task, and requires a large number of sensors. Fortunately, the emergence of increasingly inexpensive and capable Unmanned Aerial Vehicles (UAVs) enables efficient monitoring of extended areas with a limited number of sensors. In this chapter, we focus on model-based state estimation when UAVs equipped with mobile sensors are used to sequentially estimate a system of high dimensionality. We compare performance of the classical Kalman filter with its variants in constructing probabilistic maps of air pollution transport. We test various sensor configurations including a combination of fixed and UAVs equipped with mobile sensors, and present an empirical analysis of the value added by mobile sensing. We describe the use of consensus Kalman filter for cooperative estimation in the case of a team of UAVs, formulate an information-aware path planning algorithm for UAVs, and illustrate the importance of UAV path planning on simulation data.

Keyword

Model-based estimation • Distributed parameter systems • Unmanned aerial vehicle • Mobile sensors • Kalman filter • Particle filtering

---

H. Iqbal (✉) · K. Qian · C. Claudel
The University of Texas at Austin, Austin, TX 78712, USA
e-mail: hassan.iqbal@utexas.edu

K. Qian
e-mail: kunqian@utexas.edu

C. Claudel
e-mail: christian.claudel@utexas.edu

M. Abdelkader and A. Koubaa (eds.), *Unmanned Aerial Vehicles Applications: Challenges and Trends*, Synthesis Lectures on Intelligent Technologies,
https://doi.org/10.1007/978-3-031-32037-8_7

1 Introduction

Many important chemical, physical and biological systems have dual spatial and temporal aspects. These systems, referred to as Distributed Parameter Systems (DPS), are mathematically modelled by partial differential equations (PDEs). The control and estimation of DPS is usually more complex than that of Lumped Parameter Systems, that is, systems for which the state is a time-dependent vector. Example applications of DPS estimation and control include but not limited to oil-spill estimation, air or water pollution monitoring, urban flash flood monitoring, weather prediction, soil moisture estimation, fire and smoke spread estimation, wildfire monitoring.

Fixed and mobile sensors can be deployed to monitor the state of a DPS. Mobile sensing of a DPS can sometimes provide more information than fixed sensing given a fixed number of total sensing devices. For example, sensors that identify harmful levels of pollutant concentrations are expensive. It becomes impractical to cover a large region with many such sensors at fixed locations in order to get a reliable estimate of state. Instead, a set of these sensors may be mounted on a team of UAVs that route to various locations in the region to map the pollutant distribution. Said that, the computation of mobile sensor trajectories further increases the complexity of the problem. State-estimation by employing sensors-equipped mobile robots—either UAVs, ground or underwater robots—is studied extensively in controls and robotics domains [10, 17, 18, 50].

Generally, identifying the state of a DPS is theoretically impossible with a finite number of sensors (owing to the infinite-dimensional nature of the state), which motivates the development of state observers. An extensive theory and several techniques are developed for estimating the parameters and state of a dynamic system given noisy and incomplete measurements, and a model that describes the underlying physical process. In this chapter, we compare the performance of different estimation schemes when used in conjunction with distributed parameter systems and mobile sensing. Note that while we use as running example in which UAVs or ground robots are to estimate the concentration of a pollutant gas, the background and solutions presented in this chapter are also applicable to DPS described by other linear PDEs. A block diagram of the proposed UAV based monitoring for pollutant transport is shown in Fig. 1.

Subsequently, we assess qualitatively and quantitatively the value added by mobile sensors, with respect to fixed sensors. We show that in case of a large number fixed sensors with sufficient sampling frequency, the observer converges to the true state of the system i.e within tight uncertainty bounds, the true concentration map of the pollutant in the region. Through numerical experiments, we illustrate the value that a mobile sensor adds into the estimation process in terms of reduction in state estimation error. As each UAV can only measure the environment within a limited sensing range, depending on its real-time position and orientation, we need to devise a path planner. We explain the process to compute in real-time the control inputs for all involved UAVs such that the generated trajectories allow them to take measurements on the unknown air environment, which provides as much infor-

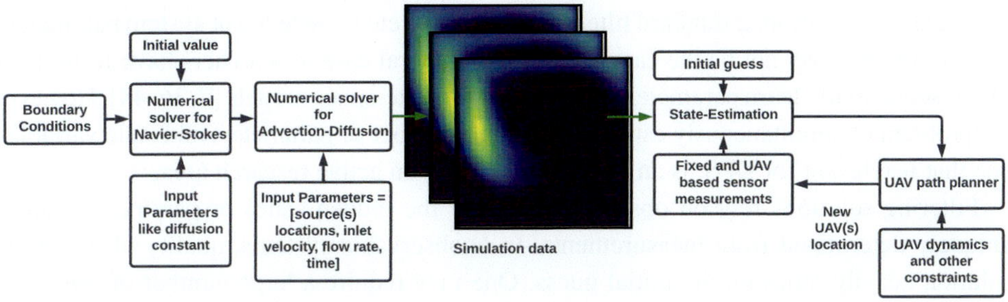

Fig. 1 Block diagram of the proposed UAV based monitoring for pollutant transport

mation as possible. Lastly, we empirically show the effect of unknown or incorrect model parameters, the underlying wind flow in the case of our running example, on the quality of state estimate of the pollutant concentration map.

2 Literature Review

The state and parameter estimation problems in distributed parameter systems have some similarities with standard state estimation. Approaches relying on Kalman Filter [29], Extended Kalman Filter, Ensemble Kalman Filter [14], particle filter [15] and their variants are available. Kalman filter (KF) gives exact solution for linear Gaussian systems by recursively updating posterior density. Extended Kalman Filter (EKF) is essentially a Kalman Filter applied to a linearlized model. EKF does not provide guarantees for nonlinear systems, though it has shown to work in systems with unimodal symmetric posteriors. The Ensemble Kalman Filter (EnKF) propagates an ensemble of samples through state-space and has shown to work reliably in non-linear systems with Gaussian noise distribution. Particle Filter is a popular filtering approach for estimation of non-linear and non-Gaussian systems. However, in large dimensional systems, such as those observed as distributed parameter system, base version of particle filter, also referred as the bootstrap particle filter, often fail to converge at all due to impoverishment of the particles' population. This is exacerbated if particles are not initialized properly.

In recent years, researchers have worked on efficient particle filters, such as those that employ optimal sampling [16], equivalent weights [1], constrained sampling [25], and probability truncation, to estimate relatively large dimensional systems. A filtering technique called Rao-Blackwellized Particle Filter (RBPF) has also been extensively studied and applied to various real-life cases. In RBPF, linear part of the high dimensional state is marginalized and solved via KF, while particles in PF are used to estimate the non-linear states only. This way a high dimensional state may carefully be factorized into states with smaller dimensions and solved efficiently. We refer the reader to [16] for an introductory survey on classic filtering techniques and their variants.

The above-mentioned standard filters assume complete knowledge of system parameters. In real life, this may not be the case. For a more general case, researchers have formulated PFs to sequentially learn parameter distributions for state-space models [7, 36, 44]. However, the problem of simultaneously estimating a large number of parameters and state variables, one that is relevant for DPS, is an open problem and an active research area.

Filtering solutions rely on observability i.e. if the hidden states or parameters can be uniquely determined from measurements. In unobservable systems, quality of numerical solution heavily relies on the initial guess. One may require a large number of sensors at fixed locations to design a consistent estimator for state and parameters, depending on the system. Hence, it is of interest that a relatively smaller number of mobile sensing devices can instead be used that autonomously reposition themselves within the spatial domain as per guiding principle that either full observability condition is always imposed if possible [26], or that mobile sensors traverse to information rich regions thereby maximizing observability [22]. UAVs are an example of mobile sensors. The collaboration among mobile sensors require communication. Researchers have considered varying connectivity topology to achieve task assignment and encoded communication constraints in the path planning optimization program [32] to achieve realistic path plans for a team of mobile sensors.

Various metrics are employed to optimize path plans for mobile sensors, that broadly can be classified as information based, input energy based, and a combination of both. As engineers, we would like to gather as much information as possible given a fixed number of mobile sensors while using the least fuel cost. Indeed, the problem of active information gathering has been broadly studied in the literature [2–5, 23, 24, 30, 35, 42, 49, 51, 52]. A combination of optimization tools, covariance or information matrices of Kalman filter, upper confidence bound (UCB) in multi-armed bandits is used to optimally explore the spatial domain while exploiting current knowledge [13].

3 Problem Formulation

3.1 Modelling

In this chapter, we consider the case of pollutant transport in air. The transport of the pollutant can be modelled by an advection-diffusion equation [9], which is linear in the concentration state, but has a distributed set of uncertain parameters i.e. the underlying wind velocity field. If y is the concentration of the pollutant gas, then the evolution of the gas field as governed by the advection-diffusion equation is given as

$$\frac{\partial y}{\partial t} = -(\boldsymbol{u}.\nabla)y + \gamma \nabla^2 y + S \tag{1}$$

where the first term on the right hand side represents the advection and the second term represents diffusion with γ being the coefficient of the diffusion. S term represents the gas sources. The underlying wind velocity field $\boldsymbol{u}$, in which the pollutant traverses, can safely be assumed to respect physics of an incompressible fluid flow i.e volume of fluid in any sub-region is constant over time. This can mathematically be represented as

$$\nabla.\boldsymbol{u} = 0 \tag{2}$$

Further, we assume that air is homogeneous i.e. the density of air is constant in space. These two assumptions together imply that density of air is constant in both time and space. In order to compute the underlying wind velocity field, we use Navier–Stokes equations [11]. We use the vorticity-streamfunction formulation [21] of the Navier–Stokes equations as it allows us to skip the pressure correction step i.e. inherently meets the incompressibility constraint (2). The spatial domain is discretized in form of a Cartesian grid. Steamfunctions, ψ, are lines parallel to the flow defined as follows:

$$\boldsymbol{u} = \nabla^{\perp}\psi \equiv \left(\frac{\partial \psi}{\partial y}, -\frac{\partial \psi}{\partial x}\right) \equiv (u, v) \tag{3}$$

where u, v are horizontal and vertical velocities respectively. Vorticity, ω, at any point in Cartesian grid is defined as twice the angular velocity and defined as

$$\omega \equiv \nabla \times \boldsymbol{u} = \frac{\partial v}{\partial x} - \frac{\partial u}{\partial y} = -\frac{\partial^2 \psi}{\partial x^2} - \frac{\partial^2 \psi}{\partial y^2} \equiv -\nabla^2 \psi \tag{4}$$

We discretize the domain using second-order centered finite differences, and implicitly march the governing equations forward in time. The following vorticity-steamfunction equations can easily be derived [21] to describe the flow of an incompressible fluid:

$$\frac{\partial \omega}{\partial t} + (\nabla^T \psi).\nabla\omega = \frac{1}{Re}\nabla^2\omega \tag{5}$$

$$\nabla^2\psi = -\omega \tag{6}$$

Equation 6 enforces incompressibility. Suitable boundary conditions may be chosen to reflect various scenarios. In this chapter, we consider the case of wind flow in an indoor environment with four walls on each side as shown in the Fig. 2. Note that the estimation process presented in the chapter is also applicable to scenarios with different boundary conditions or linear partial differential equations as long as the problem is formulated accordingly. For the indoor environment, we assume Dirichlet boundary conditions on ψ and Thom's boundary conditions on ω i.e. in rectangular domain Ω, $\psi = 0$ on the entire $\partial\Omega$ which ensures that the walls are a single streamline, i.e. that the wall-normal velocities are zero. Moreover, Thom's boundary conditions on ω can be derived with Taylor's series expansion for the streamfunction at a point adjacent to a wall and is given by:

$$\omega_b = 2\left(\frac{\psi_b - \psi_a}{\nabla n^2} + \frac{u_t}{\nabla n}\right) \tag{7}$$

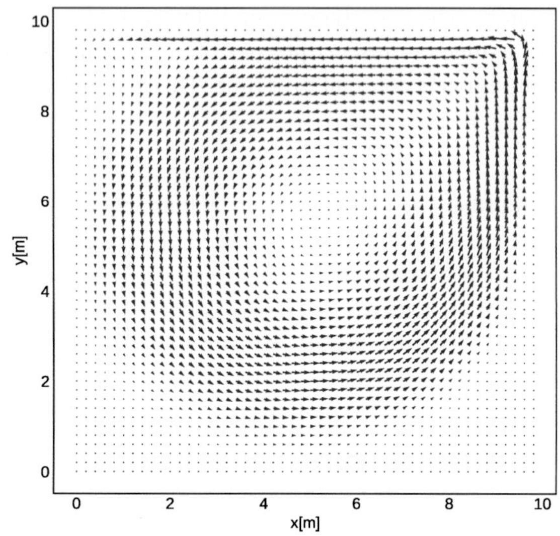

Fig. 2 Steady-state wind velocity field used for numerical experiments

where subscript ‘a’ representing the wall-adjacent point and the subscript b representing the point at the wall. ‘n’ is a coordinate representing the wall-normal direction, and Δn is the spatial discretization in the direction normal to the wall. Readers interested in derivation of Thom’s formula and brief overview of its variants may refer to [48].

By employing the above-mentioned boundary conditions, the resultant steady-state velocity field is computed using Eqs. (5) and (6) and shown in Fig. 2. A pollutant is leaked at time $t = 0$ and the pollutant concentrations at various times is shown in Fig. 3. This system may correspond, for instance, to a dynamically changing methane gas distribution in an indoor environment with potentially multiple moving gas leaks. Exact parameter values for numerical results shown in Figs. 3 and 2 are mentioned in Table 1 under Sect. 3.4.

3.2 State Estimation

Let us consider a sensor network with a goal to estimate the unknown pollutant concentration map in Fig. 3 at each time step. We assume that the underlying wind velocity has reached a steady state i.e. $\boldsymbol{u}$ is constant or

$$\frac{\partial \omega}{\partial t} = \frac{\partial \psi}{\partial t} = 0 \tag{8}$$

Assumption 8 makes the Eq. 1 linear with respect to the concentration state allowing us to represent concentration evolution in the form

$$\mathbf{y}_{t+1} = A_t \mathbf{y}_t + \mathbf{w}_t, \;\; \mathbf{w}_t \sim \mathcal{N}(\mathbf{0}, W_t), \tag{9}$$

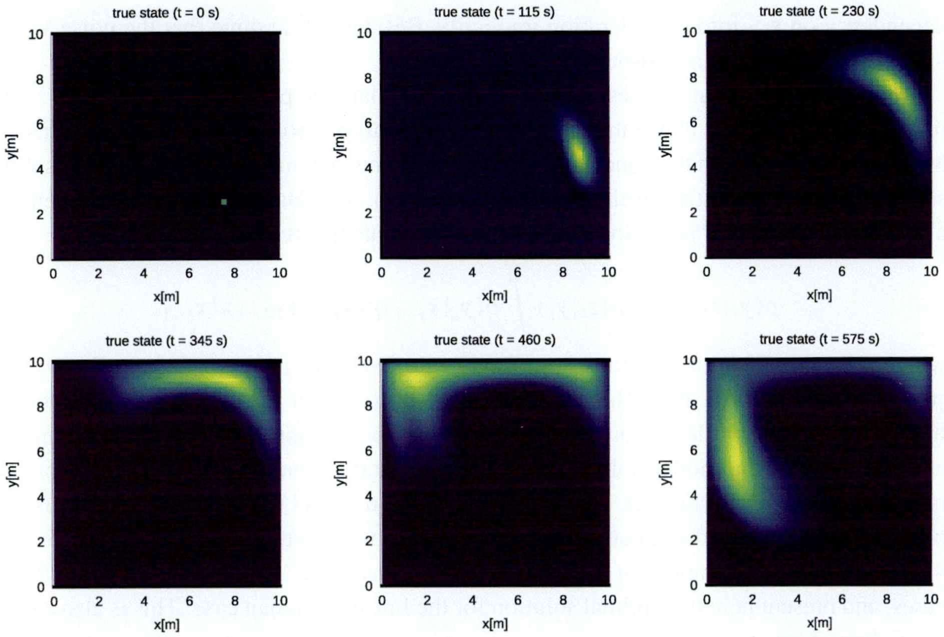

Fig. 3 Dynamic pollutant distribution at various times used for numerical experiments

where $A_t \in \mathbb{R}^{d_y \times d_y}$ is the system parameter and the process noise $\mathbf{w}_t \in \mathbb{R}^{d_y}$ follows Gaussian distribution $\mathcal{N}(\mathbf{0}, W_t)$ with zero mean and covariance $W_t \in \mathbb{R}^{d_y \times d_y}$. Here d_y is the dimension of the concentration state and depends on resolution of grid being used in approximating the DPS.

To estimate the unknown concentration state $\mathbf{y}_t$, we assume that each UAV based sensor i has the following stochastic linear measurement model,

$$\mathbf{z}_{i,t} = H_{i,t}(\mathbf{x}_{i,t})\mathbf{y}_t + \mathbf{v}_{i,t}(\mathbf{x}_{i,t}), \;\; \mathbf{v}_{i,t} \sim \mathcal{N}(\mathbf{0}, V_{i,t}(\mathbf{x}_{i,t})). \tag{10}$$

where $\mathbf{z}_{i,t} \in \mathbb{R}^{d_z}$ denotes the data measured by the i-th sensor; $H_{i,t}(\mathbf{x}_{i,t}) \in \mathbb{R}^{d_z \times d_y}$ is the sensing matrix and $\mathbf{v}_{i,t}(\mathbf{x}_{i,t}) \in \mathbb{R}^{d_z}$ is the sensing noise following Gaussian distribution with zero mean and covariance $V_{i,t}(\mathbf{x}_{i,t}) \in \mathbb{R}^{d_z \times d_z}$. Here d_z is dimension of observation and may be computed depending upon how large or small of a region a sensor can reliably measure, i.e. with reasonable signal to noise ratio, from its current location. We note that both dynamic model (9) and measurement model (10) are linearly dependent on concentration state. Hence, the estimation process of pollutant concentration becomes a linear problem.

In addition, both the sensing matrix and noise are assumed to be dependent on the sensor's state $\mathbf{x}_{i,t}$, since the measurement might be related to the sensor's position and/or orientation. Nevertheless, for the sake of presentation, shorthand notations $H_{i,t}$, $\mathbf{v}_{i,t}$ and $V_{i,t}$ are later used to denote the sensing matrix, noise and covariance, respectively, where we take the dependency on $\mathbf{x}_{i,t}$ into consideration implicitly. Besides, we assume that the noise $\mathbf{v}_{i,t}$ is uncorrelated for different sensors.

The goal of Bayesian state estimator is to approximate the posterior density, $p(y_t|z_t)$ at time t over all possible concentration states y_t using all previous sensor measurements z_t. More specifically, given the concentration measurements obtained from UAV based mobile sensors, $p(\mathbf{z}|\mathbf{y})$, we estimate the posterior $p(\mathbf{y_t}|\mathbf{z_{1:t}})$. The Bayes filter provides general algorithm for computer posterior from all measurements in a recursive manner [45].

$$p(\mathbf{y}_t|z_{1:t}) = \eta p(z_t|\mathbf{y}_t) \int p(\mathbf{y}_t|\mathbf{y}_{t-1}) p(\mathbf{y}_{t-1}|z_{1:t-1}) d\mathbf{y}_{t-1} \tag{11}$$

where η is normalization constant from Bayes rule. We have $p(\mathbf{y}_{t-1}|\mathbf{y}_t)$ from Eq. (9) and $p(z_t|\mathbf{y}_t)$ from Eq. (10). Note that in general, the posterior in Eq. (11) is computationally intractable to solve in closed form. However, under the assumptions that the measurements are independent, the model hold Markov property, state dynamics (9) and measurement model (10) is linear-Gaussian, a closed form solution of (11) is obtained to give us the minimum mean squared estimate of the state and its covariance. We refer the reader to [28, 45] for theoretical derivations of standard filters that can handle non-linear and non-Gaussian cases, and present here the optimal solution for the linear-Gaussian case. This is also called the Kalman filter.

$$\begin{aligned}
\bar{y_t} &= A_t y_{t-1} + B_t u_t \\
\bar{\Sigma}_t &= A_t \Sigma_{t-1} A_t^T + W_t \\
K_t &= \bar{\Sigma}_t C_t^T [C_t \bar{\Sigma}_t C_t^T + V_t]^{-1} \\
y_t &= \bar{y_t} + K[z_t - C_t \bar{y_t}] \\
\Sigma_t &= [I - K_t C_t] \bar{\Sigma}_t
\end{aligned} \tag{12}$$

where K_t is the Kalman gain, $\bar{y_t}$ and $\bar{\Sigma}_t$ are the predicted state and covariance at time t. Note that in our case, we assume no control. Hence, $B_t u_t = 0$ for all t. Lastly, y_t and Σ_t are final state estimates after incorporating the measurement update.

3.3 Kalman Consensus Filter

Let us assume a total number of $\mathcal{I}$ UAVs are placed inside the spatial domain of Fig. 2 and their initial position is known. For simplicity, we assume no UAV position uncertainty in this work. The UAVs can communicate with each other via wireless platforms such as Bluetooth, WiFi etc. We assume no delay in communication. The UAVs are equipped with a pollutant gas sensor whose signal-to-noise ratio is known. Let us consider a centralized perspective in which local information tuple $(\mathbf{y}_t^i, \mathbf{z}_t^i)$ can be collected from UAVs and utilized globally.

Given that the sensing noise is assumed to be uncorrelated for distinct sensors, the measurement model for the network can be simply written as

$$\mathbf{z}_t^i = H_t^i(x_t^i)\mathbf{y}_t + \mathbf{v}_t^i, \tag{13}$$

where the measurement matrix $H_t^i(x_t^i)$ represents measurements from ith UAV at time t as a function of current pose of the UAV. H_t stacks all local measurement matrices as a collective global one and is given as

$$H_t = [H^1(\mathbf{x}_t^1)^\top, H^2(\mathbf{x}_t^2)^\top, \ldots, H^I(\mathbf{x}_t^I)^\top]^\top \in \mathbb{R}^{d_z \times d_y} \tag{14}$$

Moreover, the global measurement noise represented by

$$\mathbf{v}^t = [(\mathbf{v}_t^1)^\top, (\mathbf{v}_t^2)^\top, \ldots, (\mathbf{v}_t^I)^\top]^\top \in \mathbb{R}^{d_z} \tag{15}$$

is assumed to have zero-mean Gaussian noise with covariance expressed as

$$V := \text{diag}\{V^1, V^2, \ldots, V^I\} \in \mathbb{R}^{d_z \times d_z}. \tag{16}$$

Subsequently, the centralized Kalman filter for estimating the mean $\hat{\mathbf{y}}_t \in \mathbb{R}^{d_y}$ and covariance $\Sigma_t \in \mathbb{R}^{d_y \times d_y}$ performs the following recursions,

$$\Sigma_{t+1} = A_{t+1}(\Sigma_t^{-1} + I_t)^{-1}A_{t+1}^T; \tag{17a}$$

$$\hat{\mathbf{y}}_{t+1} = A_{t+1}(\hat{\mathbf{y}}_t + (\Sigma_t^{-1} + I_t)^{-1}(\mathbf{i}_t - I_t\hat{\mathbf{y}}_t)), \tag{17b}$$

where the two variables $I_t := (H_t)^\top V^{-1} H_t \in \mathbb{R}^{d_y \times d_y}$ and $\mathbf{i}_t := (H_t)^\top V^{-1}\mathbf{z}_t \in \mathbb{R}^{d_y}$, often referred to as the new information, incorporate the measurements into the updates. The i_t can be understood as measurement information from each sensor in the network at step t. I_t corresponds to the measurement noise term at step t. Further information on Kalman consensus filter can be found in [40].

It is also worth mentioning that the Kalman consensus filter (17) readily estimates the unknown environment in an on-line manner, i.e., the team of UAVs move to new positions, obtains the new measurements, and updates their estimations. However, we should note that such an on-line procedure is performed in a centralized way, since the new information I^k and $\mathbf{i}^k$ are involved with the data obtained/maintained by all UAVs. For simplicity purposes, we skip devising a distributed scheme to run the Kalman consensus filter (17), and refer the interested reader to many existing works, e.g., [37–39].

3.4 Information Based Guidance

Observability properties of a state-estimator affects its performance. More specifically, a filter may not be consistent, i.e. estimation errors may not be zero-mean or true covariance matrix may be larger than the one calculated by filter, if the system has poor observability characteristics. As mentioned in Sect. 2, in the case of linear system, ideally we would like to ensure that while computing paths for UAVs to monitor the pollutant transport, full observability conditions are imposed. This can be done by ensuring that rank of Observability Gramian remains equal to the state dimension i.e.

$$\text{rank}(W(t_0, t_f)) = d_y \tag{18}$$

where $W(t_0, t_f)$ is Observability Gramian for some time t_f and is defined as

$$W(t_0, t_f) = \int_{t_0}^{t_f} \Phi^T(t, t_0) C_{1:n}^T(x_t) C_{1:n}(x_t) \Phi(t, t_0) dt, \tag{19}$$

where $\Phi(t, t_0) \in \mathcal{R}^{d_y \times d_y}$ is state transition matrix.

Full observability condition may not be guaranteed to meet if environment dynamics are such that it requires a prohibitively large number of sensors. In this case, the best possible solution is to compute a combination of control inputs for UAVs such that they bring sensors to locations that maximize observability characteristics of the system or reduce uncertainty of state-estimate. This can be done by maximizing an information based metric in an optimization program for path planning [22] given the UAV dynamics.

This solution has been employed in non-linear systems as well [26], although the observability conditions are different for non-linear system; Observability Gramian of a linearized non-linear system provides information about the observability of the original nonlinear system i.e. if the linear time-varying system is observable, then the nonlinear system is locally observable about the nominal trajectory [33]. Optimizing UAV paths so they route to information-rich areas in order to to reduce uncertainty is also called informative path-planning.

Information theory [43] provides ways to measure the uncertainty associated with a system. The entropy can be used to describe the uncertainty on what the state of a system is. The reduction of entropy through measurement can be used as a way to measure information gain [19]. Similarly, we can use information associated with the covariance matrix itself, for example the log of determinant of the covariance matrix, which is also used in several articles [2, 27, 42, 47] to measure the uncertainty of systems. From the estimation point of view, the Cramer Rao bound [46] is the lower bound of error for nonlinear estimators. In many other articles related to sensor management [8, 20, 41], Kullback Divergence (KL Divergence) [34] or Renyi Divergence are popular choices. In this chapter, we elaborate a particular case of employing information gain in path planning for UAVs.

A centralized active information gathering problem has been widely studied in the literature, e.g., [2, 35]. An early attempt [2] reduces the original problem with stochasticity into the following deterministic optimization,

$$\underset{\{\mathbf{u}_{1:n,t}\}_{t\in\mathcal{T}}}{\text{maximize}} \quad \log\det(\Sigma_T^{\mathcal{I}}), \tag{20a}$$

$$\text{subject to} \quad \mathbf{x}_{i,t+1} = f_i(\mathbf{x}_{i,t}, \mathbf{u}_{i,t}), \tag{20b}$$

$$\Sigma_{t+1} = \left(A_t(\Sigma_t)^{-1}A_t^\top + W_t\right)^{-1} + I_t, \tag{20c}$$

$$\mathbf{u}_{i,t} \in \mathcal{U}_{i,t}, \quad i \in \mathcal{I},\ t \in \mathcal{T}. \tag{20d}$$

Note that the constraint (20c) is due to the information form of Kalman filter [12]. Equation (20a) represents motion model for UAVs. Observability condition is implicitly imposed by assuming that all the control actions in the action space $\mathcal{U}$ satisfy Eq. (18).

The online sensor placement optimization programs such as above are usually NP-hard and submodularity property is exploited to efficiently achieve near-optimal observation locations. In order to solve the above formulation (20), we refer the reader to the FVI algorithm developed in [35] that constructed a search tree of all possible control sequences and then found the optimal one which gives the maximum objective function value. Krause and Guestrin [31] also presents solution methods to similar optimization programs.

4 Results

This section presents results from pollutant transport simulation experiments in the scenario of a high concentration of pollutant gas injected at $t = 0$ s into a constant velocity field. The velocity field is computed with a finite-difference based numerical solver for vorticity-streamfunction formulation of the Navier–Stokes Eqs. (5) and (6), with the initial velocity boundary conditions of zero on each wall except the north wall which has an initial non-zero velocity. The numerical solver is run until convergence to steady-state velocity field is reached within a residual error of less than 1×10^{-6}. The different parameters used in the simulation study are shown in Table 1. The resultant velocity field and dynamic distribution of pollutant gas is shown in Figs. 2 and 3 respectively.

4.1 Qualitative Comparison

In Fig. 4, four snapshots are shown for the estimation process of density field using four fixed sensors at equally distant locations in the spatial domain. The same figure illustrates in third and fourth columns respectively, the state estimation process with only one mobile sensor, and a combination of four fixed and one UAV based mobile sensor. The first column in Fig. 4 represents evolution of the true state. A process noise is added to the true state, and an observation noise is added when making observations, representing the signal to noise ratio of pollutant gas concentration sensors. It can be seen that a moderately large dimensional state of $N_x \times N_y$ can be reliably estimated using Kalman filter with only four

Table 1 The parameter choices for the experimental study

Parameter	Unit	Value	Meaning
(L_x, L_y)	m	(10, 10)	Domain size
(N_x, N_y)	–	(50, 50)	Spatial mesh shape
CFL	–	0.5	Courant number
Δx	m	L_x/N_x	Spatial resolution
Δt	s	$CFL * \Delta x$	Time step
v	–	0.1	Observation noise parameter
q	–	0.1	Process noise parameter
u_n, u_s, u_e, u_w	m/s	$(1, 0, 0, 0)$	Initial boundary conditions
$c\rho$	–	1.0	Product of specific heat and mass density
eps	–	1.0	Porosity
Re	–	100	Reynold's number
$\Delta_{t,meas}$	s	$5 * \Delta t$	Time gap for measurement update
$\Delta_{s,meas}$	–	25	Observation subsample (4 fixed sensors)

sensors at fixed locations. Secondly, it can be observed that the UAV based mobile sensor on its own isn't able to detect gas concentration until the UAV physically moves closer to the gas source or high pollutant concentration areas areas. Once the UAV has detected the gas, the Kalman filter is able to reliably estimate the state evolution even as the UAV moves along a random path.

Note that in this case, we employ a point-wise observation matrix dependent on state of the mobile sensor. Hence, the 'jitter' in the measurement, reflective of observation noise, can be seen at times $t = 350$ s and $t = 525$ s in the third and fourth columns. This can be improved if the point-wise observation value is assigned as an average value to the whole observation radius around the UAV. Linear and Gaussian observation matrices are alternates that provide smoother estimates.

4.2 Performance of Kalman filter

Root-mean-square error (RMSE) measures difference between values predicted by the state estimator and the true state values. It is defined as

$$\text{RMSE} = \sqrt{\frac{\Sigma_{i=1}^{N}(y_i - \hat{y}_i)^2}{N}} \tag{21}$$

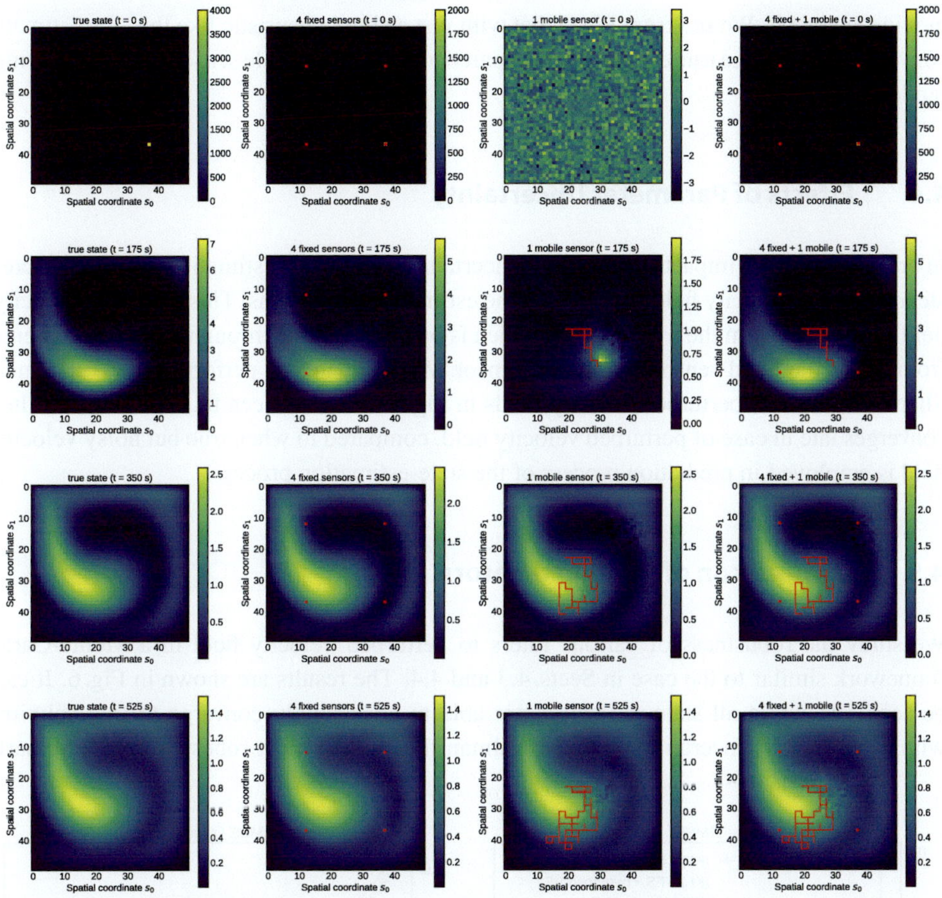

Fig. 4 Snapshots from online estimation of pollutant concentration map. Red dots and lines show locations of fixed sensors and random path trajectory of UAV respectively

where y_i is the true value, $\hat{y}_i$ is the estimated value and N is dimension of state. Figure 5 shows how the root mean squared error (RMSE) of state estimate. We run each scenario $N = 10$ times as a Monte-Carlo study to illustrate consistent results.

4.3 Effects of Guidance

We compare the performance of assigning random path points to UAV versus a simple heuristic based path planner. In this particular case, our guidance heuristic is simply the UAV based mobile sensor computing gas concentration gradient at its current location, and moving against it. We see from Fig. 5 that with just a simple heuristic like this, the estimate converges faster to the true estimation, compared to when the mobile sensor is only executing random walk.

4.4 Effects of Parameter Uncertainty

In order to study the impact of parameter uncertainty in the state-estimation process, we create a perturbed velocity field and use it in the estimation algorithms. This perturbed velocity field is different from the true velocity, which is only observable through the measurements from either the fixed sensors or mobile sensor. We compare the performances of Kalman filter with true and perturbed velocity fields in Fig. 5. It can be seen that the Kalman filter converges late in case of perturbed velocity field, compared to when true but noisy velocity field is employed in prediction process of the state-estimation process.

4.5 Comparison of State-Estimators

We study the robustness of various filters to perturbed velocity field in a Monte-Carlo framework similar to the case in Sects. 4.3 and 4.4. The results are shown in Fig. 6. It can be seen that while all state-estimators are able to consistently converge to true solution within reasonable uncertainty bounds, Kalman filter is the most robust filter. We remark

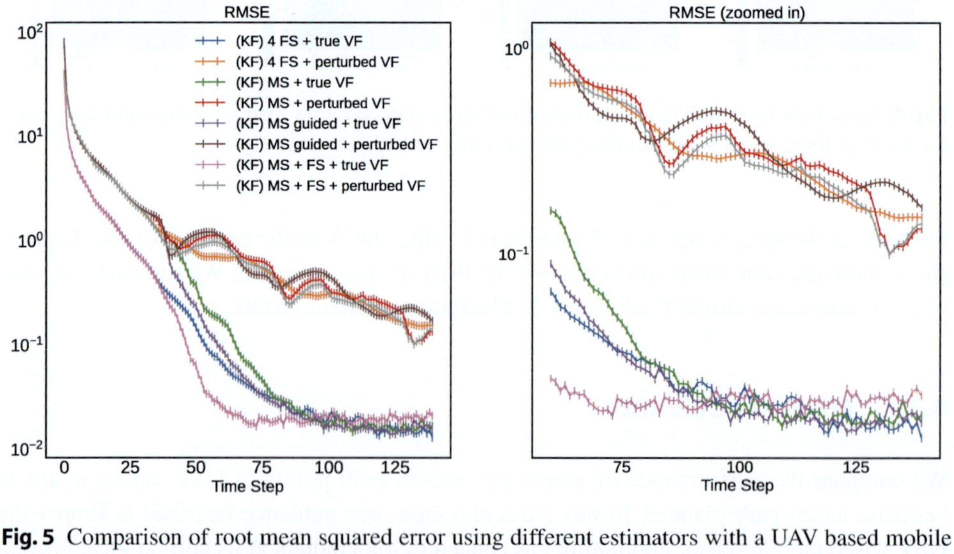

Fig. 5 Comparison of root mean squared error using different estimators with a UAV based mobile sensor in semi-log scale. Here, KF = Kalman Filter, MS = UAV based Mobile Sensor, VF = Underlying wind Velocity Field

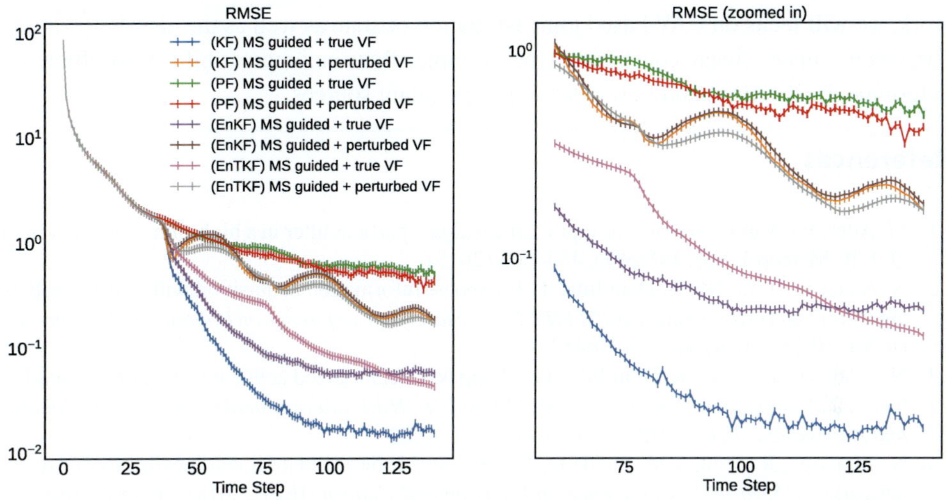

Fig. 6 Comparison of root mean squared error using different estimators with a mobile sensor in semi-log scale. Here, MS = UAV based Mobile Sensor, VF = Underlying wind Velocity Field

that it is only relevant for this particular case study in which assumption (8) makes the problem linear-Gaussian. For the most general case of non-linear non-Gaussian, Kalman filter becomes inconsistent. Furthermore, it can be seen that the performance of particle filter with perturbed velocity field is slightly better than with true velocity field. This is because introducing more noise into the system than the actual case, addresses the weights degeneracy issue of particles. Interested readers may refer [36] to learn more about weight degeneracy in particle filters. In our study, Ensemble Transform Kalman Filter (EnTKF) is the best performing filter when perturbed velocity field is used in the estimation step. Readers interested to learn about EnTKF are referred to [6].

5 Conclusion

In this chapter, we consider the case of UAVs-based monitoring of gas pollutant transport in an indoor environment, given an underlying steady-state wind velocity field . The pollutant transport is modeled via Advection-Diffusion equation. We compare several schemes for estimating the state of pollutant gas concentration evolving over space and time. In the case of gas pollutant monitoring with a leak, we show that adding UAV-based mobile sensors can improve the quality of estimates, provided that their paths are adequately determined. In the case when number of pollutant sensors is smaller than minimum required to meet observability condition (18), UAV allows us to reduce uncertainty in state-estimation. We show that parameter uncertainties can have considerable effect on the quality of the estimation, particularly for computationally intensive schemes such as the particle filtering. In future work, we will focus on UAV based joint estimation of state as well as the distributed model parameters, under linear constraints (for example, flow incompressibility), to efficiently determine the locations and flow-rates of gas pollutant sources.

References

1. M. Ades, P.J. Van Leeuwen, The equivalent-weights particle filter in a high-dimensional system. Q. J. R. Meteorol. Soc. **141**(687), 484–503 (2015)
2. N. Atanasov, J. Le Ny, K. Daniilidis, G.J. Pappas, Information acquisition with sensing robots: algorithms and error bounds, in *2014 IEEE International Conference on Robotics and Automation (ICRA)* (IEEE, 2014), pp. 6447–6454
3. N. Atanasov, J. Le Ny, K. Daniilidis, G.J. Pappas, Decentralized active information acquisition: theory and application to multi-robot SLAM, in *2015 IEEE International Conference on Robotics and Automation (ICRA)* (IEEE, 2015), pp. 4775–4782
4. N. Atanasov, R. Tron, V.M. Preciado, G.J. Pappas, Joint estimation and localization in sensor networks, in *53rd IEEE Conference on Decision and Control* (IEEE, 2014), pp. 6875–6882
5. G. Best, O.M. Cliff, T. Patten, R.R. Mettu, R. Fitch, Dec-MCTS: decentralized planning for multi-robot active perception. Int. J. Robot. Res. **38**(2–3), 316–337 (2019)
6. C.H. Bishop, B.J. Etherton, S.J. Majumdar, Adaptive sampling with the ensemble transform kalman filter. Part i: theoretical aspects. Mon. Weather Rev. **129**(3), 420–436 (2001)
7. C.M. Carvalho, M.S. Johannes, H.F. Lopes, N.G. Polson, Particle learning and smoothing. Stat. Sci. **25**(1), 88–106 (2010)
8. E.K.P. Chong, C.M. Kreucher, A.O. Hero, Partially observable markov decision process approximations for adaptive sensing. Discr. Event Dyn. Syst. **19**(3), 377–422 (2009)
9. J. Clairambault, *Reaction-Diffusion-Advection Equation* (Springer, New York, New York, 2013), p. 1817
10. M.A. Demetriou, Emulating a mobile spatially distributed sensor by mobile pointwise sensors in state estimation of partial differential equations via spatial interpolation, in *2017 American Control Conference (ACC)* (IEEE, 2017), pp. 3243–3248
11. P.G. Drazin, N. Riley, *The Navier-Stokes Equations: A Classification of Flows and Exact Solutions*. London Mathematical Society Lecture Note Series (Cambridge University Press, 2006)
12. D. Bin, K. Qian, C. Claudel, D. Sun, Parallelized active information gathering using multisensor network for environment monitoring. IEEE Trans. Control Syst. Technol. **30**(2), 625–638 (2022)
13. B. Du, K. Qian, H. Iqbal, C. Claudel, D. Sun, Multi-robot dynamical source seeking in unknown environments, in *2021 IEEE International Conference on Robotics and Automation (ICRA)* (IEEE, 2021), pp. 9036–9042
14. G. Evensen, The ensemble kalman filter: theoretical formulation and practical implementation. Ocean Dyn. **53**(4), 343–367 (2003)
15. N.J. Gordon, D.J. Salmond, A.F.M. Smith, Novel approach to nonlinear/non-gaussian bayesian state estimation, in *IEE Proceedings F (Radar and Signal Processing)*, vol. 140 (IET, 1993), pp. 107–113
16. F. Gustafsson, Particle filter theory and practice with positioning applications. IEEE Aerospace Electron. Syst. Mag. **25**(7), 53–82 (2010)
17. H. Hajimirsadeghi, Cooperative state estimation for mobile sensors with optimal path planning
18. J. Haugen, E.I. Grøtli, L. Imsland, State estimation of ice thickness distribution using mobile sensors, in *2012 IEEE International Conference on Control Applications* (IEEE, 2012), pp. 336–343

19. A.O. Hero, D. Cochran, Sensor management: past, present, and future. IEEE Sensors J. **11**(12), 3064–3075 (2011)
20. A.O. Hero, D. Castañón, D. Cochran, K. Kastella, *Foundations and Applications of Sensor Management* (Springer Science & Business Media, 2007)
21. E.J. Hinch, *Streamfunction-Vorticity Formulation*. Cambridge Texts in Applied Mathematics. (Cambridge University Press, 2020), pp. 9–34
22. B.T. Hinson, Observability-based guidance and sensor placement. Ph.D. Thesis (2014)
23. G.A. Hollinger, G.S. Sukhatme, Sampling-based motion planning for robotic information gathering, in *Robotics: Science and Systems*, vol. 3 (Citeseer, 2013)
24. G.A. Hollinger, G.S. Sukhatme, Sampling-based robotic information gathering algorithms. Int. J. Robot. Res. **33**(9), 1271–1287 (2014)
25. H. Chongyang, Y. Liang, X. Wang, X. Linfeng, A particle filter via constrained sampling for nonlinear dynamic systems. Int. J. Robust Nonlinear Control **30**(13), 4944–4959 (2020)
26. G.P. Huang, A.I. Mourikis, S.I. Roumeliotis, Observability-based rules for designing consistent ekf slam estimators. Int. J. Robot. Res. **29**(5), 502–528 (2010)
27. S.T. Jawaid, S.L. Smith, Submodularity and greedy algorithms in sensor scheduling for linear dynamical systems. Automatica **61**, 282–288 (2015)
28. A.H. Jazwinski, *Stochastic Processes and Filtering Theory* (Courier Corporation, 2007)
29. R.E. Kalman, A new approach to linear filtering and prediction problems. Trans. ASME–J. Basic Eng. 82(Series D), 35–45 (1960)
30. Y. Kantaros, B. Schlotfeldt, N. Atanasov, G.J. Pappas, Asymptotically optimal planning for non-myopic multi-robot information gathering, in *Proceedings of Robotics: Science and Systems (RSS), Freiburg, Germany* (2019), pp. 22–26
31. A. Krause, C. Guestrin, Near-optimal observation selection using submodular functions. AAAI **7**, 1650–1654 (2007)
32. A. Krause, C. Guestrin, A. Gupta, J. Kleinberg, Near-optimal sensor placements: maximizing information while minimizing communication cost, in *Proceedings of the 5th International Conference on Information Processing in Sensor Networks* (2006), pp. 2–10
33. A.J. Krener, K. Ide, Measures of unobservability, in *Proceedings of the 48h IEEE Conference on Decision and Control (CDC) Held Jointly with 2009 28th Chinese Control Conference* (IEEE, 2009), pp. 6401–6406
34. S. Kullback, R.A. Leibler, On information and sufficiency. Ann. Math. Stat. **22**(1), 79–86 (1951)
35. J. Le Ny, G.J. Pappas, On trajectory optimization for active sensing in gaussian process models. In *Proceedings of the 48h IEEE Conference on Decision and Control (CDC) Held Jointly with 2009 28th Chinese Control Conference* (IEEE, 2009), pp. 6286–6292
36. J. Liu, M. West, Combined parameter and state estimation in simulation-based filtering, in *Sequential Monte Carlo Methods in Practice* (Springer, 2001), pp. 197–223
37. R. Olfati-Saber, Distributed kalman filter with embedded consensus filters, in *Proceedings of the 44th IEEE Conference on Decision and Control* (IEEE, 2005), pp. 8179–8184
38. R. Olfati-Saber, Distributed kalman filtering for sensor networks, in *Proceedings of the 46th IEEE Conference on Decision and Control* (IEEE, 2007), pp. 5492–5498
39. R. Olfati-Saber, J. Shamma, Consensus filters for sensor networks and distributed sensor fusion, in *Proceedings of the 44th IEEE Conference on Decision and Control* (IEEE, 2005), pp. 6698–6703
40. R. Olfati-Saber, Kalman-consensus filter: optimality, stability, and performance, in *Proceedings of the 48h IEEE Conference on Decision and Control (CDC) Held Jointly with 2009 28th Chinese Control Conference* (IEEE, 2009), pp. 7036–7042
41. B. Ristic, B.-N. Vo, Sensor control for multi-object state-space estimation using random finite sets. Automatica **46**(11), 1812–1818 (2010)

42. B. Schlotfeldt, D. Thakur, N. Atanasov, V. Kumar, G.J. Pappas, Anytime planning for decentralized multirobot active information gathering. IEEE Robot. Autom. Lett. **3**(2), 1025–1032 (2018)
43. C.E. Shannon, A mathematical theory of communication. Bell Syst. Techn. J. **27**(3), 379–423 (1948)
44. G. Storvik, Particle filters for state-space models with the presence of unknown static parameters. IEEE Trans. Signal Process. **50**(2), 281–289 (2002)
45. S. Thrun, Probabilistic robotics. Commun. ACM **45**(3), 52–57 (2002)
46. P. Tichavsky, C.H. Muravchik, A. Nehorai, Posterior cramér-rao bounds for discrete-time nonlinear filtering. IEEE Trans. Signal Process. **46**(5), 1386–1396 (1998)
47. M.P. Vitus, W. Zhang, A. Abate, J. Hu, C.J. Tomlin, On efficient sensor scheduling for linear dynamical systems. Automatica **48**(10), 2482–2493 (2012)
48. E. Weinan, J.-G. Liu, Vorticity boundary condition and related issues for finite difference schemes. J. Comput. Phys. **124**(2), 368–382 (1996)
49. J.L. Williams, Information theoretic sensor management. Ph.D. Thesis, Massachusetts Institute of Technology (2007)
50. W. Huai-Ning, X.-W. Zhang, An enhanced state estimation for linear parabolic pde systems with mobile sensors. Int. J. Adapt. Control Signal Process. **33**(10), 1552–1566 (2019)
51. L. Zhou, V. Tzoumas, G.J. Pappas, P. Tokekar, Resilient active target tracking with multiple robots. IEEE Robot. Autom. Lett. **4**(1), 129–136 (2018)
52. L. Zhou, V. Tzoumas, G.J. Pappas, P. Tokekar, Distributed attack-robust submodular maximization for multi-robot planning (2019). arXiv:1910.01208

Hassan Iqbal is a Ph.D. Student with the Department of Civil, Architectural, and Environmental Engineering at UT-Austin. He received his MS degree in Civil Engineering from UT-Austin and Bachelor's degree in Electrical Engineering from National University of Sciences and Technology, Islamabad. His research interests include online state-parameter estimation of distributed parameter systems and motion planning for autonomous mobile sensors traversing these systems.

Kun Qian received the bachelor's degree from Tongji University, Shanghai, China, in 2017, and the master's degree from the University of California, Berkeley, Berkeley, CA, USA, in 2018. He received his PhD degree from the University of Texas at Austin, Austin, TX, USA. He is currently an Applied Scientist in Amazon Search Science and AI department. His research interests include multi-agent systems, sequential decision process and machine learning.

Christian Claudel is an Associate Professor of Civil, Architectural and Environmental Engineering at UT-Austin. He received the PhD degree in Electrical Engineering from UC-Berkeley in 2010, and the MS degree in Plasma Physics from Ecole Normale Superieure de Lyon in 2004. He received the Leon Chua Award from UC-Berkeley in 2010 for his work on the Mobile Millennium traffic monitoring system. His research interests include control and estimation of distributed parameter systems, wireless sensor networks and unmanned aerial vehicles.

Leveraging Drone Technology in the Accounting Profession: Current Trends, Future Opportunities, and Anticipated Challenges

Amer Qasim, Ghaleb A. Elrefae, Shorouq Eletter, and Haythem A. Bany Salameh

Abstract

Emerging technologies are envisioned to advance the accounting profession to a new level. The emergence of big data, artificial intelligence, and robotic process automation has profoundly impacted traditional accounting and auditor tasks and procedures. This chapter focuses on a new technology that has recently emerged in business operations: Unmanned Aerial Vehicles (UAVs) or "drones". This chapter is intended to highlight the role of drones in current accounting practices as well as future opportunities for exploiting the unique capabilities of drones in the accounting profession. In this domain, the potential of utilizing drones in the accounting profession in the areas related to inventory management, revenue recognition, cash flow management, and risk management is presented. However, applying this technology in this domain faces several implementation challenges, which are highlighted and discussed in this chapter. The significance of this chapter stems from the fact that very limited research has been conducted in this area in the accounting literature. This chapter provides a visionary framework that links the asymmetric information dilemma with the utilization of drones in the accounting profession.

A. Qasim · G. A. Elrefae · S. Eletter
College of Business, Al Ain University, Abu Dhabi, UAE
e-mail: amer.qasim@aau.ac.ae

G. A. Elrefae
e-mail: ghalebelrefae@aau.ac.ae

S. Eletter
e-mail: shorouq.eletter@aau.ac.ae

H. A. Bany Salameh (✉)
College of Engineering, Al Ain University, Al Ain, UAE
e-mail: haythem.banysalameh@aau.ac.ae

Telecommunications Engineering Department, Yarmouk University, Irbid, Jordan

M. Abdelkader and A. Koubaa (eds.), *Unmanned Aerial Vehicles Applications: Challenges and Trends*, Synthesis Lectures on Intelligent Technologies,
https://doi.org/10.1007/978-3-031-32037-8_8

1 Introduction

The emergence of the 4th industrial revolution in the twentieth century has affected how businesses operate. The digital transformation made fundamental shifts in how business operations occur. Enterprises are using technology to improve their business processes, increase productivity, and improve efficiency and effectiveness. Revolutionary technologies like the Internet of Things (IoT), blockchain, artificial intelligence, data analytics, etc., are the drivers for such transformation. Driven by business needs, organizations can operate better and communicate effectively. The study by [1] stated that digital transformation had enhanced customer experience and centricity. Key technology has played a vital and integral role in business operations. Enterprises across industries implement intelligent, innovative acting processes to build a competitive advantage. The implementation improved the company's innovativeness, quality, business operations, resource savings, cost reduction, horizontal and vertical integration, and product customization [2]. Numerous technology-driven ideas (e.g. Artificial Intelligent technology, robotics, business analytics) substituted existing operations in many organizations to increase productivity and competitiveness, which enhanced the efficiency and operational capability of an organization [3, 4]. The IoT is one of the advances in technology that enable communication between different devices. Organizations adopted IoT to improve business processes [5]. This chapter focuses on a new technology that has recently emerged in business operations: Unmanned Aerial Vehicles (UAVs) or "drones". This chapter intends to highlight drones' role in the accounting profession. In addition, the chapter is designed to provide current practices and future opportunities for utilizing the unique features of drones in the accounting profession. In this case, the potential of utilizing drones in the accounting profession in the areas related to inventory management, revenue recognition, cash flow management, and risk management is presented. However, applying this technology in this domain faces several implementation challenges, which are highlighted and discussed in this chapter. The importance of this work stems from the fact that very limited research has been conducted in this area in the accounting literature. This chapter provides a visionary framework that links the asymmetric information dilemma with the utilization of drones in the accounting profession.

It is important to note that using drones in the accounting profession does not suggest replacing the existing methods used by accountants. Instead, this chapter argues that drone technology can be used as a supportive tool for accountants and auditors to carry out some repetitive accounting tasks or any other task that can be automated. Other artificial intelligence (AI) and robotic-enabled technologies are also expected to complement each other to facilitate accountants' tasks. Integrating drone technology with traditional accounting methods has great potential to improve efficiency and accuracy, as well as relax the physical presence requirements of the accountants during asset measurement and assessment procedures.

2 The Emergence of New Technologies in Accounting Profession

Business transformation continues to accelerate as technology evolves. However, the scope of technology that has led to business transformation is enormous, including enterprise resource planning (ERP), financial software, accounting software, customer relationship management, AI, blockchain (BC), big data, data analytics, drone technology, etc. The IoT is the backbone of such transformation. Such emerging technologies have reshaped the landscape of the organization and their impact on the business process, efficiency, effectiveness, performance, etc., as well as the talent capability of employees.

The growing influence of industry 4.0 drivers requires that educational institutions and professional training equip accountants with interdisciplinary skills and competencies to manage the missing knowledge and expertise [6]. AI applications, robotics, and machine learning have been used for the tasks of public accountants. However, AI adoption at a large scale is promising to create substantial changes on the horizon [7]. The survival of business organizations is highly linked to employees' ability to adapt to technological transformation. The future labor force needs to acquire soft skills to satisfy the dynamic and varying demands of the labor market. The workplace develops through innovative tasks that employees perform [8].

Advancing the accounting domain with AI/Machine learning (ML)-based drone systems is becoming an outstanding educational issue. New professional qualifications are required to meet the research stream's needs and employers' expectations. Accounting educators are calling to integrate AI tools drone technology, and data analytics into accounting curricula to equip accounting graduates with the technology-based skill set and competencies to be prepared for the future accounting profession [9]. Accountants usually collect data (structured, semi-structured, or unstructured), analyze data, process information, measure it, and report or communicate the information with stakeholders. In addition, they have to keep accurate and timely financial records for the organization, perform periodic auditing, and prepare reports. The development of ICT and the availability of a large amount of data, called Big data, might inspire accountants to capture more new insights about business activities. Big data analytics (BDA), including software, hardware technologies, procedures, and tools, are instruments accountants can use to utilize Big data and come up with big insights. Recently, cutting-edge methods of performing audits have included the utilization of drone technology along with other AI mechanisms. Incorporating more data in real-time operations is possible thanks to the versatility of drones and AI in data collection, storage, and transmission. Therefore, it is anticipated that AI-based drone systems will greatly impact the business profession in general and the accounting profession in specific.

Auditors can use Big data and BDA to improve financial reporting, detect fraud, identify operational risks, and boost risk controls. Auditors can take advantage of the implementation of Big by increasing insufficiency, availability of appropriate evidence, automating transaction testing, and increasing audit quality by capturing more information about the various clients' processes [11]. Lack of adequate knowledge among accountants and external auditors is the main challenge in the accounting industry nowadays. The ongoing advancement in technology surged the challenge. Education institutions, as well as accounting and auditing professionals, need to consolidate changes in the accounting and auditing profession and develop graduates' skills. Qualified external auditors can utilize Big data better to increase the richness of evidence sources for revealing misstatements and fraud [10]. The utilization of Big data in auditing opened opportunities and challenges. To excel in their profession, they need to acquire the relevant professional skills. Big data and BDA are an opportunity for accountants to create value. The problem-driven analysis is a fundamental activity within the profession that they master probably. Their familiarity with processing structured data will facilitate the shift to working with unstructured data as well. Therefore, they will possess knowledge of business fundamentals. BDA complements an accountant's skills and knowledge. The accounting profession is witnessing a change in task structures due to BDA. Thus, accountants are encouraged to leverage their current skills along with afresh revised ones. This will enable them to take the lead in problem-driven structured and unstructured data analysis and to help scientists in the exploratory analysis [12].

3 The Concept of Drone Technology and Its Diverse Business Applications

UAVs referred to as drones, are increasingly gaining popularity due to the huge decrease in their form factors and cost, as well as the recent advances in their sensing and computing capabilities. UAVs can be seen as aircraft that can autonomously fly with no human pilot. UAVs are considered a key future technology to effectively address inefficient operations in different business sectors by providing flexible access to the operating environment. Advances in UAV technology have contributed to advancing the military, civilian, and business sectors. UAVs can effectively collect information about various phenomena in environments that are hard to access or hard to be noticed by humans [13–15].

The use of UAVs in several business applications and services has attracted significant attention from both military, civilian, and industry sectors in the last few years. The interest in UAVs is mainly attributed to their ability to enable efficient resource utilization, allow flexible operation, and provide remote-access solutions in scenarios where the human being present is not feasible. To be specific, the self-optimization and autonomous operation of UAVs are exceptionally valuable in operating scenarios where human access is not possible or supporting timely and efficient services, and applications are needed. The flexible capability of UAVs allows for utilizing different technologies and adapting their operating/flying parameters based on the surrounding operating environment and application requirements. Furthermore, due to their flexible movement and flying nature, UAVs are more resource-efficient and robust than their non-UAV counterparts against harsh operating conditions. Specifically, UAVs have been crucial to the operations of many enterprises, military, and governmental organizations in recent years. However, the massive-scale adoption and usage of UAVs are still in the early stages, even though UAVs have already broken through the inflexible old-style working environment in different business environments and industries [16–18].

3.1 UAV Operating Environment and Business Applications

The recent advances in several enabling technologies (e.g., UAV onboard computational intelligence, HD cameras, and autonomous functionalities) have resulted in utilizing the UAV systems in more business applications and services [19, 20]. Depending on the operating scenarios, UAVs introduce hardware and operational constraints in terms of battery capacity, UAV size, UAV speed, UAV, localization, computational capability, and storage. In the anticipated dynamics and the hybrid nature of the UAV systems, the operation is extremely complex due to the high number of tunable operating parameters. Based on the operating environment, UAV systems can be categorized into indoor and outdoor UAV systems [21]. Figures 1 and 2 respectively depict illustrative examples of indoor and outdoor UAV systems. Indoor UAV systems are designed to provide services, applications, and business solutions within indoor environments such as shopping malls, factories, hospitals, and offices, among others. On the other hand, outdoor UAV systems offer services in outdoor business environments. The different operating conditions of indoor scenarios (e.g., the height of the local obstacle, human workers/machines on the floor, with respect to the

Fig. 1 An example of an indoor UAV system operating in a shopping mall

Fig. 2 Indoor UAV environment

height of the UAV) make the design and applications of indoor UAV systems fundamentally different from outdoor UAV systems. Specifically, while most outdoor UAV business applications and services heavily rely on the satellite-based Global Positioning System (GPS) localization system, a non-GPS localization solution is required for UAVs' deployment into complex indoor environments [22].

Compared to outdoor deployment scenarios, the indoor density of obstacles is much higher because of the existing stationary objects such as pillars and good shelves. Human movements and activities are also more frequent, which has a huge influence on the UAVs' operation. Compared to the outdoors, the closed, dense nature of the indoor environment results in less flying flexibility and strict requirements for collision avoidance. Hence, flight during task execution for the UAVs must be carefully planned in non-colliding paths. Specifically, the damage risks to UAVs and objects (e.g. human workers or machines on the shop floor) are much higher in indoor environments. Thus, indoor UAV systems have more physical constraints and require more intelligent and accurate control [23].

The offered autonomy and flexibility of UAVs, along with their efficient data collection capabilities, make them attractive in several indoor and outdoor applications [24, 25]. UAV-based systems can offer a large number of applications and services in indoor and outdoor scenarios. Hundreds of indoor and outdoor applications and services of UAVs are underway (implemented or envisioned) because of the increasing interest from industries, businesses, and academia in UAVs [26, 27]. Figure 3 summarizes key outdoor and indoor UAV applications. Specifically, the key outdoor UAV applications include Aerial photography (i.e., for Journalism and filming, Accounting-based applications, Financial Auditing, Remote sensing, Infrastructure inspection/monitoring, Road safety, Payload transportation, Precision agriculture, Surveillance and security (e.g., border control), Search-and-rescue missions, Disaster Management, Target Detection and Tracking, Express shipping and delivery, Geographic mapping of inaccessible locations, Safety inspections for Buildings and bridges, Wildlife Monitoring, Pollution monitoring, Storm tracking, and forecasting hurricanes and tornadoes [28, 29]. On the other hand, indoor UAV applications include Indoor Search-and-rescue missions, Manufacturing Industry, Payload transportation, Indoor Infrastructure Inspection, In-door Surveillance, and security, Indoor Intrusion detection and tracking, Indoor Vertical Farming, Target Detection, and Tracking, Express shipping and delivery, On-demand Indoor Internet Access (hot-spot coverage), Safety inspections for Buildings, Navigation inside a closed area, Entertainment and announcements, and Pandemic response, among others [30].

3.2 AI-Aided Drone Systems

AI-driven UAVs are expected to be the next revolutionary paradigm in UAV technology. Integrating AI in UAV systems promises to offer new applications and services at low cost, low risk, and efficient operation (e.g., smart infrastructure monitoring, smart inspection for the level of task completion, and task-driven cash-flow, among others) [15]. Specifically, Intelligence plays a key role in enabling efficient indoor and outdoor UAV system operation. In general, integrating intelligence in UAV operations is a key requirement for efficient operations. The increased complexity of UAV operating conditions resulted in intractable control of the different monitoring and management elements in the UAV systems. In addition, Machine-learning provides a systematic approach to extracting valuable information from UAV-collected data. Based on such information, the UAV allows us to automatically identify possible correlations and provide conclusions that cannot be noticed or reported by human experts [31, 32]. ML enables UAVs to intelligently monitor their operating environment and exploit the collected (experienced) data for adapting their operating strategies. Specifically, ML allows the UAVs to learn, predict, and adapt based on the dynamic of their operating environment (e.g., including location dynamics, application requirements, and flaying patterns, among others) [14, 30]. ML takes as input data from interacting with the operating environment (collected by the UAVs) and, through the application of different learning techniques, i.e., unsupervised/ supervised, deep or reinforcement learning, enables the UAV system to adapt to their operating parameters in an autonomous and dynamic manner. This allows the UAVs to proactively optimize system efficiency and the likelihood of satisfying different application and service requirements [30].

Fig. 3 Application and service scenarios for indoor and outdoor UAV systems

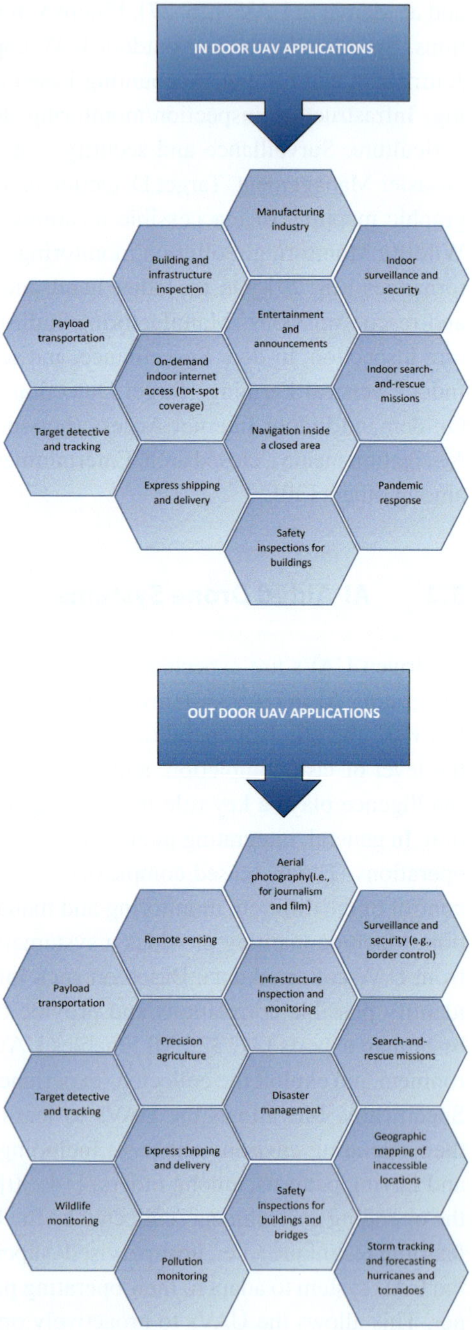

4 Integrating Drone Technology in Accounting Profession

Drones (UAVs) are usually equipped with portable cameras that can offer very high spatial resolution data. Drones take videos and images to enable several complex business processes in real-time. Drones have been proposed for use in inspecting cell towers and other large structures for defects, with significant developments taking place across industries including real estate, insurance, agriculture (farming), film industry, etc. [33]. Drones could be a financially feasible way to help people in hard-to-reach areas and provide access to healthcare and health-related services [34]. Auditors rely on in-person physical evidence collection mechanisms like inventory observations, verification, document collection, plant, property, and equipment evaluation, discussions, interviews, comments, etc. [35]. The aerial views gave the audit team more high-quality information [36]. Compared to manual techniques, drones can reach difficult-to-reach locations and quickly collect massive amounts of high-quality data compared to conventional methods. Moreover, drone use is believed to result in genuine efficiency gains in conducting audits. An innovative study in [37] explored the potential use of drone technology for remote auditing in real estate accounting. Using the completion-percentage approach to calculate revenue from long-term contracts, the drone will inspect the site to evaluate and report the construction progress. This cutting-edge technology will improve auditory duties while saving time and money and enhancing site security.

5 The Applications of Drone Technology in the Accounting Profession

New technologies have had a huge impact on the accounting profession in various aspects including accounting analytics and continuous auditing [38]. These technologies include AI, robotic process automation, data analytics, and blockchain. Drones have recently been recognized as a potential tool for acquiring data in the form of images and videos, and the term "Dronification" has been coined to describe the process of using drones in business operations [39]. Drones can be equipped with high-tech sensing devices that utilize Internet-of-Things (IoT) technology and are suitable for various activities, including long-term construction projects [40]. The benefits of using drones in business operations include cost, time, and effort savings, increased efficiency, and enhanced safety, making drones useful in several aspects, such as environmental appraisals and large-scale construction projects (i.e., highways and bridges).

In the accounting profession, UAVs could be utilized to automate and expedite several routine and repetitive accounting services and activities. The use of drones in inventory management systems has delivered practical solutions, according to studies conducted in this field. It is anticipated that the accounting profession will be vastly affected by the recent advances in robotic and AI-based technologies and their utilization in remote-controlled drones in the future [39]. This transformation is expected to be characterized by the increasing use of robots, AI, and machine learning in the profession.

5.1 Using UAVs as an Inventory Management Tool

In the case of inventory management, [39] indicated that certain auditing firms whose clients have large mineral deposits have experimented with using drones to fly over areas with natural resources and take videos and images to assist the auditor in the process of estimating the capacity of the natural resource. Based on the captured pictures and videos of the monitored natural resource, the auditor can then utilize the captured videos and images by the drones to conduct the required analysis for natural reserve estimations. Keeping in mind that the natural resources are situated on the ground, the auditor (or the licensed third party) is capable of flying the drones over the designated area to collect the needed images and videos to measure the natural resource estimates. One of the examples that can be highlighted in this regard is the project conducted by the international auditing firm PricewaterhouseCoopers (PwC), which started the first experimental inventory count audit in 2017 using drones. PwC started the process by flying drones to capture images and videos of a coal mine in the UK. The captured images and videos have then been utilized to estimate the coal's volume in the mine, and the coal's value is estimated using the estimated coal volume. Based on that, several advantages are assumed when using this technology for auditing natural resources in remote areas, such as time and effort efficiency, the accuracy of measurements, and the achieved safety benefits. Furthermore, [39] explained that drones could be employed in several construction project operations, beginning with the construction site survey and continuing until the surveys are needed at the completion of the construction projects. In this regard, traditional site surveys that usually take up to several weeks to complete can be completed in the order of hours if drones (with cameras and advanced capture software) are employed. Other examples include the audit of an oil rig based on onboard sensing devices that link thermal imaging measurements with gas leakage readings. This linkage reduces the time needed to complete the required mission (i.e., a few days are required instead of a few weeks needed when using traditional inspection activities). Similarly, drones are employed by auditors when auditing companies with agricultural operations. In this regard, the technology can save and process pictures and videos from huge and remote agricultural regions. In addition, by applying the right technology and software, farmers can also determine nutrient deficiencies in the air. Also, the firefighters could exploit the technology to follow the bushfires' progress. The authors in [41] utilized drones and RPA accounting systems to conduct inventory counting for livestock. Results demonstrated a reduction in the count time by more than 95% as well as error rates falling by more than 75%. As a result, [41] concluded that using this technology to count inventory improved the three main audit quality dimensions, these are: effectiveness, efficiency, and documentation quality.

5.2 Using UAVs as a Tool for Assessing Long-Term Projects' Completion

The employment of drones in several business operations can be utilized by relating it to the accounting profession by assessing the applicability of this technology to support emerging applications (or technologies) to assess recognizing revenues [40, 42]. In this regard, and from a financial accounting standpoint, the method utilized to determine the revenue in long-term construction projects is called "the percentage of completion method." The completion percentage of a certain construction project is calculated by dividing the current billings of the project by the overall expected cost of the long-term project. The envisioned application in this chapter, proposed by [37], is to determine the project's completion percentage based on the actual inspection provided by a drone. In this regard, the ERP system is provided with pictures and videos captured by the drone and blackprints for the project. ERP then compares the blackprints and images to determine the completion percentage for the projects. The work in [37] explained that this proposal has a visionary nature; the study studies the use of innovative drone technology to perform site inspections to evaluate and screen the progress of the construction project's completion. Consequently, the proposed framework is purely based on a theoretical and conceptual basis. It highlights the idea of using drone technology to facilitate the auditor's tasks regarding evidence collection by making the auditing process more efficient and practical, particularly during lockdowns. This is significant for the following two reasons:

1. It adds to previous research on the use of emerging technologies in the accounting profession by advocating the use of drones in long-term construction projects.
2. The utilization of drones to assess the auditor during the audit engagement will be a useful tool to be considered, specifically when the physical presence of auditors to collect the data necessary for the audit work is not feasible (for example, in the recent lockdown and social distancing guidelines related to the COVID-19 pandemic's outbreak).

5.2.1 Utilizing UAV Technology in Long-Term Construction Projects

Drones are widely used in the construction and real estate industries, both professionally and in scientific research. For example, [42] explained that the attention toward utilizing drones in the field of civil engineering had increased progressively in recent periods. They explained how drones are used as supplementary inspection tools for bridge inspections. In this regard, it is explained that when applying this technology for bridge field inspections, benefits related to image quality and damage identification capabilities are significantly noticed at a remarkably less costly compared to traditional inspection mechanisms. Drones can capture data from areas of the bridge that can be perceived as hazardous if data collection is conducted in person, which will, in return, preserve individuals' safety and protect their lives from any dangerous and risky events. In addition, drones can collect enhanced-quality data, adding value to the quality of construction site operations. On the other hand, several challenges are documented in previous research that may encounter the application of this technology in the construction site, including [42]:

1. wind speed in the designated area during the flying time,
2. overexposure to sunlight that may impair the quality of captured images,
3. dark areas and areas with shades and low brightness in certain areas in the construction site
4. government-restricted drone flying areas and
5. objects identified as obstacles during the flying time.

Nevertheless, employing drones in construction site inspections is revolutionizing conventional inspection methods. It is expected to be more efficient and effective in recognizing the diverse types of damages on specific elements in the construction site [42]. As mentioned previously, the accounting profession's operations are witnessing significant transformations led by the emerging technological development in big data, robotic process automation, AI, and blockchain technologies [43]. This has inspired academics and accounting professionals to anticipate that these new technologies will not only modernize but also revolutionize existing accounting methods. One of these technologies currently being explored in the accounting profession is robotic process automation (RPA). The authors in [39] argued that the purpose of employing robots in repetitive accounting activities and procedures is not to replace human accountants but to discover innovative methods to assist and support accountants in their day-to-day routine and repetitive activities, which will have a favorable impact in terms of cost savings and accuracy. However, in comparison to the numerous studies in civil engineering directed at the utilization of drones in the profession, limited literature has examined this technology in accounting. To our knowledge, most recent studies in the accounting literature on utilizing drones in the accounting profession studied their use in inventory management systems (mainly to count inventory). In this domain, two works have investigated the utilization of drone technology in accounting [39, 41]. For example, the work in [39] specified a framework to design and implement an "audit drone automation" in external and internal auditing settings. The study demonstrated how drones are perceived as an appropriate tool for auditing procedures by considering the drone's capability to collect data and evidence to back-up certain assertions reached by management and assessed by auditors during the audit period. The authors mentioned that using drones is expected to improve the auditor's work in terms of observation, inspection, evaluation, and performance. Drones can be used as a tool for asset inspections and monitoring the progression of the construction project and asset confirmation. This state-of-the-art technology can collect better data quality data, which will help to improve auditor procedures and results. This chapter adds to the literature by investigating a different perspective on how drone technology could be utilized in accounting. In this domain, we present a new perspective that could be explored in the field of accounting based on exploiting emerging drone technology in the civil engineering industry and its applications.

5.2.2 Long-Term Construction Projects' Revenue Recognition: Percentage of Completion Approach

From the perspective of financial accounting, it is typical to acknowledge revenue when a sale is made. In this regard, the International Financial Reporting Standard (IFRS) No. 15 explained 5 steps to be followed by the financial accountant when recognizing revenues; these steps are as follows:

1. Acknowledging the contract.
2. Recognizing the obligations in the contract (services or products that have to be rendered).
3. Specifying the contract value.
4. Allocating the value to the identified contract performance obligations.
5. Recognizing revenue when obligations are fulfilled.

However, accounting for long-term construction contracts when calculating revenues is not a straightforward process. In this case, companies that undertake construction projects determine the revenue over time according to the project's percentage of completion. This allows them to report revenue in proportion to the project's lifespan as long as the current progress toward fulfilling the obligation can be estimated. To achieve this, the company requires to compute the degree of project completion periodically. However, asset verification and measurement in financial accounting require a physical inspection of the auditor, usually conducted in person at the construction site. Unfortunately, in some cases, it may be impossible or unsafe for the auditor to exist at the site, and geographical or travel restrictions may also pose significant challenges. To address these issues, we propose a new mechanism for conducting site inspections using drones as a substitute for the auditor's physical presence at the construction site. This remote audit model will rely on drones to provide the necessary data and insights, enabling the auditor to verify and measure the assets accurately.

5.2.3 Applying the UAV Technology in Long-Term Construction Projects

In the model depicted in Fig. 4 proposed by [37], the following elements are considered:

(1) the person that will fly the drone. In this case, it is the auditor or the third party. This person needs to have the required skills for flying the drones and capturing the needed images and videos of the construction site,
(2) the state-of-the-art drone with the required features and technologies to capture videos and images of the construction site, and
(3) the software (AI-enabled) incorporates the data required to relate the captured videos and images with the drawings of the construction site.

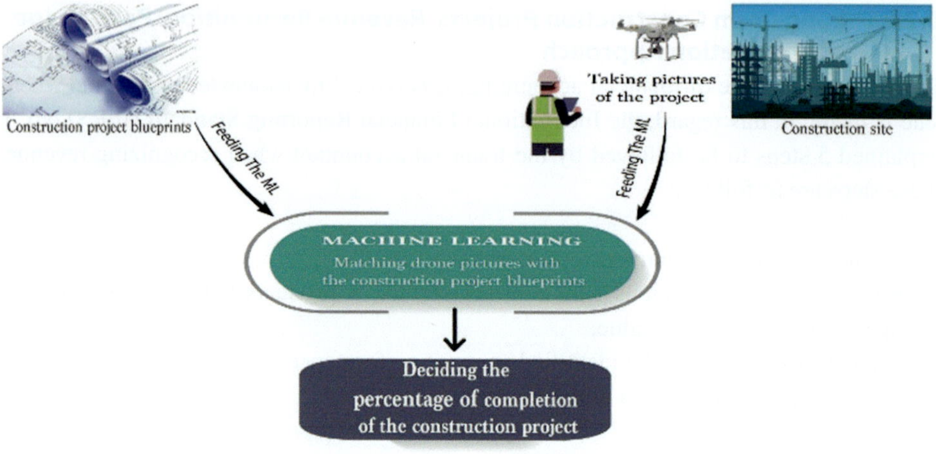

Fig. 4 Using drone technology for the recognition of revenue in construction projects. *Source* [37]

The work in [37] explained that, according to the proposed model, the auditor (or the third party) flies the drone over the specific regions in the construction site and takes the necessary images and videos needed to evaluate the completion percentage of the project. Then, the auditor ascertains the completeness of the images and videos taken. After that, the auditor feeds the AI-enabled software with the images and videos in order to compare them to the project's drawings (blackprints). After that, the outputs provided by the software will indicate (determine) the percentage of progression of the construction project. Additionally, the authors in [44] explained four steps for drone utilization in construction inspection. The 4 steps are shown in Fig. 5.

The work in [44] emphasized the required specifications for bridge inspection (p113) as: "To efficiently select a drone, the following seven considerations were studied."

- One important factor to consider when selecting a drone for site inspection is its flying time. A longer flying period is preferred for a more effective and comprehensive inspection. Therefore, the lifetime of the drone's battery is a crucial aspect to take into account.

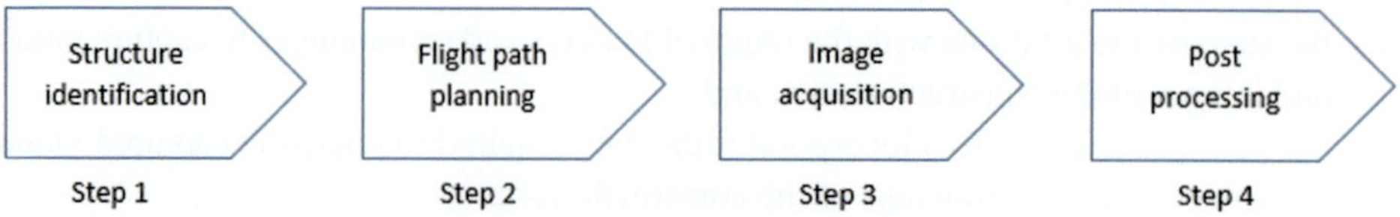

Fig. 5 Inspection workflow is divided into four steps. *Source* [44]

- The inclusion of extra high-resolution cameras on the drone can improve its capabilities to capture intricate visual content (images and videos).
- The drone's capacity to incorporate additional features like a flashlight and LED lights.
- The capability to remotely control the drone from long distances.

On the other hand, the auditor (or the accountant) should take into account the following issues when applying this technology in their work:

1. the auditing firm should approve the drone used (a certified drone)
2. ownership of the drone (is the drone owned by the auditor, the construction firm, or a third party?), and
3. the software that will link or compare the captured images with the project's drawings.

5.3 Using UAVs as a Risk Management Tool

The Covid-19 pandemic epidemic began in 2020 and has impacted how business is done. In-person activities are severely restricted due to laws governing social distance and lockdowns [37]. For some tasks, like project inspections and appraisals, being physically accessible has become a problem that organizations should overcome. Businesses have recently begun investigating the potential of artificial intelligence and robotics in numerous activities [45]. The development of robotic process automation, AI, and information and communication technologies opened the door to investigating their use in the contemporary era, which is characterized by social isolation and lockdowns [37, 46]. As a result, lockdowns and social exclusion force enterprises to consider incorporating new technology into their risk management strategies. In this regard, business continuity strategies should be stressed even if staff cannot be physically available. Additionally, when considering in-person inspections, some building sites may be deemed hazardous, or time-consuming [37]. All of these encourage the investigation of technological substitutes that eliminate the requirement for physical examination [40, 47]. This could be accomplished by using robots or drones to inspect construction projects. The implementation of drones in construction projects is covered in this chapter. This chapter's primary goal is to theoretically present the possibility of incorporating drone technology into company risk management strategies. The chapter is theoretical in nature, but more real-world trials need to be carried out. Despite their application in civil engineering projects, drones have yet to be investigated from a risk management perspective. This chapter closes this gap in the literature using a management lens.

5.3.1 Defining Risk Management

There is no widely accepted meaning of the term "Risk" in business research. However, while defining "Risk," most of the literature agreed on three fundamental concepts: probability, the future, and an undesirable consequence. To that aim, "Risk" is sometimes equated with the likelihood that an unpleasant reality (unfavorable repercussions) may develop as a result of a natural occurrence or human activity [45]. For example, lockdowns and social isolation are risk factors that firms should incorporate in their risk management strategies because they are likely, futuristic, and unpleasant. In this sense, businesses should think about the likelihood that these occurrences may occur in the future and develop plans of action to reduce losses in the event they do. Risk management can be viewed as a preventive planning procedure that a business should always take into account to ensure the reduction of any potential expected losses in the event of an emergency incident. This includes ongoing evaluations of the business's operations and preparations for the availability of appropriate people, resources, technologies, and processes [45].

5.3.2 Proposing a Risk Management Framework

Figure 6 shows the suggested risk management framework from this chapter. The framework takes into account two time periods: the first reflects regular times when there are not any emergencies or pandemic rules, and the company makes sure that it is constantly "being prepared" for any unforeseen incident during these times. The second period is during a pandemic when lockdowns and social seclusion lead to limitations that prevent engineers from going on-site at a construction site to conduct site inspections. The business must ensure that the individuals using drones for site inspections receive the essential training throughout the

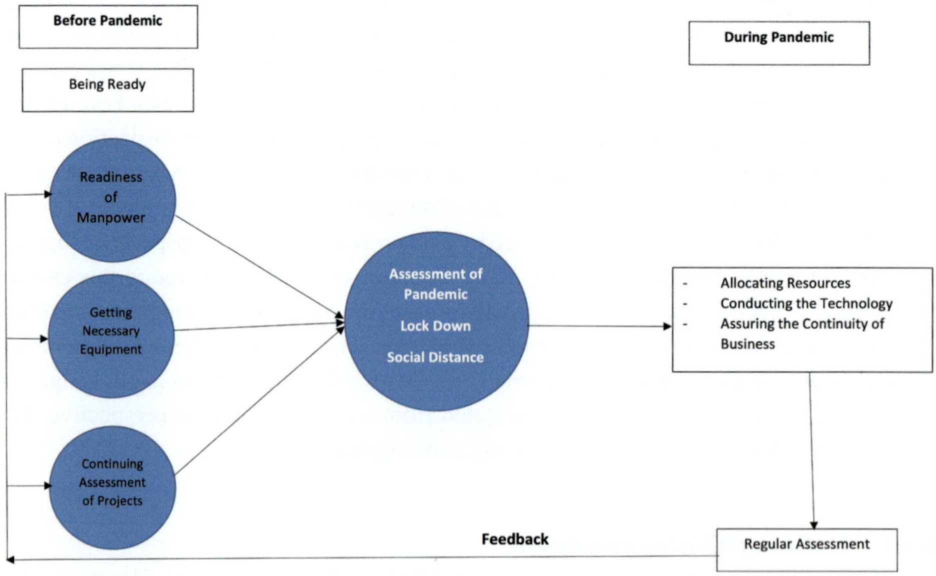

Fig. 6 Including drones in risk management plans: a framework. *Source* [50]

"always ready" phase. The drones to be deployed in emergency situations and on construction sites should be taken into account during training. Additionally, the business should now review the requirements for the drones that will be required. Engineers are prohibited from physically being present at the construction site to perform site inspection during the second phase, which is the unforeseen events of "lockdowns" and "social distance." In this instance, the business will begin deploying drones for construction project site visits. It is crucial for the business to record the experiences of those related to this phase so that they may provide feedback and continually enhance their suggestions for future equipment and training choices.

6 Challenges of Implementing Drone Technology in Accounting Profession

Despite the several advantages of drones in the accounting profession, many challenges are anticipated when implementing this technology. In addition, this technology is still in its early stages, both in terms of research and practice. Also, evidence regarding using this technology in the accounting profession needs to be more extensive, experimental, and theoretical in nature. The following are some anticipated challenges when using this technology in the accounting profession.

6.1 Drones' Ownership

More transparency regarding the ownership and use of drone photographs may increase confidence, allowing for full use of the advantages of drones in audits and advancing environmental management and compliance processes toward the achievement of their improved environmental outcomes goals. As explained earlier in this chapter, drones are used as a data collection tool for many accounting purposes. It can be used for asset verification and asset measurement in inventory management systems. In addition, it can be used for data collection when assessing the completion percentage of long-term construction projects, which will be used later for revenue recognition. However, it would be relevant to ask questions about the governance of drones. For example,

- Who should have ownership of the drone?
- Shall there be certified drones tested and approved by professional accounting bodies?

The accountant and the external auditor should agree upon these issues before implementing drones in the company's accounting system.

6.2 Suitable Drones' Technical Specifications

Prior literature indicated specific technical specifications that should be available in the drones used in the assets' inspections and identifications. For example, extended battery capacity, extra lighting fixtures on the drone, and state-of-the-art high-definition and high-resolution capabilities of cameras, the rules surrounding drone purchase, and operational costs are some important ones.

6.3 Technical Knowledge of Drone Users

Audit time savings are influenced by site size, geography, and the auditor's familiarity with the site. The person responsible for flying the drone on the construction site or the desired asset to be measured should have the necessary technical knowledge related to flying the drone and capturing the needed images and videos. In addition, the ability to correlate the company's ERP system with the captured images and videos for asset measurement purposes. In addition, companies should provide accountants with the necessary training.

6.4 Legal and Governmental Restrictions

Some restrictions might result from legal and governmental issues. For example, there might be flight restrictions in areas where it is legally prohibited to fly drones. These restrictions are inherent and cannot be controlled by the accountant. However, it is essential to understand and check if any restrictions are present regarding the flying areas and whether government-related permissions must be produced before flying the drones.

6.5 Weather Condition Challenge

Another inherent restriction that is out of the control of the user of the drone is weather conditions in terms of wind speed, storms, snow, and rain. In this regard, the person flying the drone must check if the weather conditions are acceptable for flying the drone.

7 Summary, Conclusions, and Recommendations

Business operations are undergoing significant transformations because of the growing advancements in information and communication technologies [48, 49]. Recently, remote auditing and robotic process automation have received increased attention and deployment in the accounting profession [47]. One of the robotic and AI-enabled technologies to be considered is "drone technology." This chapter discussed the utilization of drones in the accounting profession in three different areas: (1) inventory management, (2) revenue recognition in long-term construction projects, and (3) risk management tools. Based on existing literature, drones are already being deployed in different industries (for example, bridge and highway inspections) [40, 42]. However, the application of drones in the accounting and auditing professions is still in the early stages, and therefore more experimental and applied research is needed [39, 41]. This chapter discussed, in theory, the applicability of utilizing drones to capture pictures and videos to assess auditors in determining (1) the inventory levels, (2) the completion percentage in long-term construction projects, and (3) the corporate risk management plan. The implementation of lockdowns and social distancing measures has forced businesses to explore alternative methods to ensure business continuity. This chapter addresses the challenges faced by the accounting profession when these measures prevent in-person site visits. One potential solution is the use of drones to replace on-site accountants. Companies must consider using drone technology in their risk management plans. Figure 6 illustrates the proposed framework, which emphasizes the importance of being prepared, assessing risks, obtaining feedback, and continually improving policies.

Recommendations for future directions in this area should be based on experiments with this technology in real-life settings. In this regard, it is highly recommended to carry out applied research to examine whether this technology will add value to the work of the accountant and the auditor [51]. On the other hand, regulation issues need more in-depth research to organize the utilization of this technology in accounting and auditing. In this regard, the work in [46] argued that companies must establish appropriate regulations for the use of emerging technologies by adapting, in a continuous manner, their privacy and confidentiality policies, as well as establishing policies concerning data security in accordance with legal authorities' directives. The work in [39] showed that many of the advantages of using drones would depend on the used software by the drones, the drone's security measures, and the drone's technology's extended potential.

Acknowledgements This work was partly supported by ASPIRE Award for Research Excellence (AARE) 2020, Abu Dhabi, UAE, Grant number AARE20-161.

References

1. S.A. Taylor, G.L. Hunter, A.H. Zadeh, D. Delpechitre, J.H. Lim, Value propositions in a digitally transformed world. Ind. Market. Manag. (2019). https://doi.org/10.1016/j.indmarman.2019.10.004. Research, 1–33
2. B. Gajdzik, R. Wolniak, Influence of industry 4.0 projects on business operations: literature and empirical pilot studies based on case studies in Poland. J. Open Innov.: Technol. Market Compl. **8**(1), 44 (2022)
3. J. Pensri, D. Piyada, The role of value chain in enhancing competitiveness and operational efficiency of organic rice business (2021)

4. M. Zeebaree, S. Sattar, M. Aqel, A.N. Qader, D. Zebari, Impact of innovation technology in enhancing organizational management. Int. J. Sup. Chain. Manag. **9**(5), 1183 (2020)
5. M. Toshio, Modeling IoT and big data impacts to business performance, in *2021 IEEE International Conference on Industrial Engineering and Engineering Management (IEEM)* (IEEE, 2021), pp. 1127–1131
6. E. Stancheva-Todorova, The knowledge and skills profile of accountant 4.0, in *11th International Conference "Digital Transformation of the Economy and Society: Shaping the Future"*, Prilep, North Macedonia (2019)
7. K.M. Bakarich, P.E. O'Brien, The robots are coming...but aren't here yet: the use of artificial intelligence technologies in the public accounting profession. J. Emer. Technol. Account. **18**(1), 27–43 (2021)
8. E.R. Schislyaeva, O.A. Saychenko, Labor market soft skills in the context of digitalization of the economy. Soc. Sci. **11**(3), 91 (2022)
9. E. Stancheva-Todorova, B. Bogdanova, Enhancing investors' decision-making-An interdisciplinary AI-based case study for accounting students, in *AIP Conference Proceedings*, vol. 2333, No. 1 (AIP Publishing LLC, 2021), p. 150007
10. D. Balios, P. Kotsilaras, N. Eriotis, D. Vasiliou, Big data, data analytics and external auditing. J. Modern Account. Audit. **16**(5), 211–219 (2020)
11. A. Eltweri, A. Faccia, N. Sawan, The Importance of Big Data Visualisations for Auditors' Decisions, in *2021 5th International Conference on Software and e-Business (ICSEB)* (2021), pp. 7–12
12. G. Richins, A. Stapleton, T.C. Stratopoulos, C. Wong, Big data analytics: opportunity or threat for the accounting profession? J. Inf. Syst. **31**(3), 63–79 (2017)
13. G. Aiello, F. Hopps, D. Santisi, M. Venticinque, The employment of unmanned aerial vehicles for analyzing and mitigating disaster risks in industrial sites. IEEE Trans. Eng. Manag. **67**(3), 519–530 (2020)
14. A. Masadeh, M. Alhafnawi, H.A.B. Salameh, A. Musa, Y. Jararweh, Reinforcement learning-based security/safety UAV system for intrusion detection under dynamic and uncertain target movement. IEEE Trans. Eng. Manag. 1–11 (2022)
15. H. Shakhatreh, A. Khreishah, J. Chakareski, H.B. Salameh, I. Khalil, On the continuous coverage problem for a swarm of UAVs, in *2016 IEEE 37th Sarnoff Symposium* (2016), pp. 130–135
16. A.M. Hayajneh et al., Performance analysis of drone assisted multiple antenna backscatter iot sensor network, in *2021 8th International Conference on Internet of Things: Systems, Management and Security (IOTSMS)* (2021), pp. 1–7
17. K.P. Valavanis, G.J. Vachtsevanos, *Handbook of Unmanned Aerial Vehicles* (2014)
18. H.B. Salameh, M. Alhafnawi, A. Masadeh, Y. Jararweh, Federated reinforcement learning approach for detecting uncertain deceptive target using autonomous dual UAV system. Inf. Process. & Manag. **60**(2) (2023)
19. H.B. Salameh, R. Al-Maaitah, H. Al-Obiedollah, A. Al-Ajlouni, Energy-efficient power-controlled resource allocation for MIMO-based cognitive-enabled B5G/6G indoor-flying networks. IEEE Access **10**, 106828–106840 (2022)
20. O. Bouachir, M. Aloqaily, I.A. Ridhawi, O. Alfandi, H.B. Salameh, Uav-assisted vehicular communication for densely crowded environments, in *NOMS 2020 - 2020 IEEE/IFIP Network Operations and Management Symposium* (2020), pp. 1–4
21. H.B. Salameh, A. Masadeh, G.E. Refae, Intelligent drone-base-station placement for improved revenue in b5g/6g systems under uncertain fluctuated demands. IEEE Access **10**, 106740–106749 (2022)
22. M. Alhafnawi, A. Masadeh, H.B. Salameh, M. Ayyash, R. El-Khazali, H. Elgala, A review of indoor uav-based tracking systems: classification, status, and challenges, in *2022 International*

Arab Conference on Information Technology (ACIT), Abu Dhabi, United Arab Emirates (2022), pp. 1–6
23. M. Khosravi, S. Enayati, H. Saeedi, H. Pishro-Nik, Multi-purpose drones for coverage and transport applications. IEEE Trans. Wireless Commun. **20**(6), 3974–3987 (2021)
24. J.-M. Li, C.-W. Chen, T.-H. Cheng, Motion prediction and robust tracking of a dynamic and temporarily-occluded target by an unmanned aerial vehicle. IEEE Trans. Control Syst. Technol. **29**(4), 1623–1635 (2020)
25. B. Yang, X. Cao, C. Yuen, L. Qian, Offloading optimization in edge computing for deep learning enabled target tracking by internet-of-UAVs. IEEE Internet Things J. (2020)
26. S. Śledź, M.W. Ewertowski, J. Piekarczyk, Applications of unmanned aerial vehicle (uav) surveys and structure from motion photogrammetry in glacial and periglacial geomorphology. Geomorphology **378**, 107620 (2021)
27. F. Zhang, H. Liu, Z. Ma, Y. Yang, X. Wan, Study of uav application in wireless sensor networks, in *2020 3rd International Conference on Mechanical, Electronics, Computer, and Industrial Technology (MECnIT)* (2020), pp. 343–348
28. F. Outay, H.A. Mengash, M. Adnan, Applications of unmanned aerial vehicle (uav) in road safety, traffic and highway infrastructure management: recent advances and challenges. Trans. Res. Part A: Policy Pract. **141**, 116–129 (2020)
29. W. Jin, J. Yang, Y. Fang, W. Feng, Research on UAVplication and deployment of uav in emergency response, in *2020 IEEE 10th International Conference on Electronics Information and Emergency Communication (ICEIEC)* (2020), pp. 277–280
30. M. Al-Hefnawi, A. Masadeh, H.B. Salameh, A. Musa, Reinforcement learning method for autonomous UAVs monitoring an uncertain target, in *2021 Eighth International Conference on Social Network Analysis, Management and Security (SNAMS)* (2021), pp. 1–5
31. F.F. Lizzio, E. Capello, G. Guglieri, A review of consensus-based multi-agent UAV applications, in *International Conference on Unmanned Aircraft Systems (ICUAS)* (2021), pp. 1548–1557
32. P. Bithas et al., A survey on machine-learning techniques for UAV-based communications. Sensors **19**(23), 5170 (2019)
33. D. Appelbaum, R.A. Nehmer, Using drones in internal and external audits: an exploratory framework. J. Emer. Technol. Account. **14**(1), 99–113 (2017)
34. B. Hiebert, E. Nouvet, V. Jeyabalan, L. Donelle, The application of drones in healthcare and health-related services in north america: a scoping review. Drones **4**(3), 30 (2020)
35. D. Appelbaum, S. Budnik, M. Vasarhelyi, Auditing and accounting during and after the COVID-19 crisis. CPA J. **90**(6), 14–19 (2020)
36. X. Lucock, V. Westbrooke, Trusting in the "eye in the sky"? Farmers' and auditors' perceptions of drone use in environmental auditing. Sustainability **13**(23), 13208 (2021)
37. A. Qasim, G.A. El Refae, H. Issa, S. Eletter, The impact of drone technology on the accounting profession: the case of revenue recognition in long-term construction contracts, in *2021 22nd International Arab Conference on Information Technology (ACIT)* (IEEE, 2021), pp. 1–4
38. H. Issa, T. Sun, M.A. Vasarhelyi, Research ideas for artificial Intelligence in auditing: the formalization of audit and workforce supplementation. J. Emer. Technol. Account. **13**(2), 1–20 (2016)
39. D. Appelbaum, R. Nehmer, Using drones in internal and external audits: an exploratory framework. J. Emerg. Technol. Account. **14**(1), 99–114 (2017)
40. D. Oksana, M. Volodymyr, S. Oleg, Accounting automation in agroindustrial enterprises using drones (UAVs), in *the 11th International Conference on Advanced Information Technologies (ACIT)* (2021)

41. M.H. Christ, S.A. Emett, S.L. Summers, D.A. Wood, Prepare for takeoff: improving asset measurement and audit quality with drone-enabled inventory audit procedures. Rev. Account. Studies (2021)
42. J. Seo, L. Duque, J. Wacker, Drone-enabled bridge inspection methodology and application. Autom. Construct. **94**, 112–126 (2018)
43. A. Al-Chahadah, G. El Refae, A. Qasim, The use of data mining techniques in accounting and finance as a corporate strategic tool: an empirical investigation on banks operating in emerging economies. Int. J. Econ. Bus. Res. **15**(4), 442–452
44. E. Ciampa, L. De Vito, M.R. Pecce, Practical issues on the use of drones for construction inspection. J. Phys.: Conf. Seriese. **1249**, 1–11 (2019)
45. R. Ortwin, Three decades of risk research: accomplishments and new challenges. J. Risk Res. **1**(1), 49–71 (1998). https://doi.org/10.1080/136698798377321
46. M.B. Burns, A. Igou, "Alexa, write an audit opinion": adopting intelligent virtual assistants in accounting workplaces. J. Emerg. Technol. Account. **16**(1), 81–92 (2019)
47. A. Qasim, F. Kharbat, Blockchain technology, business data analytics, and artificial intelligence: use in the accounting profession and ideas for inclusion into the accounting curriculum. J. Emerg. Technol. Account. **17**(1), 107–117 (2020)
48. A. Al- Chahadah, A. Qasim, G. El Refae, Financial inclusion indicators and their effect on corporate profitability. AAU J. Bus. Law. **4**(2), 08–23 (2020)
49. A. Qasim, H.. El. Issa, G. Refae, A. Sannella, A model to integrate data analytics in the undergraduate accounting curriculum. J. Emerg. Technol. Account. **17**(2), 31–44 (2020)
50. A. Qasim, G.A. El Refae, S. Eletter, The application of drone technology in construction projects: risk management approach, in *2022 Cesare Conference Publications* (2022). ISSN: 2788-6204
51. M.J. Boyle, *The Drone Age: How Drone Technology will Change War and Peace* (Oxford University Press, USA, 2020)

Amer Qasim is a Professor of Accounting at the College of Business, Al Ain University—UAE. He received his Ph.D. in Accounting from the University of Aberdeen—UK. His research interests include Accounting Education, AI technologies in Accounting, Data analytics in accounting, Corporate Governance, and Financial Reporting.

Ghaleb A. Elrefae is a Professor of Financial Economics and Accounting at the College of Business at Al Ain University, UAE. He has published in various journals, books, and journal proceedings. He served as a Guest Editor and has recently become a chief-editor and an editorial member of several economics, business, and information technology journals. His research interests include university corporate governance, risk management, financial contract theory, theory of the firm, industrial organization, business analytics, and Artificial Intelligence.

Shorouq Fathi Eletter is an associate professor in business management at Al Ain University. She holds a Ph.D. in Business Intelligence and Operation research. Her research interests are business intelligence, deep learning, big data, credit risk management, quality management, Knowledge management, management, marketing research, sentiment analysis, data mining, text analysis, and others.

Haythem A. Bany Salameh received his Ph.D. degree in Electrical and Computer Engineering from the University of Arizona, USA, in 2009. He is currently the Dean of Scientific Research and Graduate Studies at Al Ain University, Al Ain, UAE. He is also with Yarmouk University (YU), Irbid, Jordan. He also holds a visiting professor position at Staffordshire University, UK. He was a recipient of the Jordan Science Research Support Foundation (SRSF) Prestigious Award for Distinguished Research in ICT in 2015, the Best Researcher Award for Scientific Colleges in YU in 2016, and the SRSF Award for Creativity and Technological Scientific Innovation in 2017. He authored and co-authored more than 160 highly esteemed research articles. He is among World's Top 2% Scientists List for various disciplines, compiled by Stanford University in 2020, 2021, and 2022. His research interests include wireless networking, with an emphasis on dynamic spectrum access, cognitive radio networking, drone communication, Internet-of-Things, and security. Prof. Bany Salameh has served and continues to serve on the Technical Program Committee of many international conferences and as a reviewer for many international conferences and journals. He supervised and examined more than 100 master's theses at most Jordanian universities. He is an IEEE Senior Member class of 2016.

Security and Communication

Security Challenges for UAV Systems Communications: Potential Attacks and Countermeasures

Wojdan BinSaeedan, Atheer Aldawsari, Laila Alhussain, Lama Alrushud, and Lujain Alfawzan

Abstract

Unmanned aerial vehicles (UAVs) are aircraft that are guided either remotely by humans using a remote control or autonomously by an intelligent computer-based autopilot system. Recently, UAVs have been used to perform essential roles in critical sectors, such as the entertainment or industrial sector; these roles include responding to security or emergency situations, controlling natural disasters, delivering medicines. However, UAVs have vulnerabilities that criminals can use to exploit the UAV system, abuse it, and perform numerous types of attacks. In this chapter, the security challenges for UAV system communications are discussed, and their security vulnerabilities are addressed. Details of potential threats and attacks exploiting UAV system communications and their effect on individuals' lives, privacy and society are discussed. Throughout the chapter, countermeasures to prevent and detect these attacks are demonstrated.

1 Introduction

The unmanned aerial system (UAS) contains unmanned aerial vehicles (UAVs), a ground control station (GCS), and a communication protocol, as shown in Fig. 1. The GCS is responsible for monitoring and controlling UAVs during flight, while the communication protocol transfers information between the UAV and GCS. The communication protocols are categorized into radio frequency line of sight (LOS) and beyond line of sight (BLOS). LOS is

W. BinSaeedan (✉) · A. Aldawsari · L. Alhussain · L. Alrushud · L. Alfawzan
Department of Computer Science, College of Computer and Information Sciences, Imam Mohammad Ibn Saud Islamic University (IMSIU), RAMA4158, Riyadh 11432, Saudi Arabia
e-mail: wmasaeedan@imamu.edu.sa

M. Abdelkader and A. Koubaa (eds.), *Unmanned Aerial Vehicles Applications: Challenges and Trends*, Synthesis Lectures on Intelligent Technologies,
https://doi.org/10.1007/978-3-031-32037-8_9

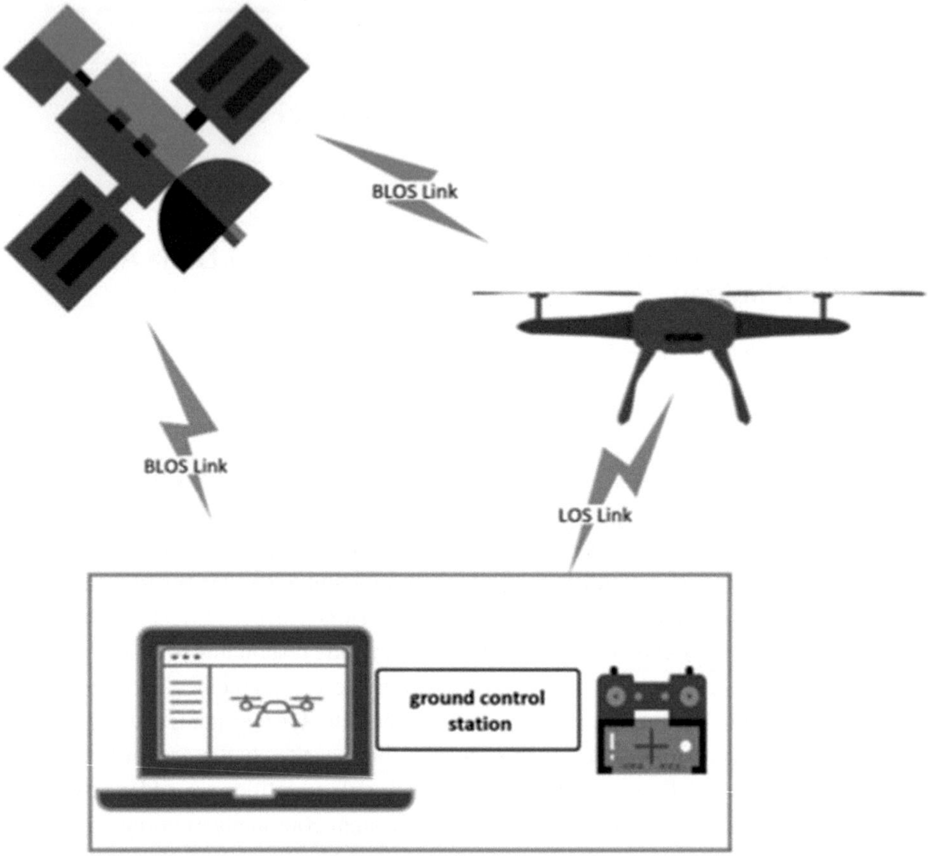

Fig. 1 The basic architecture of the UAS Components

short-range limited wireless communication, operated via direct radio waves, whereas BLOS exceeds the limits of radio via satellite communications. Therefore, securing the communication protocols of the UAS and computing powerful defence measures are important for improving UAS security.

UAVs perform critical tasks in different sectors, such as defence, surveillance, and delivery. It was first used in the military sector by the U.S. navy back in 1918 to attack warships in World War I [1], and it has been used widely since to perform several other tasks, such as providing Internet access in uncovered areas, aerial photography, delivering medicines and controlling natural disasters [2].

However, one of the greatest challenges that affect UAVs is their communication capabilities. Moreover, employing UAVs in such tasks with this limitation might lead to serious losses since UAVs are exposed to cyberattacks such as GPS spoofing, jamming, and eavesdropping that threaten their system and reveal critical information as it transfers through communication protocols.

In an example of this attack, militants of a country took advantage of unencrypted GPS signals and used SkyGrabber, a cheap software available on the Internet, to capture live video feeds that the GCS receives from the UAV. This provided the militants with battlefield advantages. Moreover, this attack happened repeatedly, and it was only discovered later when American soldiers arrested a militant and found files of intercepted UAV video feed on his laptop [3, 4].

This chapter is motivated by the unique UAV communication challenges the UAS deal with that we discussed earlier. Most studies in the field of UAV security have not focused on where these cyberattacks occur. In this chapter, the security challenges of the UAS are analysed, and an overview of the potential threats against it are provided. The security requirements that are affected by such attacks are also addressed, and new defence and countermeasures for securing the UAS against cyber-attacks are covered. The chapter concludes with suggestions for future research opportunities to enhance the security of the UAS.

The UAS vulnerabilities are first explained in Sect. 2. In Sect. 3, a general overview of security threats that exploit UAS vulnerabilities and the side effects caused by these cyberattacks are provided. In Sect. 4, different existing countermeasures that secure UAV applications against damage by these threats are discussed. The chapter concludes by providing future research opportunities and summarizing the security of UAS.

2 Vulnerabllities

The number of cyber threats is increasing due to vulnerabilities of UAS entities and the communication data links between them. These data links have weak capabilities because UAVs communicate with one another via wireless communications that are often unencrypted and unauthenticated. The communication protocols between the UAS entities vary and have different capabilities, such as UranusLink, UAVCAN, and MAVLink.

The UranusLink protocol is an open-source lightweight packet-oriented communication protocol that was designed for robotic applications; a stable version has not yet been proposed. On the other hand, UAVCAN is a lightweight open-source communication protocol that is designed for small overhead aerial vehicles and small data flow. The UAVCAN and UranusLink protocols are not widely used because they use unscalable protocols and support a limited number of programming languages [5].

MAVLink is the most widely used communication protocol and is an open-source scalable protocol that supports multiple programming languages. MAVLink allows UAS entities to communicate in both directions. Even though the MAVLink communication protocol is widely accepted and used, it is unencrypted and lacks security measures that protect messages transferring through it, which makes it vulnerable to security attacks targeting the UAS [6].

Fig. 2 Illustration of a cyber-attack on a UAS communication

Additionally, most civilian UAVs depend on an equipped global positioning system (GPS) to obtain guidance, navigation, control, and positioning data from a geosynchronous satellite [7]. This reliance on GPS could lead to serious cyberattacks such as GPS spoofing, since civilian GPS is unencrypted and unauthenticated and is thus easy to fabricate, or GPS Jamming, since civilian GPS signals have low power at approximately 130 dBm [2].

With these unsecured wireless communication data links, the UAV and its data feeds are vulnerable to various threats that could lead to serious consequences, such as loss of confidentiality if the attacker steals the UAV data, loss of authentication if the attacker adjusts the data or fakes new data rather than stealing it, or loss of availability if the attacker gains full control of the UAV or damages one of its entities. Figure 2 illustrates how an attacker takes advantage of the vulnerabilities discussed in this section.

3 Cyber-Attacks Against UAVs

The UAS is vulnerable to security threats that target the UAV, GCS, and communication data links in between. These vulnerabilities make the UAS an easy target for attackers to actively disrupt UAV operations. Moreover, cyber-attacks tend to affect the privacy, confidentiality, integrity, availability, and authenticity of the UAS. Losing these security requirements leads

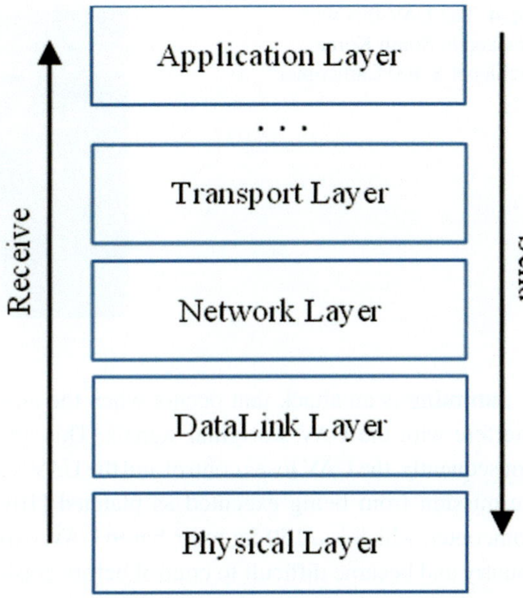

Fig. 3 Communication network layers

the attacker to access the network and control the UAV. These attacks accrue at different network layers, as shown in Fig. 3. In this section, we discuss security threats against the UAS entities according to where the attack is most likely to occur, beginning with the physical layer and progressing through the data link layer, network layer, transport layer, and application layer.

3.1 Cyber-Attacks Targeting the Physical Layer

The physical layer is the lowest layer of the network and is the fundamental layer that supports higher-level functions. In particular, the physical layer is responsible for controlling how bitstream data are sent and received. Attacks that can occur on this layer include GPS spoofing, jamming, and eavesdropping.

GPS Spoofing is an attack that occurs because UAVs rely on GPS signals for navigation, guidance, control, and positioning, which are not encrypted in civilian UAVs. The attackers exploit this feature by generating malicious signals and sending them at a higher power than legitimate signals; the UAV then receives these malicious signals as authentic signals. In 2014, a GPS spoofing attack was carried out by sending false GPS data to the UAV, causing it to land in the hands of militants instead of its home base [8]. Thus, this attack will change the behaviours of UAVs and gain control over the UAV. This attack affects the availability and integrity of the UAV system so the attackers can easily capture, change, and steal data [9].

Fig. 4 The UAV that was attacked in South Korea "Schiebel S-100 Camcopter" [11]

Jamming is an attack that occurs when the attacker sends radio frequency signals that interfere with the UAV's original signals. This affects the integrity of the UAV's system. Consequently, the UAV loses control and the UAV's operators cannot reach it, which prevents the mission from being executed as planned [10]. On May 10, 2012, a Schiebel S-100 Camcopter, which is a 150-kg rotor-based UAV, experienced a jamming attack by a militant country and became difficult to control before crashing into its own GCS in Incheon, South Korea. This incident killed the Schiebel engineer and severely injured others. Afterwards, investigators found that the attack started on April 28, 2012. The Schiebel S-100 Camcopter is shown in Fig. 4.

An **eavesdropping** attack is the act of intercepting, modifying, or deleting information exchanged through network communication; thus, the attacker intercepts communications and obtains information. This attack occurs when the information between the UAV and its communication links is transmitted. In which the confidentiality of the system is compromised [12, 13].

A **keylogging** attack is a malware attack which is a type of software that can be used to monitor or track everything done in the system. This type of attack is considered dangerous since antivirus systems generally cannot detect it. This virus can stay in the compromised system as long as the attacker needed. since this virus is considered a legitimate program it can share the resources of the system, in which Keylogging has the ability to track or monitor most activities done in the system, including the screen as well as the operations done for files, and the usage of the internet [14, 15]. In 2011 at Creech Air Force Base in the United States, the ground control unit was compromised by a keylogging virus. This virus allowed remote access by creating a backdoor in the GCS, enabling the keylogger to monitor the keyboard strokes of UAV operators who manage the UAV fleet. Therefore, such attack results in compromising the authentication and confidentiality of the UAV system, which may cause operators to lose control of sensitive data or the UAV itself [16].

3.2 Cyber-Attacks Targeting the Data Link Layer

The data link consists of two layers, the logical link layer (LLC) and the media access control layer (MAC). The data link layer is responsible of organizing and transmitting data.

A **denial-of-service (DoS)** attack is an attack that occurs in wireless networks. DoS attack prevents the UAV from operating normally, especially when UAVs receive requests from users. DoS attacks accomplish this by flooding the network with a large amount of meaningless network traffic. Which affects the network availability [12].

3.3 Cyber-Attacks Targeting the Network Layer

The basic purpose of the network layer is to deliver packets from a source to a destination across multiple intermediate networks. The network layer depends on the process performed by each physical layer and the data link layer, in which the network layer selects the suitable routes to send the data packets to the transport layer.

A **black hole** is an attack that exploits routing protocols, which can be performed by one independent malicious node as (single black hole) or by gathering many malicious nodes as in (collaborative black hole). This malicious node tries to attract all data packets by convincing the valid nodes that a better route to a specific destination exists. For instance, the attacker can access the data while packet is being transferred through the black hole, by injecting fake control packets into the network. If these packets are handled, a route is created, which leads the data packets to be delivered to the black hole instead of the destination node. The black hole attack is used to launch a directed pull attack which entirely alter the routing data. A variation of this attack is the **grey hole** attack, where the attacker selects which packets to drop [17, 18] to thus compromise the availability of data in the UAV system [15].

A **replay**, which is similar to DoS attacks, intercepts data transmissions. By exploiting communication link vulnerabilities, attackers either delay the sent data or retransmit altered data via broadcast or direct links without decrypting the captured message [18]. Address Resolution Protocol (ARP) requests are replayed, providing a significant amount of information crack the encryption key, which mostly occur when the communication data links are secure. Performing such attack on small UAVs might lead to de-authentication, reverse engineering, DOS, data tampering, hijacking, and system halting [17].

Packet sniffing is an attack that enables unauthorized users to illegally gain access to data by using packet-sniffing software to eavesdrop on network traffic. When a UAV is used to intercept data, it violates confidentiality, which can have serious consequences [10].

In a **man in the middle (MiTM)** attack, attackers intercepts the communication data link between the UAV and the GCS, in which becomes “in the middle” of the communication, usually without the knowledge of the operators. The attackers will control communications between UAVs and GCS and will secretly collect confidential data and use it to behave as although they are legitimate operators [19]. A MiTM attack thus violates authentication, confidentiality, and integrity. UAVs are especially vulnerable to such attack, due to the constant broadcast of UAVs movement and flight information, which makes them an easy targets [15].

Adversarial attacks are generated by small perturbations of data input that have a statistically significant impact on artificial intelligence (AI) agent decisions. This is possible by knowing the parameters of the AI agent being attacked. The methods used to generate this attack are the fast gradient sign method (FGSM) [20], projected gradient descent (PGD) [21], and Carlini and Wagner (C and W) [22]).

A **Sybil** attack is an attack that creates new identities or impersonates existing ones to give the attacker multiple identities. This attack can vitiate various operations of these networks, such as data aggregation, misbehaviour detection, and routing mechanisms [23]. Sybil attacks affect confidentiality, availability, and integrity.

3.4 Cyber-Attacks Targeting the Transport Layer

Transport layer is the fourth layer of the five-layer architecture. The main responsibility of the transport layer is to provide communication between the network layer and the application layer. The layer's basic role is to accept data from the previous layer, split it into smaller units, and then send it to the next layer.

An **IP spoofing** attack occurs when the UAV firewall configuration is set to give permission to a set of fixed IP addresses, and the attacker impersonates the genuine IP address with another false IP address. This provides the attacker with access to the UAV system to perform other attacks, such as MiTM or DoS. IP spoofing is an active attack, and it affects the UAV system integrity [24].

3.5 Cyber-Attacks Targeting the Application Layer

The outer layer of the five-layer architecture is the application layer. The application layer manages communications between network applications and provides several ways to manipulate data (information), which enables any type of user to easily access the network. For communication or data transmission, this layer depends on the lower layers of the five-layer architecture.

In **Distributed denial-of-service (DDoS)** attack, the attacker sends requests across several devices, thus making the identification of the attacker in the network more difficult. DDoS attacks may occur on UASs to prevent them from receiving their original control signal, which leads to false transmissions of control messages, sensor information, and mission information [12]. Consequently, the mission cannot be successfully carried out since the

UAV can't be in a steady condition. A DDoS attack violates the availability of the system. According to [25], this attack can be performed through either by Wi-Fi jamming attack or by a de-authentication attack.

4 Countermeasures

In this section, different existing countermeasures that secure UAVs against damage from the threats and attacks discussed in the previous section are discussed.

4.1 Cryptographic Approaches

Cryptographic schemes are implemented to secure UAV feeds, whether they are text, images, or videos, transferred over unencrypted or insecure data links. As discussed in the vulnerabilities section, the GPS signals, Wi-Fi connection, and most communication links are unencrypted. Applying encryption algorithms to UAV feeds that pass through these communication protocols will ensure the privacy, integrity, availability, and confidentiality of this information.

Yoon et al. [26] proposed an asymmetric authentication scheme to determine the authenticity of the information the UAV receives. Asymmetric encryption is based on two keys: a public key and a private key. The UAV uses a public key to encrypt a random stream and send it to the GCS; it uses a private key to decrypt the encrypted random stream, re-encrypt it using the public key, and resend it to the UAV. The UAV then indicates whether the UAV is under threat by comparing the received stream with its data. If the communication link is safe, the UAV will receive the command; otherwise, a threat will be detected, and the communication data link will be disconnected and no longer receive the command. This scheme requires high bandwidth in the case of extensive data.

Steinmann et al. [27] provide an asymmetric encryption scheme where the key incessantly and randomly changes. Their scheme ensures the security of the keys and the information. Encryption is performed using the public key and sent with the hash code, and this information is then decrypted using the private key. A threat is detected if an alert arises by comparing the received hash code and calculated hash code.

Lin et al. [28] proposed lightweight symmetric encryption to encrypt the transferred feed information. Asymmetric encryption is only based on the key that encrypts and decrypts the transferred information. If the attackers have access to the data, the symmetric encryption scheme will prevent them from reading the information.

Sharma et al. [29] used a functional encryption approach to secure the UAV networks against threats. The authors encrypt the data feed using functional encryption, and the decryption of the encrypted data shows only particular information since there is no need to decrypt the whole message.

Additionally, [30] allowed the UAV to receive and execute the commands only if they have a valid digital signature, which proves that the command is from an authenticated GCS; otherwise, no command will be performed, and the UAV will immediately return to the take-off position and return-to-launch (RTL) mode. Their lightweight protocol secures the UAV against MiTM attacks. Unfortunately, traditional protocols cannot be efficient in real-time applications, and they have a high processing time, high computational cost, and issues with key distributions due to the overhead and high transfer rate.

4.2 Blockchain Approaches

Blockchain is a shared ledger that uses cryptographic techniques to secure shared data. Along with enhancing UAV security and transparency, and guarantee the integrity of the data.

Ge et al. [31] presented a distributed blockchain techniques to ensure communications protection through data collection and transmission. And alleviate the possibility of attacks from a maliciously manipulated UAV. The common blockchain approach demands substantial bandwidth overhead and expensive processing, which can overwhelm the small memory and limited resources of the UAV. Instead, a lightweight blockchain approach is proposed to mitigate processing and bandwidth costs while maintaining the privacy and the security. Moreover, multiple transaction types in the blockchain architecture are identified to describe numerous data accesses, and to guarantee the reliability of the decentralized network, a novel reputation-based consensus protocol has been developed. The proposed architecture makes it possible for UAV-based applications to acquire sensor data that is managed in a reliable way, as well as to collect data in real time to meet the requirements of integrity, confidentiality, and availability.

Ch et al. [32] introduced a blockchain solution that provides an improved privacy and data storage with unique properties, such as transparency, security, and effective distribution methods. A cloud platform is used to stores all the instructions, UAV responses, authentication and integrity information. To ensure the privacy of the data being stored, the cloud is fully secured with pentatope-based elliptic curve cryptography and SHA encryption protocol. The data is then moved to an Ethereum-based public blockchain to allow smooth transactions. Results show that the proposed methodology protected data from multiple cyber-attacks.

García-Magariño et al. [33] introduced a blockchain solution using asymmetric encryption to enable the detection of false information produced by UAVs under attack. The attacked UAV has two options: raise an alert as false alarm or ignore true observations. Even if the UAV fails to make valid observations, its yet considered a dependable system since other UAVs will most likely be able to identify it. Because the proposed solution uses a secure signing system. The experiments showed that the system worked efficiently even if a minority of UAVs were compromised.

Khan et al. [34] suggested using a machine learning (ML) approach based on a blockchain technique and as a case study, they decided a collaborative intrusion detection. Their solution was conducted in two stages. In the first stage, ML was used independently at each node. And in the second phase, the ML models by shared other blockchain network participants are stacked. ML stacking positively affected the accuracy of the classifiers.

4.3 Rule-Based Approaches

In the rule-based approach, the engineer must identify a set of rules, monitor the behaviours of UAVs, and then compare it against these rules to detect threats. However, in recent years, fewer published studies have proposed a rule-based approach than the other approaches presented.

Sedjelmaci et al. [35] proposed a rule-based approach to distinguish between unauthenticated and authenticated signals. In their approach, they compare a defined set of rules based on a specific threshold limit and signal value. If the value of the transmitted signals is under the threshold, it indicates that there is no threat to the UAV; if the value exceeds the threshold value, a threat will be detected. Employing their approach allows the UAV to detect GPS spoofing and jamming attacks that might lose the availability and confidentiality of the UAV and its data. Although their approach allows the UAV to distinguish between authenticated and unauthenticated signals, it might be unable to detect some threats due to the fixed threshold value.

The work of Lee et al. [36] is based on a set of rules that detect cyberattacks threatening the availability, confidentiality, and integrity of the UAV. As soon as the threat is detected, a defence method is launched based on the rule that reached the limit. Since the number of rules is high, the processing time can be long, which could delay the detection process. However, the false negative rate is reduced. In [37], a framework is proposed to detect cyber-physical threats against small UASs. In this study, the authors defined a set of rules based on an investigation done on the sensors of UAVs and determined which sensor is more vulnerable to cyber-physical threats that damage the integrity of the UAS.

However, this approach may have limitations regarding detection and require efforts to maintain its performance. The rule-based approach alone is unable to detect threats that are complex or new and have not been covered by the defined rules; thus, these rules should be updated over time to detect such threats.

4.4 Supervised Learning Approach

Supervised learning is an ML approach. This approach tends to use labelled samples to train algorithms and predict unknown cases with the trained model. Therefore, the training method is the main factor in detecting attacks.

Manesh et al. [38] applied an artificial neural network (ANN) technique to secure the UAV system from GPS spoofing attacks. Several GPS signal parameters, for example, the Doppler shift, pseudorange, and signal-to-noise ratio (SNR) are used as features fed into the neural network. Spoofing signals were detected with high accuracy in the results.

Greco et al. [39] suggested a supervised learning framework to secure the UAV network from jamming attacks. Two different ML techniques were used, which are decision tree (DT) and multilayer perceptron (MLP). The framework depends on analysing the network parameters, namely, throughput, received signal strength indicator, and packet delivery rate. The framework includes two main steps: First takes a series of packets and extracts a sequence of features, and the second begins when each feature it detects is assigned a value, regardless of whether jamming is present.

Fenga et al. [40] proposed a two steps GPS spoofing attacks detection method by using GA-XGBoost using an inertial measurement unit (IMU) and GPS. The attacks in this study were generated using a spoofing device that was modified to transmit GPS signals to the UAV, which is confused between the real and the fake signal. The experiments show that the method performed an accuracy of 96.3%.

Shafique et al. [41] developed a new method to secure UAVs from GPS signal spoofing attacks by using GPS signal characteristics. The method classifies the spoofed and authentic signals once the UAV receives a GPS signal is received. Different algorithms were used, such as naive Bayes (NB), linear regression (LR), DT, random forest (RF), and support vector machines (SVM), along with K-fold analyses to choose a proper classification algorithm. The authors found that the SVM with a polynomial kernel has the highest accuracy at 98%.

To improve security by detecting attackers, Manesh et al. [42] studied the efficiency of using ML-based classifiers to detect jamming attacks on ADS-B receivers. Several supervised ML techniques, including ANN, SVM, K-nearest neighbour (KNN), and DT, are examined and compared. Features such as the energy statistic, bit error rate, and bad packet ratio are utilized to train these models and to differentiate between the received jamming signals and the authentic signals.

4.5 Unsupervised Learning Approach

Unsupervised learning is a type of ML technique that analyses unlabelled datasets by discovering patterns or clustering the data. This learning technique solves one of the most severe problems in UAV IDS research: the unavailability of a well labelled dataset. Unsupervised ML detection techniques are robust against known and unknown cyberattacks, which is important since not all attacks are discovered by researchers. One reported incident was when ScanEagle, which is a fixed-wing UAV designed by Boeing, was attacked by an unknown cyber-attacker, causing the UAV to fall into the hands of the attackers. Unfortunately, there is a downside to unsupervised learning-detection techniques because the rate of false alarms is high.

Park et al. [43] proposed an IDS with unsupervised learning, and this model has the ability to detect the abnormal status of the UAV despite the attack type. A one-class linear autoencoder was trained only with normal flight data with no attacks in the mission to check different reconstruction losses for the normal flight and the abnormal flight. The reconstruction loss was used to recognize an intrusion on the UAV, and the model produced considerably higher reconstruction loss with the abnormal flight than with the normal flight. Regarding feature extraction, two rules were established: hardware generality and sensor stability. Feature engineering was also performed to overcome the different scales and periods of each feature. Although the model was trained with benign data, the model has the ability to detect GPS spoofing and DoS attacks patterns during the validation phase.

Panice et al. [44] presented an approach to detect GPS spoofing attacks on UAVs. This approach is based on a state estimate study that uses the SVM as a tool to detect abnormalities; it is combined with a GPS-inertial navigation system (INS) hybrid positioning by analysing the error distribution between the INS and GPS. This research detects abnormal signs and intrusions by adjusting parameters such as window length, threshold, and attack duration to enhance system performance and predict the next sensor value and location information based on a given mission. The authors categorised UAV status into two categories using state estimation as a key feature: normal status and abnormal status. They evaluate and investigate the UAV behavior in an abnormal status by using a simulator to produce data from a post-mission that simulates false GPS data injection. Numerous scenarios were tested and showed that the highest TP rate was 95%.

Whelan et al. [45] suggested a novelty-based IDS that overcomes the lack of labelled UAV IDS datasets problem by using unsupervised ML classifiers: one-class support vector machine (OC-SVM), autoencoder neural network, and local outlier factor (LOF). These are generated per sensor to detect any attacks to the UAV, even the undiscovered attacks, and identify the sensor being targeted. The classifiers are trained only on normal behaviour, and any other behaviour is considered potentially malicious. The results indicate that the autoencoder scored 94.81%F1, and showed the best performance; the OC-SVM scored 81.17% F1, and LOF showed poor performance scoring 58.93% F1.

4.6 Reinforcement Machine Learning Approaches

In addition to supervised and unsupervised learning, reinforcement learning involves an agent learning directly from its environment by creating its own learning experiences. Through the trial-and-error approach, the agent learns what decision is correct and what is incorrect. Thus, it is used in detecting and classifying attacks that occur against UAVs.

Bouhamed et al. [46] suggested a cost-effective intrusion prevention and detection (IPD) system through a distributed architecture, where a swarm of UAVs is employed. The suggested framework includes two main processes: applying a DRL module on the UAVs and implementing offline periodic learning adapted to the GCS. A deep DQN algorithm was adopted to build a multiclass classifier to detect intrusions using a labelled dataset. The learning environment in reinforcement learning was replaced with a sampling function of stored normal and intrusion cases. In addition to sampling the training dataset, the proposed offline periodic-learning approach ensures that the IPD system is always up to date by adding new knowledge to the already trained IPD system without forgetting what has already been learned. The proposed model was tested by two different reward functions: simple reward and customized reward. The second function enables the IPD system module to detect minor classes, and it outperforms the first function and other ML algorithms with a 99.7% accuracy and has a 0.95 precision, 0.97 recall, and 0.96 F1-scores.

Xiao et al. [47] proposed DQN to establish an allocation strategy against attacks on UAVs considering the optimal safe rate and the secrecy capacity. Their method creates an interaction between the attacker and the UAV transmission using a smart attack game based on the Nash equilibrium (NE) and prospect theory (PT) to reach the best response. Their implementation of the power-allocation strategy applied to the UAV shows a safe rate equal to 93% and an improvement of the secrecy capacity by 16% compared with other Q-learning algorithms.

Elnaggar and Bezzo [48] proposed inverse reinforcement learning (IRL) to predict sensor spoofing attacks in an autonomous vehicle. The IRL predicts the intention of the attacker to spoof the sensor and determines the set of sensors that were spoofed; it then recovers the system by using the Bayesian inverse reinforcement learning (BIRL) approach. The BIRL approach can determine the compromised sensors and the system can be recovered by observing and comparing measurement data and control inputs. Moreover, the authors used optimal Markov decision processes (MDPs) and improved the security of the system. Three scenarios for a case study on UAV navigation in a stochastic environment were simulated.

Praveena et al. [49] suggested a deep reinforcement learning technique that is optimized by using the black window optimization (DRL-BWO) algorithm, DRL. This is an integration of reinforcement learning and a deep neural network, which involves for intrusion detection an improved reinforcement-learning-based deep belief network (DBN). In which in UAV networks the DBN model is used to determine the existence of intrusions. The BWO technique is used to determine the appropriate DRL hyperparameter settings, which improve the performance. The results of extensive experiments showed that the simulation values were 98%, 99%, 98%, and 98% in terms of the precision, recall, F-measure, and accuracy, respectively.

Fei et al. [50] proposed a reinforcement-learning-assisted flight recovery approach against cyber-physical attacks targeting sensors and motors, in which reinforcement learning assists fault-tolerant control by maintaining UAV control under various attacks. The reinforcement-learning-based policy performed better than traditional FTC approaches, which are intended to detect and recover from specific faults or attacks. Therefore, the proposed policy runs in tandem with the flight controller of the UAVs; while the UAV is operating normally, the policy creates no or minimal control command adjustments and has no effect on the operation. When the fault situation happens or the UAV is under attack, the policy takes the state inputs and generates a suitable actuator command modification with little to no latency to mitigate the fault or attack condition and maintains UAV position and path tracking. The results shown that the reinforcement-learning-based policy can keep the vehicle stable in the face of attacks.

Wang et al. [51] proposed a reinforcement-learning-based energy-efficient antijamming relay (REAR) scheme that allows UAVs to enhance relay power and strategy policies without knowing the UAV network and channel model. Additionally, for the UAV relay the hot-booting method was used to accelerate the learning process. The results of the simulation for the network consisting of five UAVs against a random jammer shows an improvement on the network utilities, such as by boosting communication reliability and lowering energy usage.

Table 1 highlights a taxonomy of the common cyberattacks, the security requirements affected and the methods proposed to defend the UAS against such attacks.

Table 1 Compilation of the security requirement affected by cyber-attacks and the prevention countermeasures to these cyber-attacks

Cyber-attack	Security requirements affected	Countermeasures	
GPS spoofing	Availability integrity	Rule-based approach	[35]
		Supervised learning	[38, 40, 41]
		Unsupervised learning	[43, 44]
		Reinforcement learning	[47, 48, 50]
Jamming	Integrity	Rule-Based approach	[35, 36]
		Supervised learning	[39, 42]
		Unsupervised learning	[45]
		Reinforcement learning	[47, 51]
Eavesdropping	Confidentiality	Cryptographic approach	[28]
		Reinforcement learning	[47]
DoS	Availability	Cryptographic approach	[26]
		Blockchain approach	[34]
		Unsupervised learning	[43]
		Reinforcement learning	[46, 49]
Black hole	Availability	Rule-based approach	[35]
Grey hole	Availability	Rule-based approach	[35]
MiTM	Authentication Confidentiality Integrity	Cryptographic approach	[29, 30]
		Blockchain approach	[33]
DDOS	Availability	Reinforcement learning	[46]

5 Future Research Opportunities

Research on UAVs, which have been referred to as revolutionary, has been rapidly increasing due to their importance. New research possibilities arise as more issues are being discovered, and recent solutions are showing limitations. After analysing the existing countermeasures and addressing their limitations, we will discuss further research possibilities in this section.

First, it is essential to secure the wireless communication protocol of UAVs, which transfers the information between the UAV and GCS. The challenge in securing these protocols is to achieve high security without affecting the network throughput. Therefore, developing authorization and authentication schemes can secure UAV flight from attacker access. Shafique et al. [52] suggested a combination of encryption and watermark algorithms to secure Wi-Fi passwords. This combination reduces the probability of the attacker gaining control of the UAV by cracking the password. Another suggestion for improvement is to incorporate the algorithm with different transforms to secure transmission [53].

Second, intrusion detection system (IDS) solutions are effective and necessary to improve the detection of cyberattacks in UAVs and to protect UAV communications. Existing solutions that use only one approach, such as anomaly based, signature-based, or specification-based approaches, must be enhanced because they lack the ability to detect or perform well. To boost the effectiveness of IDS and to boost detection rates, the system must be lightweight for monitoring UAV communications and detecting attacks by utilizing hybrid detection approaches, which are a mixture of various approaches that will fill in the gaps in each approach.

Third, as discussed earlier, the cost of encrypting civilian UAV communication protocols is high. Therefore, we recommend that security be conducted at the software level; building a UAV with more security in mind will reduce the costs of losing the UAV and implementing a new scheme with a high cost, such as encryption. Fourth, regarding UAV swarms, the network of the swarms does not completely meet security requirements, which makes it vulnerable to cyber-attacks. Thus, we recommend implementing a scheme to secure these networks.

Additionally, the adversarial reinforcement-learning method is used for intrusion detection and includes the environment behaviour for the learning process. In this approach, a study in [54] suggested a model uses a reinforcement learning for intrusion detection. A neural network was used by the agent and environment to train the model. This reinforcement-learning-based method makes a significant contribution to detecting network security intrusions.

Finally, further work is required to establish the viability of how cyberattacks could occur in different OSI layers, since cyberattacks may overlap in different layers; however, fewer papers involve such topics.

6 Conclusion

UAVs communicate with other UAS entities through insecure communication protocols, which are vulnerable to cyber threats. Thus, to ensure a sufficient level of security for a UAS, the UAV and the feeds that transfer between the UAS entities through the communication protocol must fully meet the security requirements of the UAS. In this chapter, we discussed the security challenges and vulnerabilities in UAS entities and covered the side effects of the security threats to them . It sets out the significant growth of countermeasures carried out against these cyber threats in past years. We conclude by suggesting further works that could assess the long-term solution of security in these systems.

References

1. R.K. Barnhart, D.M. Marshall, E. Shappee, *Introduction to Unmanned Aircraft Systems* (CRC Press, 2021). (Cited on pages 1 and 8.)
2. S. Arteaga, L. Hernández, G. Pérez, A. Orozco, L. Villalba, Analysis of the GPS spoofing vulnerability in the drone 3DR solo. IEEE Access. **7**, 51782–51789 (2019)
3. R. Guo, B. Wang, J. Weng, Vulnerabilities and attacks of UAV cyber physical systems, in*Proceedings of the 2020 International Conference on Computing, Networks and Internet Of Things* (2020), pp. 8–12. https://doi.org/10.1145/3398329.3398331
4. K. Hartmann, C. Steup, The vulnerability of UAVs to cyber attacks-an approach to the risk assessment, in *2013 5th International Conference on Cyber Conflict (CYCON 2013)* (2013), pp. 1–23
5. N. Khan, N. Jhanjhi, S. Brohi, A. Nayyar, Emerging use of UAV's: secure communication protocol issues and challenges, in *Drones In Smart-cities* (2020), pp. 37–55
6. N. Khan, N. Jhanjhi, S. Brohi, A. Almazroi, A. Ali, A secure communication protocol for unmanned aerial vehicles. CMC-Comput. Mater. Continua. **70**, 601–618 (2022)
7. A. Tang, A review on cybersecurity vulnerabilities for urban air mobility, in *AIAA Scitech 2021 Forum* (2021), pp. 0773
8. R. Altawy, A. Youssef, Security, privacy, and safety aspects of civilian drones: a survey. ACM Trans. Cyber-Phys. Syst. **1**(11) (2016). https://doi.org/10.1145/3001836
9. F. Tlili, L. Fourati, S. Ayed, B. Ouni, Investigation on vulnerabilities, threats and attacks prohibiting UAVs charging and depleting UAVs batteries: Assessments & countermeasures. Ad Hoc Netw. **129**, 102805 (2022)
10. G. Loukas, *Cyber-Physical Attacks: A Growing Invisible Threat* (Butterworth-Heinemann, 2015)
11. Schiebel Corporation, 2016 Schiebel. Camcopter® S-100, (n.d.). https://schiebel.net/products/camcopter-s-100/
12. W. Saad, M. Bennis, M. Mozaffari, X. Lin, *Wireless Communications and Networking for Unmanned Aerial Vehicles* (Cambridge University Press, 2020)
13. R. Kumar, R. Sharma, P.K. Pattnaik, *Multimedia Technologies in the Internet of Things Environment*, vol. 2 (Springer Singapore Pte, Limited, 2021)
14. M.R. Manesh, N. Kaabouch, Cyber-attacks on unmanned aerial system networks: detection, countermeasure, and future research directions. Comput. Secur. **85**, 386–401 (2019)

15. F. Tlili, L.C. Fourati, S. Ayed, B. Ouni, Investigation on vulnerabilities, threats and attacks prohibiting UAVs charging and depleting UAVs batteries: assessments & countermeasures. Ad Hoc Netw. **129**, 102805 (2022)
16. A.A. Tarr, J.A. Tarr, M. Thompson, J. Ellis (eds.), *Drone Law and Policy: Global Development* (Risks, Regulation and Insurance (Routledge, 2021)
17. R. Guo, B. Wang, J. Weng, Vulnerabilities and attacks of UAV cyber physical systems, in *Proceedings of the 2020 International Conference on Computing, Networks and Internet of Things*. (Apr 2020), pp. 8–12
18. D. He, S. Chan, M. Guizani, Drone-assisted public safety networks: The security aspect. IEEE Commun. Mag. **55**(8), 218–223 (2017)
19. R.M. Haris, S. Al-Maadeed, Integrating blockchain technology in 5G enabled IoT: A review, in *2020 IEEE International Conference on Informatics, IoT, and Enabling Technologies (ICIoT)* (IEEE, 2020), pp. 367–371. (February)
20. T. Hickling, N. Aouf, P. Spencer, Robust adversarial attacks detection based on explainable deep reinforcement learning for UAV guidance and planning (2022). arXiv:2206.02670
21. A. Madry, A. Makelov, L. Schmidt, D. Tsipras, A. Vladu, Towards deep learning models resistant to adversarial attacks (2017). arXiv:1706.06083
22. N. Carlini, G. Katz, C. Barrett, D. Dill, Provably minimally-distorted adversarial examples (2017). arXiv:1709.10207
23. A. Gupta, D. Sukheja, A. Tiwari, Impact of sybil attack and security threat in mobile adhoc network. Int. J. Comput. Appl. **124**(9) (2015)
24. K. Jain, K. Khoshelham, X. Zhu, A. Tiwari, in *Proceedings of UASG* (2019)
25. D.P. Shepard, J.A. Bhatti, T.E. Humphreys, A.A. Fansler, Evaluation of smart grid and civilian UAV vulnerability to GPS spoofing attacks, in *Proceedings of the 25th International Technical Meeting of the Satellite Division of the Institute of Navigation (ION GNSS 2012)* (2012), pp. 3591–3605. (Sept 2012)
26. K. Yoon, D. Park, Y. Yim, K. Kim, S.K. Yang, M. Robinson, Security authentication system using encrypted channel on uav network, in *2017 First IEEE International Conference on Robotic Computing (IRC)*(IEEE, 2017), pp. 393–398. (Apr 2017)
27. J.A. Steinmann, R.F. Babiceanu, R. Seker, Uas security: Encryption key negotiation for partitioned data, in *2016 Integrated Communications Navigation and Surveillance (ICNS)* (IEEE, 2016), pp. 1E4-1. (Apr 2016)
28. C. Lin, D. He, N. Kumar, K.K.R. Choo, A. Vinel, X. Huang, Security and privacy for the internet of drones: challenges and solutions. IEEE Commun. Mag. **56**(1), 64–69 (2018)
29. D. Sharma, A. Rashid, S. Gupta, S.K. Gupta, A functional encryption technique in UAV integrated HetNet: a proposed model. Int. J. Simul. Syst. Sci. Technol. **20**(S1), 7–1 (2019)
30. Y. Li, C. Pu, Lightweight digital signature solution to defend micro aerial vehicles against man-in-the-middle attack, in *2020 IEEE 23rd International Conference on Computational Science and Engineering (CSE)* (IEEE, 2020), pp. 92–97. (Dec 2020)
31. C. Ge, X. Ma, Z. Liu, A semi-autonomous distributed blockchain-based framework for UAVs system. J. Syst. Archit. **107**, 101728 (2020)
32. R. Ch, G. Srivastava, T.R. Gadekallu, P.K.R. Maddikunta, S. Bhattacharya, Security and privacy of UAV data using blockchain technology. J. Inf. Secur. Appl. **55**, 102670 (2020)
33. I. García-Magariño, R. Lacuesta, M. Rajarajan, J. Lloret, Security in networks of unmanned aerial vehicles for surveillance with an agent-based approach inspired by the principles of blockchain. Ad Hoc Netw. **86**, 72–82 (2019)
34. A.A. Khan, M.M. Khan, K.M. Khan, J. Arshad, F. Ahmad, A blockchain-based decentralized machine learning framework for collaborative intrusion detection within UAVs. Comput. Netw. **196**, 108217 (2021)

35. H. Sedjelmaci, S.M. Senouci, N. Ansari, A hierarchical detection and response system to enhance security against lethal cyber-attacks in UAV networks. IEEE Trans. Syst. Man Cybern. Syst. **48**(9), 1594–1606 (2017)
36. H. Lee, S. Eom, J. Park, I. Lee, UAV-aided secure communications with cooperative jamming. IEEE Trans. Veh. Technol. **67**(10), 9385–9392 (2018)
37. D. Muniraj, M. Farhood, A framework for detection of sensor attacks on small unmanned aircraft systems, in *2017 International Conference on Unmanned Aircraft Systems (ICUAS)* (IEEE, 2017), pp. 1189–1198). (June 2017)
38. M.R. Manesh, J. Kenney, W.C. Hu, V.K. Devabhaktuni, N. Kaabouch, Detection of GPS spoofing attacks on unmanned aerial systems, in *2019 16th IEEE Annual Consumer Communications & Networking Conference (CCNC)* (IEEE, 2019), pp. 1–6. (Jan 2019)
39. C. Greco, P. Pace, S. Basagni, G. Fortino, Jamming detection at the edge of drone networks using multi-layer perceptrons and decision trees. Appl. Soft Comput. **111**, 107806 (2021)
40. Z. Feng, N. Guan, M. Lv, W. Liu, Q. Deng, X. Liu, W. Yi, Efficient drone hijacking detection using two-step GA-XGBoost. J. Syst. Archit. **103**, 101694 (2020)
41. A. Shafique, A. Mehmood, M. Elhadef, Detecting signal spoofing attack in uavs using machine learning models. IEEE Access **9**, 93803–93815 (2021)
42. M.R. Manesh, M.S. Velashani, E. Ghribi, N. Kaabouch, Performance comparison of machine learning algorithms in detecting jamming attacks on ADS-B devices, in *2019 IEEE International Conference on Electro Information Technology (EIT)* (IEEE, 2019), pp. 200–206. (May 2019)
43. K. Park, E. Park, H. Kim, Unsupervised intrusion detection system for unmanned aerial vehicle with less labeling effort, in *International Conference On Information Security Applications* (2020), pp. 45–58
44. G. Panice, S. Luongo, G. Gigante, D. Pascarella, C. Di Benedetto, A. Vozella, A. Pescapè, A SVM-based detection approach for GPS spoofing attacks to UAV, in *2017 23rd International Conference on Automation and Computing (ICAC)* (2017), pp. 1–11
45. J. Whelan, T. Sangarapillai, O. Minawi, A. Almehmadi, K. El-Khatib, Novelty-based intrusion detection of sensor attacks on unmanned aerial vehicles, in *Proceedings of the 16th ACM Symposium on QoS and Security for Wireless and Mobile Networks* (2020), pp. 23–28
46. O. Bouhamed, O. Bouachir, M. Aloqaily, I. Al Ridhawi, Lightweight ids for uav networks: a periodic deep reinforcement learning-based approach, in *2021 IFIP/IEEE International Symposium on Integrated Network Management (IM)* (IEEE, 2021), pp. 1032–1037. (May 2021)
47. L. Xiao, C. Xie, M. Min, W. Zhuang, User-centric view of unmanned aerial vehicle transmission against smart attacks. IEEE Trans. Vehic. Technol. **67**(4), 3420–3430 (2017)
48. M. Elnaggar, N. Bezzo, An IRL approach for cyber-physical attack intention prediction and recovery, in *2018 Annual American Control Conference (ACC)* (IEEE, 2018), pp. 222–227. (June 2018)
49. V. Praveena, V. Praveena, C. Ponnusamy, A. Ihsan, R. Alroobaea, S. Yahya, M.A. Raza, Optimal deep reinforcement learning for intrusion detection in UAVs. Comput. Mater. Contin. **70**, 2639–2653 (2022)
50. F. Fei, Z. Tu, D. Xu, X. Deng, Learn-to-recover: Retrofitting uavs with reinforcement learning-assisted flight control under cyber-physical attacks, in *2020 IEEE International Conference on Robotics and Automation (ICRA)* (IEEE, 2020), pp. 7358–7364. (May 2020)
51. W. Wang, X. Lu, S. Liu, L. Xiao, B. Yang, Energy efficient relay in UAV networks against jamming: a reinforcement learning based approach, in *2020 IEEE 91st Vehicular Technology Conference (VTC2020-Spring)* (IEEE, 2020), pp. 1–5. (May 2020)
52. A. Shafique, A. Mehmood, M. Elhadef, Survey of security protocols and vulnerabilities in unmanned aerial vehicles. IEEE Access **9**, 46927–46948 (2021)

53. J. Yaacoub, H. Noura, O. Salman, A. Chehab, Security analysis of drones systems: attacks, limitations, and recommendations. Internet of Things **11**, 100218 (2020)
54. G. Caminero, M. Lopez-Martin, B. Carro, Adversarial environment reinforcement learning algorithm for intrusion detection. Comput. Netw. **159**, 96–109 (2019)

Wojdan BinSaeedan received the B.Sc. degree from Imam Mohammad Ibn Saud Islamic University (IMSIU), Saudi Arabia, the M.Sc. degree from King Saud University, and the Ph.D. degree from the Department of Computer Science, King Saud University. She is currently an Assistant Professor in the Department of Computer Sciences at College of Computer and Information Sciences, Imam Mohammad Ibn Saud Islamic University (IMSIU), Saudi Arabia. Areas of her research interests include Artificial Intelligence, Cybersecurity, and Digital Forensics.

Atheer Aldawsari received the B.Sc. degree from Imam Mohammad Ibn Saud Islamic University (IMSIU), Saudi Arabia. Area of her research interest is Artificial Intelligence and Cybersecurity.

Laila Alhussain received the B.Sc. degree from Imam Mohammad Ibn Saud Islamic University (IMSIU), Saudi Arabia. Area of her research interest is Artificial Intelligence, Cybersecurity, and Computer Vision.

Lama Alrushud received the B.Sc. degree from Imam Mohammad Ibn Saud Islamic University (IMSIU), Saudi Arabia. The area of her research interest is Artificial Intelligence.

Lujain Alfawzan received her B.Sc. degree in computer science from Imam Mohammad Ibn Saud Islamic University (IMSIU), Saudi Arabia. Area of her research interest is Cybersecurity, Machine Learning, NLP.

Security Challenges in Multi-UAV Systems Communication Network

Mimonah Al Qathrady, Sultan Almakdi, Mohammed S. Alshehri, and Samar M. Alqhtani

Abstract

Unmanned Aerial Vehicles (UAVs) are a newly emerging kind of "flying Internet of Things (IoT)" device. Multi-UAV systems consist of a collection of UAVs. They autonomously fly in free space to carry out a mission. Multi-UAV systems provide advanced services and have diverse applications. Examples include road traffic monitoring, logistics inspection and supervision, research and rescue, simultaneous localization and mapping, and providing network coverage. They are expected to be an essential and integrated component of smart cities. The primary benefit of the multi-UAV system is its ability to accomplish tasks requiring coordinated and collaborative behavior, which is dependent on communication between UAVs. However, multi-UAV systems communication networks face security and privacy challenges that hinder the UAVs' performance or may have a devastating impact on people and properties. These challenges need to be thoroughly studied by the research community. Accordingly, this chapter focuses

M. Al Qathrady (✉) · S. M. Alqhtani
Department of Information Systems, College of Computer Science and Information Systems, Najran University, King Abdulaziz Rd, Najran 61441, Saudi Arabia
e-mail: mqalqathrady@nu.edu.sa

S. M. Alqhtani
e-mail: smalqhtani@nu.edu.sa

S. Almakdi · M. S. Alshehri
Department of Computer Science, College of Computer Science and Information Systems, Najran University, King Abdulaziz Rd, Najran 61441, Saudi Arabia
e-mail: saalmakdi@nu.edu.sa

M. S. Alshehri
e-mail: msalshehry@nu.edu.sa

M. Abdelkader and A. Koubaa (eds.), *Unmanned Aerial Vehicles Applications: Challenges and Trends*, Synthesis Lectures on Intelligent Technologies,
https://doi.org/10.1007/978-3-031-32037-8_10

on multi-UAV communication system security, illustrates potential attack scenarios, and examines countermeasures and mitigation techniques to protect multi-UAV systems from malicious activities. The chapter concludes with a discussion of the multi-UAV system open issues that should be considered in future research. This chapter aims to provide the academic, industry, and research community with valuable guidelines to assist in constructing secure multi-UAV system frameworks and standards.

Keywords
Unmanned aerial vehicle • Multi-UAV systems • Communication network • Attacks on multi-UAV (IoT) networks • Attacks countermeasure • Multi-UAV systems challenges

1 Introduction

UAVs have recently gained the attention of academics due to the fact that they are faster than ground transportation such as trucks and cars, and are not subject to road traffic jams and can visit otherwise inaccessible areas [1]. Multi-UAVs are a group of UAVs that are deployed to complete a mission. Multi-UAV systems can collaborate and complete missions efficiently and economically [2], and they are utilized to work together. With the utilization of a multi-UAV system, mission operations duration can be decreased, and unlike a single UAV, a team of UAVs can simultaneously perform tasks in different geographic areas. Also, each UAV may have a unique collection of sensors that complement one another. If a single UAV is lost, the mission is halted. However, when numerous UAVs are in operation, the loss of one could be compensated in some circumstances by giving additional responsibilities to other UAVs [3].

Consequently, multi-UAV systems could be utilized for a wide range of applications. For instance, they serve as flying ad-hoc networks (FANET) to enhance wireless communications. As the primary enabler of the IoT, UAVs can naturally operate as mobile aerial users [4]. Other applications include real-time traffic information, rescue missions, the delivery of goods, flying unmanned taxis, and forest firefighting [3, 5].

A crucial problem in the cooperation and control of many UAVs is their efficient networking over a wireless medium in constantly changing environments. Four communication architectures of multi-UAV networks have been explained: centralized, ad hoc, multi-group, and multi-layer ad hoc [6]. Figure 1a depicts the centralized communication architecture where all UAVs are connected to a central entity (i.e., the ground station), and all communications between UAVs pass through the central node.

In ad hoc architecture, each UAV will participate in the transmission of data to other UAVs inside the network, and a backbone UAV acts as the ad hoc network's gateway, passing information between the ground station and the other UAVs, as shown in Fig. 1b. Multi-Group and multi-layer UAV networks combine a centralized UAV network with UAV

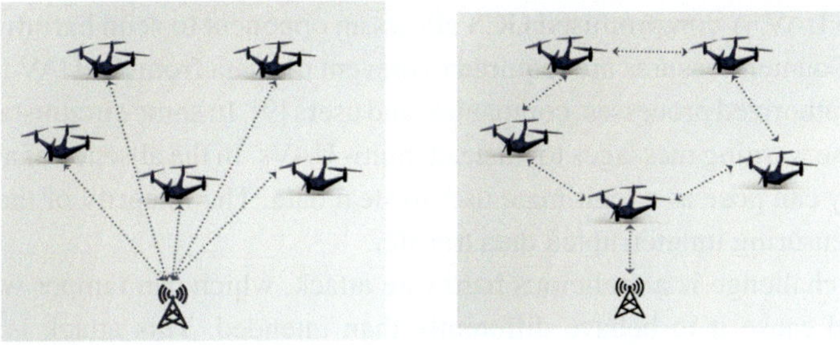

Fig. 1 Multi-UAVs communication architectures

Fig. 2 Connection channel types in Multi-UAV systems

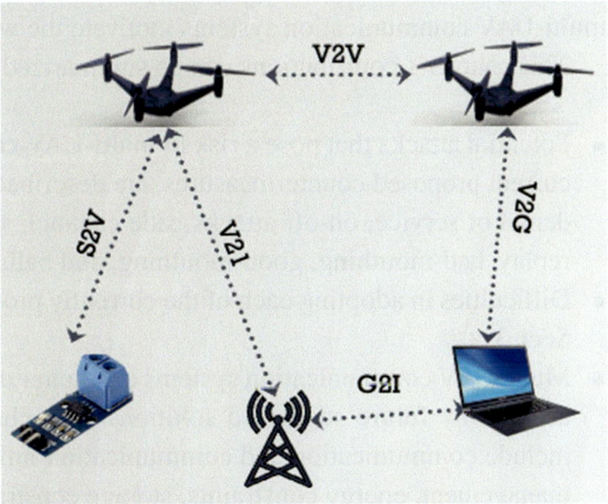

ad hoc networks, and are appropriate for mission scenarios involving a large number of UAVs with varying flight or communication characteristics. Details of these architectures can be found in [6].

There could be several channels of communication in a multi-UAVs system, as seen in Fig. 2. The most fundamental communication channels are vehicle-to-ground (V2G), vehicle-to-vehicle (V2V), vehicle-to-infrastructure (V2I), and sensor-to-vehicle (S2V). Multi-UAV systems have huge attack surfaces and vectors, especially when they are part of an IoT system.

Multi-UAVs frequently feature onboard wireless communication modules that use unencrypted or unauthenticated channels. Another significant problem with employing drones in military operations is hacking [7]. Individual privacy is jeopardized as a result of multi-UAV liability. Continuous UAV flights may reveal the trade secrets of the companies involved. Signal spoofing by hijacking can potentially harm the nature of some UAVs [8]. Ground Control System (GCS) assaults are extremely risky because the attacker can take all of the data from the UAV. A compromised GCS allows an opponent to send harmful and incorrect commands. Countermeasures are required to prevent the data from the UAV from becoming leaked to unauthorized processes, companies, and users [9]. In some circumstances, attackers broadcast false warning messages to mislead multi- UAVs. In the absence of authentications, any adversary can pose as a legitimate user to steal data. The integrity of the UAV mission is critical to ensuring uninterrupted data transfer.

A central challenge is a malicious hardware attack, which can tamper with multi-UAV hardware and cause it to behave differently than intended. This attack is carried out to steal secret data or cause the UAV's mission to fail [10]. Some attacks purposefully tamper with onboard navigational sensors, which can cause the UAV's trajectory to deviate [11]. These attacks may result in errors in transmitting precise UAV location and position to the control system. These security challenges and the devastating impacts of security threats on multi-UAV communication systems motivate the writing of this chapter.

This chapter's contributions can be summarized as follows:

- Potential attacks that pose a risk to multi-UAV communication networks, as well as their current proposed countermeasures, are described in Sect. 2. These include Sybil attacks, denial of service, on-off attacks, side channel, spoofing, message substitution, message replay, bad-mouthing, good-mouthing, and ballot stuffing attacks.
- Difficulties in adopting each of the currently proposed countermeasures are illustrated in Sect. 3.
- Multi-UAV communication systems encounter unique challenges that must be taken into account by future suggested solutions. The challenges are explained in Sect. 4. They include communication and communication infrastructure, UAV operation, UAV traffic management, energy constraints, storage constraints, and time completion issues.

2 Possible Attacks on Multi-UAV (IoT) Networks

This section explains potential security attacks facing multi-UAV communication networks and their corresponding countermeasures. Attacks include Sybil attacks, denial of service, on-off attacks, side channel, spoofing, message substitution, message replay, bad-mouthing, good-mouthing, and ballot stuffing attacks.

2.1 Sybil Attack

A Sybil attack is a classic computer science problem where in which a single attacker takes over network by pretending to be a lot of other nodes. This is a practical attack similar to DDoS attack on any sub-website. For example, when there is DDoS attack on GitHub and GitHub goes down because the attackers are sending a lot of fake requests to GitHub,

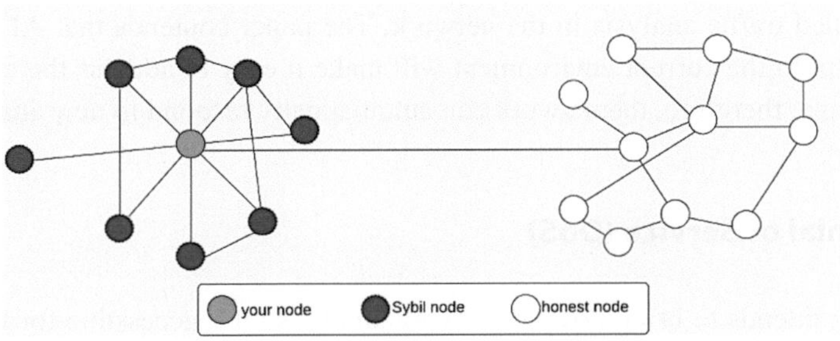

Fig. 3 Sybil attack

GitHub becomes unable to differentiate between real and unreal traffic. In this situation, Github becomes unavailable to users. This is similar to a Sybil attack, where an attacker can completely take over a pure network and render it unusable.

According to Arshad et al. [12] and Benadla and Merad-Boundia [13], IoT technology has presented ways in which devices can be interconnected to share information to improve human life. Further, they established that wireless networks are the most prevalent technologies in the 21st century. Nazir et al. [14] supported this notion as they surveyed wireless network security. They also identified those devices in the network that act as nodes. John et al. [15] argue that the distributed nature of the network makes it vulnerable to a Sybil attack, (see Fig. 3). According to Sassani Sarrafpour et al. [16], threat actors advertise several false identities or a single identity occurring in several locations, known as Sybil nodes, during the attack. When attackers introduce multiple fake peer identities to compromise information sharing in the network, this was called a Sybil attack as in Fig. 3. In addition, the attack may block legitimate users' access to a network because they can out-vote honest nodes and refuse to receive or transmit information. According to Almesaeed and Al-Salem [17], the attack can allow threat actors to control the computing power in the network, disrupting services. Tualy and Idoudi [18, 19] conducted a survey and established that the broadcast nature of vehicular networks makes them vulnerable to Sybil attacks.

The increased use of IoT devices in critical infrastructure has led to various strategies to prevent Sybil attacks. Almesaeed and Alsalem [17] stipulated that Sybil attack detection in wireless is the main security challenge since attackers can generate several identities. Khan et al. [20] identify cryptography as one of the countermeasures to the Sybil attack. They argued that cryptography could be used to verify the identity of nodes in the network, thereby eliminating counterfeit nodes. According to Khan et al. [20], this preventive measure enables the creation and management, distribution, and validation of keys for identities. Siddarth et al. [21] supported their argument by establishing that identity validation will prevent Sybil attacks because it reveals the true identity of malicious nodes. In their research, the nodes can be validated either directly or indirectly. Moreover, Siddarth et al. [21] and Swessi et al. [22] identify that the use of Artificial Intelligence (AI) in the cybersecurity field can eliminate Sybil attack risks. According to Swessi et al. [22], AI will help distinguish malicious nodes through detailed traffic analysis in the network. The paper contends that AI technology's ability to adapt to the current environment will make it easy to address the ever-changing threat landscape; therefore, the network can automatically respond to new attack vectors.

2.2 Denial of Service (DoS)

A DoS attack intends to bring down a network and make it inaccessible for intended customers [23]. In addition, it may also delay time-critical operations by preventing a machine, a server, or an entire network from responding to users' requests. As an IT professional, it is crucial to understand the history of this attack to guard against it.

Al-Hadhrami and Hussain [24] identify IoT as a rapidly growing technology in the industrial and consumer market. Despite this trend in recent years, the paper states that DoS attacks in multi-UAV networks have been a growing challenge and need to be addressed. Further, DoS attacks utilize the limited resources of IoT devices in the network, like network capacity and storage limitation, to influence the behavior of the entire network. While carrying out the attack, threat actors may use signal jammers to interfere with communication signals. According to Abughazaleh et al. [25], attackers can overwhelm the devices by sending unnecessary amounts of data to the IoT network, making it hard for the devices to distinguish between genuine and hoax communication signals. In addition, Verma et al. [26] argue that the attack may also be instigated to shut down the network making it inaccessible to other devices. According to Sinha and Sindhu [27], malicious actors can use a smurf attack, sending controlled message packets to several hosts with a spoofed IP address. Once the recipient machine responds, it will be flooded with traffic. The two researchers also identify that the attackers can use SYN flood, which occurs when attackers send requests to a target network but do not complete the connection in the three-way-handshake used by TCP to create a connection between clients and servers (see Fig. 4).

To achieve the success of multi-UAV networks, Gad et al. [28] establish that it is necessary to develop an advanced mechanism to ensure proper security levels to mitigate cyber threats in the network. Gad et al. [28], Geleano-Brajones et al. [29], and Ghali et al. [30] establish numerous strategies that can be used to protect the devices against DoS attacks. According to Gad et al. [28], individuals can prevent spoofing by checking that the network traffic has a source address consistent with the set of addresses of the available devices. The paper stipulates that artificial intelligence can automatically detect nodes using inconsistent addresses and block them from sending traffic to the network. Further, Hussain et al. [31] stipulate that using filters to destroy dial-up connections from spoofing will eliminate the risk. El Madani et al. [32] and Lyu et al. [33] establish that limiting the number of broadcasts will help mitigate the attack since attackers often send requests to multiple devices in the network, amplifying the attack. The study identifies that turning off broadcasting will disrupt the attack. According to Ghali et al. [30], individual devices on the network should be patched

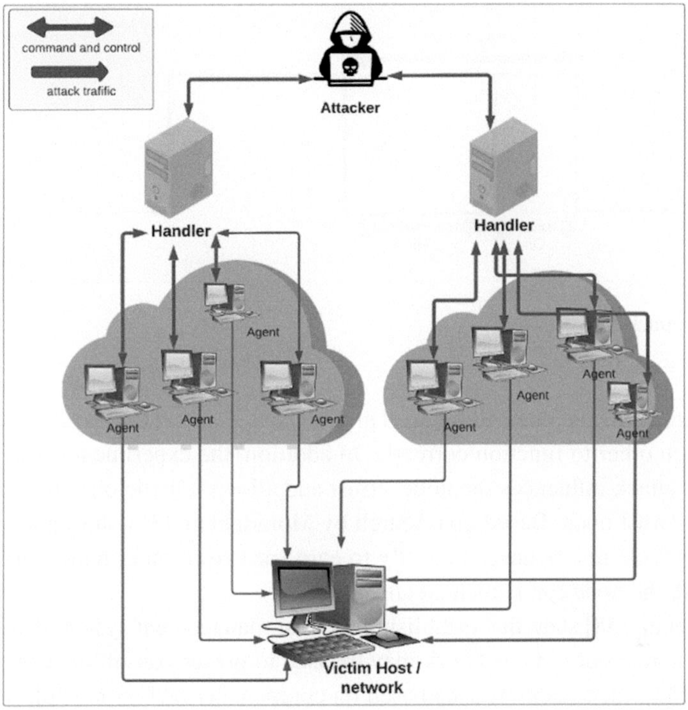

Fig. 4 Denial of service attack

to eliminate known vulnerabilities. The paper also identifies that frequent updates should be conducted to eliminate features that threat actors might use to bring down a network.

2.3 On-Off Attack

On-off attacks are major IoT trust security breaches in which malicious nodes perform random trustworthy and untrustworthy behaviors to avoid being recognized as threats, (see Fig. 5).

Nasution et al. [34] and Rehman et al. [35] conducted experiments and established that devices interact with each other in the IoT ecosystem, creating a perfect environment where heterogeneous nodes supply different services. Further, Ayed et al. [36] state that due to the interaction over wireless channels, the devices are vulnerable to malicious attacks such as on-off attacks. According to Nastion et al. [34], the on-off attack is one of IoT's trust threats. The paper states that during the attack, the malicious nodes in a network alternate between behaving well and badly, negatively affecting the network if the nodes are trusted. Therefore, the attacker operates as a sound object by sending valid trust values but then acts as a bad

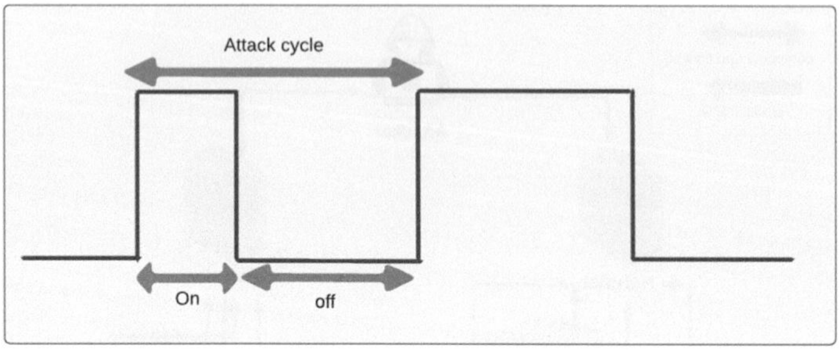

Fig. 5 On-off attack

object by sending invalid ones. Rehman et al. [35] state that in the ecosystem, various nodes must trust each other to function correctly. In addition, the experiment establishes that the sophisticated attack influences the node's trust and allows a malicious node to avoid being rated as a low-trust node. Based on research by Moradi et al. [37], during an on-off attack, the malicious node first behaves honestly to earn trust from other nodes, and once this is accomplished, the node can launch an attack.

Caminha et al. [38] state that establishing a trust management system that estimates the trust value among nodes in a network may be used to prevent on-off attacks. According to Sahoo et al. [39], the system may be based on components such as reward, severe penalty, sliding window, lenient penalty, and misbehavior frequency to identity malicious nodes. Suryani et al. [40] add that trust and reputation systems have been established as a powerful tool to defend against on-off attacks. The paper identifies that the penalty policy against the history of misbehavior of nodes in the network is a reliable factor that affects the calculation of trust values. Sahoo et al. [39] stipulate that the punishment feature helps target malicious codes that aim to launch intelligent attacks against a particular network. Zhou et al. [41] suggest that a smart trust management method can detect on-off attacks in IoT networks. However, the paper identifies that a malfunctioning node can be mismatched as an attacker based on trust values. Farea and Küçük [42] and Su et al. [43] believe that implementing artificial intelligence and an elastic slide window will eliminate false-positive cases. They identify that the AI will automatically assess nodes based on the attributed resource trust and service provider. Cananha et al. [38] and Suryani et al. [44] establish that frequent network scanning can be essential in detecting nodes installed by users in the network. In addition, the papers established that by using various scanner tools, individuals could update their device inventory to identify a malicious node before it causes harm.

2.4 Side-Channel Attack

Side-channel is a cryptographic attack in which the attacker does not compromise any data flaw between sender and receiver but steals the traffic. This attack is a severe issue for IoT devices [45].

According to Khan et al. [46], the advancement in computing technologies has presented individuals with ways of sharing massive amounts of data. Conti et al. [47] argue that the increasing reliance on data makes IoT devices vulnerable to side-channel attacks (see Fig. 6). The paper also states that the attack exploits information leakages in a network. Based on a study by Wiemers [48], leakages can result from electromagnetic signals, timing, sound, and light, among others. The paper further shows that the attack can be performed by observing, gathering, and analyzing the information leakages in the network during processing. Bandara and Kinsy [49] stipulate that threat actors use this attack to retrieve sensitive data from devices in the network; therefore, attackers can break into secure systems using a side-channel attack. Moreover, Jauvart et al. [50] argue that side-channel attacks target the implementation of standard cryptographic algorithms on devices to recover secret parameters by analyzing and measuring leaked information such as timing, electromagnetic, and power analysis. Fang et al. [51] add that the attacks can be categorized into (1) power attacks, which analyze power consumption in various node; (2) electromagnetic attacks, which exploit the electromagnetic emissions from multiple nodes during operations; (3) timing attacks, where the attacker guesses or extracts the sensitive information; and (4) cache attacks, which are performed during cache access.

Though it is challenging to eliminate side-channel attacks, they can be made unachievable. Several countermeasures have been implemented to make it hard for threat actors to perform these attacks. Fang et al. [51] establish that preventive measures against side-channel attacks fall into two categories: those that eliminate the release of leaked information and those that

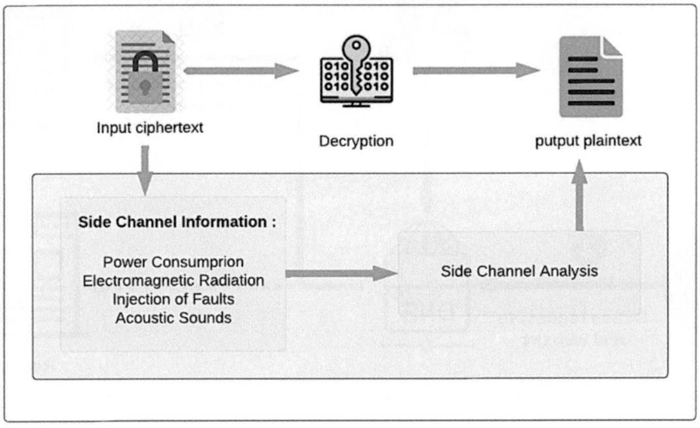

Fig. 6 Side-channel attack

eliminate the relationship between the leaked information and sensitive data. The paper stipulates that individuals can jam the emitted channel with noise in the first category. Also, Fang et al. [51] and Prates et al. [52] state that random delays can introduce to prevent timing attacks. In addition, Prates et al. [52] add that the introduction of noise causes the attackers to collect the wrong measurements. Moreover, Fang et al. [51] suggest that applying power line conditioning and filtering eliminates the risk of a network being susceptible to side-channel attacks. According to Abarzua et al. in [53], individuals can apply blinding techniques that alter the algorithm's input into an unpredictable form to eliminate information leakage. Further, Fang et al. [51] establish that the implementation of masking can eliminate the risk of an attack. The paper also adds that masking randomly splits sensitive intermediate variables into various shares. Bandara and Kinsy [49] establish that to protect the information, the device's operation should be independent of the data; the number of clock cycles should be randomized to eliminate operation timing leakages.

2.5 Spoofing Attack

In a spoofing attack, attackers pretend to be trustworthy to gain victims' trust (Fig. 7). Attackers generate fake traffic to appear sourced from a legitimate system. The method leverages various types of attacks, such as IP address spoofing, MAC address spoofing, and applications and services spoofing [54].

Abdulla et al. [55] and Sulatana et al. [56] stipulate that the potential vulnerabilities exploited wireless spoofing attacks are a growing concern in IoT, and may result in catastrophic events in critical infrastructures. Spoofing entails disguising an identity or commu-

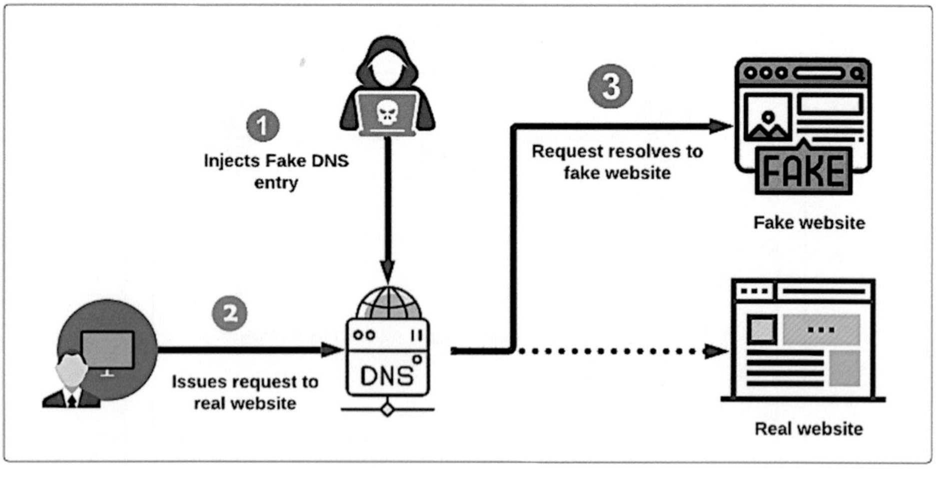

Fig. 7 Spoofing attack

nication to appear associated with a trusted node. Therefore, the attacks may take different forms, such as domain name system (DNS) server, address resolution protocol (ARP) services, or IP address spoofing in a particular IoT network. Nosouhi et al. [57] establish that the attack takes advantage of the trusted relationship by impersonating a node that the victim network knows. The paper adds that a successful spoofing attack may allow nodes in the network to share critical information with the impersonating node; therefore, attackers can harvest credentials to further their attack. Sulanta et al. [56] write that threat actors may launch a man-in-the-middle attack, a denial of service attack, or a ransomware attack. In an IP spoofing attack, threat actors send data packets from the compromised node to mask their identity. Also, Sulatana et al. [56] add that the attack vector can be used to launch a DoS attack by using the spoofed IP address to send data packets to multiple network nodes. The paper establishes that the owner of the actual IP is then flooded with responses, disrupting service. Jiang et al. [58] show that during the attack, a threat actor sends spoofed ARP messages across the network to link their MAC address to a legitimate IP address. The paper adds that successful execution will allow the attackers to steal or modify data meant for the actual IP address node. Therefore, threat actors can influence how the entire network functions by impersonating honest nodes in a network.

According to Rajashree et al. [59], because spoofing attacks can have disastrous results in a world that relies on sharing information, developing countermeasures is essential for achieving network security.Jiang et al. [58] stipulate that individuals can implement packet filtering with deep packet inspection. This approach will help analyze network IP data packets and block IPs with conflicting source information. Meghana et al. [60] add that since malicious data packets originate outside the network, it is a good strategy for eliminating spoofed IP addresses. Archana and Harini [61] establish that individuals can develop a better way of authenticating nodes in the network. The paper adds that using IP addresses for authentication increases vulnerability because IP spoofing can bypass authentication control. According to Petrovic et al. [62], the connection between nodes should be authenticated by the users or applications or through authenticity systems like IPsec and domain authentication. Mohammadnia and Slimane [63] suggested that spoofing detection software can help detect spoofing attacks, mainly ARP spoofing. Tools such as ARP monitor and NetCut can be used for this approach. Rajashree et al. [59] and Rohatgi and Goval [64] conducted a survey and established that implementing encrypted and authentication protocols to secure communication reduces the risk of spoofing attacks. Further, protocols such as transport layer security and IPSec can be used to mitigate the risks.

2.6 Message-Substitution Attack

A message substitution attack was invented by Bellare et al. [65]. This attack formalized substitution attacks to mass surveillance through symmetric encryption schemes.It attempts to replace one or more real algorithms with a subverted one to leak confidential information.

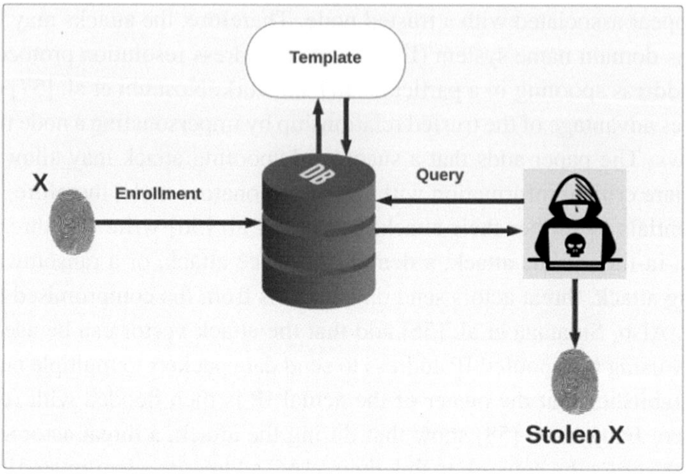

Fig. 8 Message substitution attack

Hodges and Stebila [66] state that communication is essential in an IoT network as the nodes keep sending data to one another or the server. Liu et al. [67] add that IoT networks face new threats due to the increased value of information being transmitted in the insecure channel. The most dangerous attack is the message substitution attack instigated by threat actors as shown in Fig. 8. Armour and Pottering [68] state that the attack aims to (1) replace an honest data packet in the IoT network; (2) implement a cryptographic tool with another message; (3) implement a subverted cryptographic tool that facilitates the leakage of private information being transported in the network while simultaneously generating an output that differs from the actual output. Berndt and Liśkiewicz [69] state that the attack interferes with the confidentiality of the messages since individuals without the necessary access rights can view the data, and the network lacks integrity since the information can be tampered with to produce a different output. A study by Chen et al. [70] shows that substitution attacks can affect message authentication schemes. One or more systems are replaced to reveal information to threat actors, which may remain undetected by users. The study adds that the attack may be a precursor to other attacks since the encryption or message has already been changed.

The severity of the attack requires comprehensive countermeasures to secure sensitive data in IoT networks. An et al. [71] and Jinhui et al. [67] establish one countermeasure involving industry-recommended encryption algorithms. The paper adds that encryption will make it hard for malicious actors to replace the messages or message authentication schemes. Bellare et al. [72] add that encryption will enforce confidentiality since unauthorized individuals will not access the messages. Alshemali and Kalita [73] suggest implementing intrusion detection software to help identify unauthorized individuals from accessing the network. In addition, intrusion prevention measures should be implemented to ensure the network can protect itself from similar attacks. Alshemali and Kalita [73] establish that the increased adoption of artificial intelligence can boost the automation of various processes to protect the network against message substitution attacks. According to Fischlin et al. [74], security experts can incorporate network scanning and analysis tools to establish messages that have been tampered with. This approach will be crucial in detecting data leakages within the network; therefore, it will be easy to identify the entry points of threat actors.

2.7 Message-Replay Attack

The information exchanged between sender and receiver might be useful to tempting crafty hackers to intercept it. In this attack, a malicious intruder intercepts traffic from a sender, then resends it to the intended receiver while acting as the authenticated sender [75].

Research by Chen et al. [76], and Feng et al. [77] establish an emerging trend of wireless communication technologies in IoT. The papers add that as the popularity of IoT networks grows, it is natural for security flaws to occur. One of the attacks that threat actors may use is a wireless replay attack, a low-level attack that reproduces a signal used to control some devices in the network (see Fig. 9). Na et al. [78] establish that this attack can be regarded as a man-in-the-middle attack where a malicious individual sniffs data packets or eavesdrops on data being sent in an IoT network between various nodes to intercept them and resend them under the disguise of authentic messages. According to Phimphinith et al. in [79] and Wara and Yu [80], a message replay attack is dangerous because threat actors do not have to decrypt the message that they resend and can still fool the receiving node into thinking that the message is legitimate. For example, attackers may intercept a login request to a particular resource originating from one of the nodes and then replay the message to

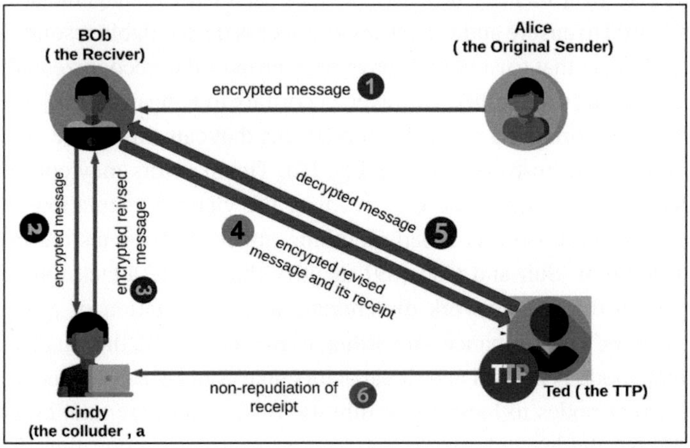

Fig. 9 Message replay attack

the intended resource, thus being granted the same access as the legitimate request. Verma and Ranga [81] add that attackers may also trick devices in the network to perform certain activities that can be precursors to a massive attack.

Agilandeeswari et al. [82] establish that replay attacks can lead to considerable damage since the integrity of the network is compromised. Therefore, various strategies have been developed to address this attack. Farha et al. [83] suggest that individuals can attach sequence numbers or timestamps to each message sent by nodes in the network. This strategy is efficient because it will allow message receivers to discard any message that bears a repeated sequence number or timestamps. According to Feng et al. [77], individuals can implement digital signatures to make it easy for receivers to authenticate that the sender is legitimate or a secure node in the network. The research also establishes that the attackers' nodes will not be able to authenticate themselves, thus blocking their actions in the network. Kurdi and Thayananthan [84] add that nodes in the network can be made to authenticate each other, and once the threat node is not authenticated, it can be blocked from sending messages. Ravanbakhsh and Nazari [85] present another countermeasure involving implementing random-session session keys to minimize replay attacks in an IoT network. The sessions are time-specific and change with time, making it hard for attackers to disguise their messages and fool receivers using an old message.

2.8 Bad-Mouthing Attack

In a bad-mouthing attack, the attackers collude to give unfavorable feedback on the targeted system, which lowers or destroys its reputation. This attack can downgrade the network's performance [86]. For example, attackers (network nodes) can provide negative information on other nodes that is false.

Bin-Yahya et al. [87] identify how IoT devices used trust management and stated that trust models are used to ensure authorized devices access the available resources. In addition, Suryani et al. [44] state that trust models have been proposed to recognize malicious objects to improve the network's reliability. However, according to Fang et al. [88], the trust models are vulnerable to bad-mouthing and collusion attacks that can exploit the system by giving deliberate false or unfair testimonies, (see Fig. 10). Threat actors may conspire to provide negative feedback on a victim to destroy or lower the victim's reputation in the network. Rehman et al. [35] establish that a bad-mouthing attack deteriorates the performance of a network. In addition, Suh and Cho [89] identify that in the attack, one or a group of malicious nodes in the IoT network disseminate a bad reputation of a legitimate node, affecting the network's performance. According to Roy et al. [90], the attack may also occur when a legitimate node is hacked and threat actors take control, sending negative reports or suggestions to other nodes to harm its credibility. Based on the reputation system, Reddy et al. [91] state that the hacked node is awarded a lower trust score; thus, it is eliminated from the network.

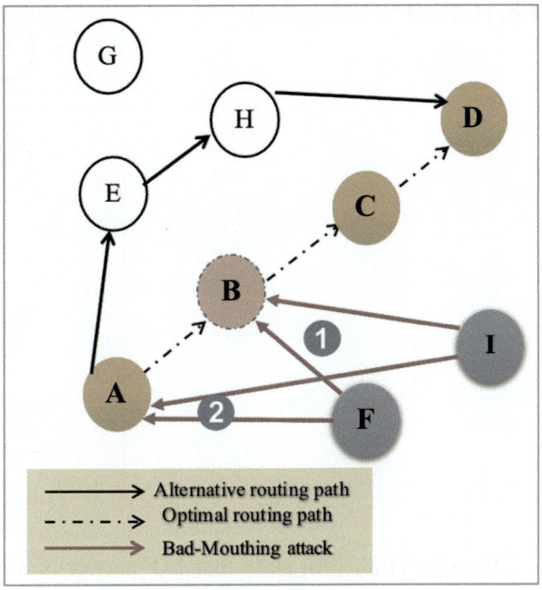

Fig. 10 Bad-Mouthing attack

To defend against insider attacks in IoT networks, trust systems implementing the idea of trust in human society may be implemented. Jiang and Liu [92] state that in the network, each node evaluates the trustworthiness of its adjacent nodes while eliminating untrustworthy nodes from the neighbor's list. Reddy et al. [91] identify that the trust system has developed into a reputation system to improve its functionality. The paper states that the reputation system evaluates a node's trustworthiness by considering adjacent nodes' evaluations. Roy et al. [90] suggest using digital signatures, which allow nodes in a network to verify their identity before evaluating adjacent nodes. According to Nangue et al. [93], digital signatures will enable nodes to filter out bad evaluations from compromised nodes. The study establishes that eliminating the bad reports will help maintain the network's integrity. Rehman et al. [35] demonstrate a prevention mechanism that implements mutual authentication algorithms, where nodes in the network can verify one anothers' identity before evaluating neighbors. According to Su et al. [94], the verification process will ensure legitimate nodes are considered in the reputation system, thus, generating a network of trustworthy devices.

2.9 Good-Mouthing Attack

As opposed to a bad-mouthing attack, one of the network nodes performs a good-mouthing attack by providing more trust information for malicious nodes to trick other nodes intentionally [95].

Fig. 11 Good-mouthing attack

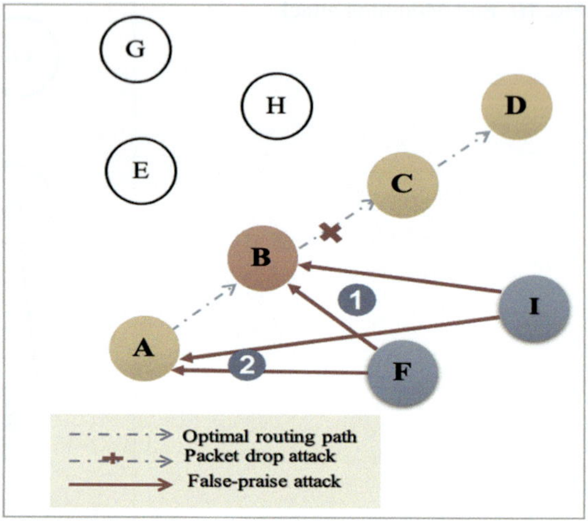

Alnasser et al. [96] state that IoT networks' trust management strategy addresses key management, safe routing, and authentication mechanisms. The paper further identifies that the system developed an efficient trust computation approach for IoT devices acting as service requesters. However, Dhaknel et al. [97] state that the trust management strategy can be vulnerable to attacks like good mouthing since the malicious nodes can provide good services by offering good advice to maintain their integrity and simultaneously compromise the network, (see Fig. 11). In addition, malicious nodes can cooperate to increase the severity of the attack. According to Kandhoul et al. [98], the nodes collude to manipulate the reputation of another malicious node by providing a good recommendation for it. Therefore, malicious nodes in the IoT attempt to exploit the trust management system to influence nodes in the network to rank another malicious node highly. According to Zaman and Sultana [99], the colluding malicious nodes provide positive testimonies to control the trust management system. The paper also identifies that nodes in the network evaluate their immediate neighbors to calculate their trust values. In addition, the nodes also use reviews from other nodes to obtain the trust value. Based on research by Zupancic et al. [100], if numerous malicious nodes collude to influence another malicious node review, it affects how legitimate nodes trust other nodes in the network. Thus, good mouthing entices honest nodes to consider malicious nodes as legitimate nodes. This attack influences the integrity of the IoT network, making it susceptible to further cyberattacks.

Similar to bad-mouthing attacks, some countermeasures apply to good-mouthing attacks. Valero et al. [101] identify that a mutual authentication strategy will be crucial in eliminating the influence malicious nodes will have on trust among nodes in the network. Wang et al. [102] add that nodes in the network can authenticate one another, preventing collusion from threat actors and verifying one another's identities before considering evaluation results. Sun and Zhao [103] identify the second countermeasure that involves implementing a trustworthy management system, where nodes evaluate the trustworthiness of adjacent nodes. The paper states that the reputation management approach can be used to improve the functionality of the network by considering the evaluation results from neighboring nodes. The third strategy to prevent good-mouthing attacks on IoT networks involves the digital signatures that will enable nodes in the network to authenticate themselves before evaluating adjacent devices. Farea and Küçük [42] identify that malicious nodes will be eliminated from making evaluations and sharing them with the other nodes, maintaining network integrity. Finally, the exponential growth of technology may introduce new attack vectors that instigate good-mouthing attacks. Therefore, Hassan [104] suggests conducting more security updates on various nodes on frequent bases to eliminate vulnerabilities that might be exploited by threat actors, compromising the integrity of IoT networks.

2.10 Ballot Stuffing Attack

Ballot stuffing happens when fake ballots (not cast by any eligible voter) are inserted into systems [105].

Nodes in an IoT network form a self-organizing network that eliminates the reliance on fixed infrastructure. Anita and Vijayalakshmi [106], and Ugur [107] state that security is essential and trust computation is implemented to improve collaboration between nodes. IoT networks may be vulnerable to reputation influence from malicious nodes. According to Jabeen et al. [108], legitimate nodes may also be influenced to act selfishly. Further, the study identifies that the presence of selfish nodes in a network enhances the reputation of their selfish peers, which may cause data loss. Khedim et al. [109] establish that selfish nodes collude in a ballot stuffing attack, as shown in Fig. 12, and propagate falsified ratings

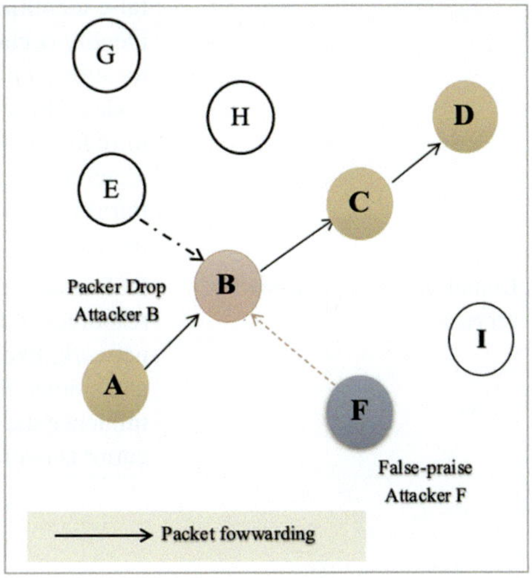

Fig. 12 Ballot-Stuffing attack

to genuine nodes. Further, the ratings interfere with how legitimate nodes interact and may block other nodes from transmitting or requesting services in the network. According to Hu et al. [110], once the network is compromised, malicious nodes can cast one more vote than the allowable, thus affecting the integrity of the network. Therefore, verifying the identity of the malicious nodes may be difficult. Hu et al. [110] state that since trust models rely on recommendations from available nodes in the network, ballot stuffing influences how devices relate to one another.

According to Jim and Gregory [111], the severity of the attack necessitates the development of comprehensive countermeasures to protect the increasing use of IoT devices. Rehman et al. [35] established that to mitigate ballot-stuffing attacks, individuals develop a trust evaluation strategy that allows trust evaluation in the network. They further argue that using low-computation calculations to calculate trust scores makes it suitable for application in most IoT networks. According to Shabut et al. [112], a non-biased trust model can be implemented, which uses the dissimilarity test. Tan et al. [113] and Suryani et al. [44] add that since the attack is trust-based, the trust evaluation method carefully considers the root cause of the attacks. Jim and Gregory [111] propose using digital signatures to authenticate individual nodes and eliminate malicious nodes in the network. Therefore, infected nodes cannot participate in decision-making. Further, Hu et al. [110] stated that collaboration between nodes should be encouraged through mutual authentication to verify one another's identities. This strategy will eliminate malicious nodes since they cannot authenticate themselves. Table 1 contains summary of attacks description, categories, and their countermeasures.

Table 1 A summary of attacks mentioned in this chapter

Attack	Category	Description	Countermeasures
Sybil attack	Passive	Threat actors advertise several false identities or a single identity occurring in several locations, known as Sybil nodes. The attackers can introduce multiple fake peer identities to compromise information sharing in the network	• Cryptography • Using artificial intelligence
Denial of service	Passive	DoS attacks utilize the limited resources of IoT devices in the network, like network capacity and storage limitation, to influence the behavior of the entire network	• Checking network traffic • Using filters to destroy dial-up connections • Limiting the number of broadcasts devices on the network should be patched

(continued)

Table 1 (continued)

Attack	Category	Description	Countermeasures
On-off attack	Active	The on-off attack is one of the IoT's trust threats. During the attack, the malicious nodes in a network alternate between behaving well and badly, negatively affecting the network if the nodes are trusted	• Establishing a trust management system • Using artificial intelligence • Frequent network scanning
Side channel attack	Passive	This attack exploits information leakages in a network. The leakages can result from electromagnetic signals, timing, sound, and light, among others	• Random delays introduction • Separating data from operations • Applying Blinding Techniques • Masking
Spoofing attack	Passive	Spoofing is the act of disguising an identity or communication to appear associated with a trusted node. Therefore, the attacks may take different forms, such as DNS server, ARP services, or IP address spoofing in a particular IoT network	• Node authentication • Packet filtering with deep packet inspection • Artificial intelligence • Intrusion detection software • Network scanning and analysis tools
Message substitution attack	Active	The attack aims to replace an honest data packet in the IoT network or implement a cryptographic tool with another message or a subverted cryptographic tool that facilitates the leakage of private information being transported in the network while simultaneously generating an output that differs from the actual output	• Encryption algorithms • Intrusion detection software • Artificial intelligence • Network scanning and analysis tools

(continued)

Table 1 (continued)

Attack	Category	Description	Countermeasures
Message Replay Attack	Passive	This attack can be regarded as a type of man-in-the-middle attack, where a malicious individual sniffs data packets or eavesdrops on data being sent in an IoT network between various nodes to intercept them and resend them disguised as authentic messages	• Sequence numbers or timestamps • Digital signatures • Random session keys
Bad-mouthing attack	Passive	In this type of attack, one or a group of malicious nodes in the IoT network disseminate a bad reputation of a legitimate node, affecting the network's performance	• Trust systems • Digital signatures • Mutual authentication algorithms
Good-mouthing attack	Passive	The nodes collude to manipulate the reputation of another malicious node by providing a good recommendation for it	• Trustworthy management system • Digital signatures • Security updates
Ballot-stuffing attack	Active	Legitimate nodes may also be influenced to act selfishly. The presence of selfish nodes in a network enhances the reputation of their selfish peers, which may cause data loss. Therefore, selfish nodes collude in a ballot-stuffing attack and propagate falsified ratings to genuine nodes	• Trust evaluation strategy • Non-biased trust models • Digital signatures

3 Challenges Facing Adopting the Previous Countermeasures in Multi-UAV (IoT) Network

This section summarizes the challenges facing adopting the aforementioned countermeasures in multi-UAV (IoT) networks. We listed the challenges based on our views after observing up-to-date solutions as shown in Table 2.

Table 2 A summary of challenges in adopting the previous countermeasures in a multi-UAV (IoT) network

Attack	Challenges
Sybil attack	Multi-UAV (IoT) network suffers from energy constraints since it required more processing power to identify the counterfeit notes [114]. In addition, artificial intelligence was not well developed enough to facilitate countermeasures and establishing artificial barriers was not cost-effective
Denial of service	Energy constraints such as analyzing network traffic were time-consuming and required more resources such as computer memory. Limiting the number of broadcasters affects the data transfer speed of the network [115]. Firewalls must be updated often to eliminate the constantly evolving threat in the sector
On-off attack	Energy constraints resulted in limited computation power to establish a comprehensive trust management system
Side channel	Random delays that were introduced created time delays [94]. Masking introduces energy constraints because data is masked to prevent unauthorized access
Spoofing	Storage and energy constraints since all data packets must be inspected to block IP addresses with conflicting source information. Technological advancement presets a new way for data packets to evade packet tracers [116]. Financial constraints since some countermeasures required premium software such as NetCut
Message substitution	There were energy constraints due to the increased encryption. In addition, there were financial constraints since it requires additional applications
Message replay	Energy constraints require more processing power to allocate sequence numbers and timestamps. In addition, using digital signatures increases the need for high processing power [117]
Bad-mouthing	Energy and time constraints due to the implementation of reputation systems. The verification steps consumes more time
Good-mouthing	Energy and time constraints due to the implementation of trust management systems and mutual authentication system
Ballot-stuffing	Time constraints due to the increased complexity in authenticating nodes in the network [118]

4 Multi-UAV Systems Challenges

Future proposed security countermeasures should be built while taking into account that Multi-UAV systems faced specific kind of challenges. These challenges include communication and communication infrastructure, UAVs operation, UAV traffic management, energy constraints, storage constraints and time completion issues.

4.1 Communication and Communication Infrastructure

Communication is a significant challenge for multi-UAVs swarms. Communication difficulties fall into two categories: assuring specific properties (for example maintaining a fully linked network) and avoiding bottlenecks [119]. One of the essential basic functions of a swarm is communication. Multi-UAVs is frequently offered as enablers Because of their good mobility and increased speed implementation, aerial wireless transmission frameworks are being considered as a supplement to or replacement for established communication infrastructure [120]. UAVs assume an unstable or degraded communication network. This could be due to the environment's unstable or degraded communication networks (indoors or in so-called urban canyons) or deteriorated infrastructure. The civilian drone market is growing, posing risks to people's, property, and privacy rights. Data communication must be secured using secure communication methods [121]. Communication is one of the most difficult design challenges for multi-UAV systems. Ad hoc networks between UAVs are classified as a FANET, separate network, and other ad hoc network types based on mobility, computational power, localization, and topology change.

4.2 UAV Operation

Multi-UAVs operations present several issues, such as testing and mimicking swarms. Multi-UAVs operates under different operational conditions than conventional piloted aircraft [7]. Concerns have also been raised about how these devices work and the use of wearable sensors. The environment in which the device is meant to operate will influence the choice of device and hardware [122]. Multi-UAVs in uncertain settings necessitate quick and adaptive path planning. The more complex an application, the more specific the environment requirements may become. Multi-UAV is increasingly capable of contributing to problem-solving and data processing activities because of their onboard processing power [123]. This advancement also makes them potentially more autonomous, as algorithms like alight path computations can be carried out by the devices themselves [124].

The challenge with software is deciding whether to use off-the-shelf software or invest in designing and implementing custom applications. The literature is brimming with unique ideas, techniques, and implementations for solving a wide range of UAV challenges. The state of Multi-UAVs and their integration into the National Airspace System (NAS) has not been approved by the appropriate authorities. The ability of Multi-UAVs to adhere to the Federal Aviation Regulations (FARs) right of way regulations remains their greatest challenge. These conditions apply when airplanes are flying in adverse weather, and they include a 'see and avoid' clause. This means that it is on to the pilot to grant his or her other pilots the right of way [125]. Although there is a lot of active research to help overcome this problem, the' see and avoid' requirement remains a huge difficulty for UAVs. According to the FAR's right-of-way rule, "Where weather conditions permit, regardless of whether an operation is carried out and under aircraft or visual flight rules, each operating an aircraft shall ensure alertness to see and avoid other aircraft" [126]. Because most UAVs use sensors, this rule provides a significant issue. Sensors struggle to match a human's 'see and avoid' capabilities when it comes to observing contradicting air traffic and ground obstructions. The powers of a real human person have yet to be matched by current technology [127]. There is continuing research to find a technical remedy to this issue. Currently, sensor payloads are used to compensate for the' see and avoid' limitation. Despite the absence of a pilot on board, "a sensor payload detects and prevents collisions during air operations" [128]. Nonetheless, the Federal Aviation Administration (FAA) has not yet certified any sense-and-avoid technology capable of replacing human presence or the employment of chase planes.

4.3 UAVs Traffic Management

Multi-rotor UAVs can fly for 15–20 min before needing to recharge. We suggest an off-line routing algorithm for UAC swarms tasked with providing uninterrupted monitoring coverage over a region. Heuristics are required to tackle such difficulties to improve dispatch and scheduling efficiency. A city-wide traffic management system must be developed to accommodate the growing traffic sharing the same airways as UAVs in metropolitan settings [129]. A cloud-based, independent traffic control system with sensing devices to keep the city safe and collision-avoidance control systems should be addressed to solve the challenge.

4.4 Energy Constraints

Multi-UAVs are progressively being used for surveillance tasks like environmental monitoring and disaster risk management [130]. This paper introduces a surveillance problem with limited energy and soft priorities using UAVs. This surveillance problem is a generalized form of implementation cases ranging from international event surveillance to crime scene or disaster site monitoring [130]. An insertion heuristic combined with an effective energy negotiation method can plan routes for real-world scenarios that are currently impossible to solve by general potential solutions. The technique can be used in changing environments where re-planning is needed during the mission. UAVs present fundamental challenges for persistence due to wireless communication and limited battery capacity [130]. The asynchronous route optimization algorithm ensures that all UAVs can achieve the base station at all times. A multi-hop or single link connects each UAV to a base station. Multi-access edge computing (MEC) has emerged as a promising solution for energy devices. UAVs provide mobile devices with computing and relay services.

4.5 Storage Constraints

Multi-UAVs are used to obtain sensed data from the sensors in sensor networks (SNs). The SNs' interaction ranges may overlap in a dense network, and the UAVs do not need to fly over them one by one. However, completing the data collection task with a single UAV with limited energy storage while ensuring information freshness is extremely difficult. Before the widespread adoption of multi-UAV-based heterogeneous flying ad hoc networks, numerous challenges must be overcome [131]. The patriarchal UAV network's stability control ensures efficient drone collaboration. Some points can be identified as data storage and collection points (CPs), where UAVs can collect data from multiple SNs simultaneously [131]. The problem of multi-UAV-assisted data collection can be divided into three parts. The graph theory underpins the SN-CP association between the CPs and their nearest neighbors. The optimal flight trajectory of every UAV is determined to reduce the SNs' maximum age of information under the constraints of limited energy capacity [131].

4.6 Time Completion Issues

Path planning is a critical component of multi-UAVs intelligent control technology. The current A* algorithm has a significant memory overhead during strategic planning and a slow search speed. In a sophisticated three-dimensional environment, it cannot meet the path planning requirements. A better algorithm can generate feasible paths and maximize the strategy to the optimal solution in time [131]. Multiple rotary-wing UAVs are used in a monitoring scenario to obtain data from sensors with relatively homogeneous Poisson Point Process distributions [131]. Sensor nodes have traditionally used multi-hop data transmission to send monitored data to a base station. As a result, a sensor should send data from other sensors and retransmit data from other sensors [131]. As a result, the sensor's battery may discharge quickly, and the multi-hop communication link may be lost. Sensor nodes can send collected data directly to UAVs as cellular data collectors.

4.7 So-Called Legal Challenges

UAVs uses and availability in the wild are growing much faster than awareness programs of potential issues, as well as relevant legislation to address these issues Different countries will impose various restrictions on the use of UAVs in the foreseeable future, and such policies may be revised and altered regularly [132]. UAV use in public airspace entails a number several societal challenges and opportunities There are no guarantees when it comes to UAV legal rules because their widespread adoption outpaces governmental, social, and legal capacity to deal with problems like privacy and conflict with promotional airspace [133].

5 Conclusion

Multi-UAVs systems consist of a collection of UAVs that fly autonomously in free space to accomplish the mission. Several applications and uses of Multi-UAVs have been developed in modern life over the last few years, resulting in huge demand for them. The applications range from surveillance and reconnaissance for security objectives, monitoring, and inspection of activities taking place in various locations, to damaging usage by terrorists to steal secret information and cause destruction. Multi-UAVs communication networks face security and privacy challenges that impede Multi-UAV performance or may have catastrophic consequences for people and property.

This chapter has explained potential security attacks facing Multi-UAV communication networks and their corresponding countermeasures. Attacks include Sybil attack, denial of service, on-off attack, side channel, spoofing, message substitution, message replay, bad mouthing, good mouthing, and ballot stuffing attacks. Also, the chapter clarified the difficulties encountered by the countermeasures. Finally, the challenges posed by Multi-UAV system are described in detail. These challenges must be taken into account when designing future security solutions. Multi-UAVs communication network security is a very challenging topic, and future research is needed to alleviate the risk. Future proposed security countermeasures should be built while taking into account that Multi-UAVs systems faced specific kinds of challenges as it has been described in this chapter. These challenges include communication and communication infrastructure, UAV operation, UAV traffic management, energy constraints, storage Constraints, and time completion issues.

References

1. B. Li, Z. Fei, Y. Zhang, M. Guizani, Secure uav communication networks over 5g. IEEE Wirel. Commun. **26**(5), 114–120 (2019)
2. A. Rovira-Sugranes, A. Razi, F. Afghah, J. Chakareski, A review of ai-enabled routing protocols for uav networks: Trends, challenges, and future outlook. Ad Hoc Netw. **130**, 102790 (2022)
3. G. Skorobogatov, C. Barrado, E. Salamí, Multiple uav systems: a survey. Unmanned Syst. **8**(02), 149–169 (2020)
4. S.A.H. Mohsan, M.A. Khan, F. Noor, I. Ullah, M.H. Alsharif, Towards the unmanned aerial vehicles (uavs): a comprehensive review. Drones **6**(6), 147 (2022)
5. H. Shakhatreh, A.H. Sawalmeh, A. Al-Fuqaha, Z. Dou, E. Almaita, I. Khalil, N.S. Othman, A. Khreishah, M. Guizani, Unmanned aerial vehicles (uavs): a survey on civil applications and key research challenges. IEEE Access **7**, 48572–48634 (2019)
6. J. Li, Y. Zhou, L. Lamont, Communication architectures and protocols for networking unmanned aerial vehicles, in *2013 IEEE Globecom Workshops (GC Wkshps)* (IEEE, 2013), pp. 1415–1420
7. J. Zhang, J. Xing, Cooperative task assignment of multi-uav system. Chin. J. Aeronaut. **33**(11), 2825–2827 (2020). https://doi.org/10.1016/j.cja.2020.02.009. (SI: Emerging Technologies of Unmanned Aerial Vehicles)

8. X. Zheng, F. Wang, Z. Li, A multi-uav cooperative route planning methodology for 3d fine-resolution building model reconstruction. ISPRS J. Photogramm. Remote. Sens. **146**, 483–494 (2018)
9. A. Tahir, J. Böling, M.-H. Haghbayan, H.T. Toivonen, J. Plosila, Swarms of unmanned aerial vehicles-a survey. J. Ind. Inf. Integr. **16**, 100106 (2019)
10. J. Zhao, F. Gao, G. Ding, T. Zhang, W. Jia, A. Nallanathan, Integrating communications and control for uav systems: opportunities and challenges. IEEE Access **6**, 67519–67527 (2018)
11. R. Shrestha, I. Oh, S. Kim, A survey on operation concept, advancements, and challenging issues of urban air traffic management. Front. Future Transp. **2**, 626935 (2021)
12. A. Arshad, Z.M. Hanapi, S. Subramaniam, R. Latip, A survey of sybil attack countermeasures in iot-based wireless sensor networks. PeerJ Comput. Sci. **7**, 673 (2021)
13. S. Benadla, O.R. Merad-Boudia, The impact of sybil attacks on vehicular fog networks, in *2021 International Conference on Recent Advances in Mathematics and Informatics (ICRAMI)* (IEEE, 2021), pp. 1–6
14. R. Nazir, K. Kumar, S. David, M. Ali, et al., Survey on wireless network security, in *Archives of Computational Methods in Engineering* (2021), pp. 1–20
15. R. John, J.P. Cherian, J.J. Kizhakkethottam, A survey of techniques to prevent sybil attacks, in *2015 International Conference on Soft-Computing and Networks Security (ICSNS)* (2015), pp. 1–6. https://doi.org/10.1109/ICSNS.2015.7292385
16. B.A. Sassani Sarrafpour, A. Alomirah, S. Pang, S. Sarrafpour, Coding observer nodes for sybil attacks detection in mobile wireless sensor networks, in *2021 IEEE 19th International Conference on Embedded and Ubiquitous Computing (EUC)* (2021), pp. 87–94. https://doi.org/10.1109/EUC53437.2021.00021
17. A. Reham, A.-S. Eman, Sybil attack detection scheme based on channel profile and power regulations in wireless sensor networks. Wirel. Netw. **28**, 1–20 (2022)
18. H.B. Tulay, C.E. Koksal, Robust sybil attack detection in vehicular networks, in *2021 IEEE 94th Vehicular Technology Conference (VTC2021-Fall)* (2021), pp. 1–7. https://doi.org/10.1109/VTC2021-Fall52928.2021.9625060
19. D. Swessi, H. Idoudi, A survey on internet-of-things security: Threats and emerging countermeasures. Wirel. Pers. Commun. **124**, 1557–1592 (2022)
20. A. Khan, M. Ahmad, S. Jedran, Authorization schemes for internet of things: requirements, weaknesses, future challenges and trends. Complex Intell. Syst. (2022)
21. D. Siddarth, S. Ivliev, S. Siri, P. Berman, Who watches the watchmen? A review of subjective approaches for sybil-resistance in proof of personhood protocols. Front. Blockchain **3**, 590171 (2020)
22. D. Swessi, H. Idoudi, A survey on internet-of-things security: threats and emerging countermeasures. Wirel. Pers. Commun. **124**(2), 1557–1592 (2022)
23. G. Carl, G. Kesidis, R.R. Brooks, S. Rai, Denial-of-service attack-detection techniques. IEEE Internet Comput. **10**(1), 82–89 (2006). https://doi.org/10.1109/MIC.2006.5
24. Y. Al-Hadhrami, F.K. Hussain, Ddos attacks in iot networks: a comprehensive systematic literature review. World Wide Web **24**(3), 971–1001 (2021)
25. N. Abughazaleh, R. Bin, M. Btish, Dos attacks in iot systems and proposed solutions. Int. J. Comput. Appl. **176**(33), 16–19 (2020)
26. A. Verma, R. Saha, N. Kumar, G. Kumar, et al., A detailed survey of denial of service for iot and multimedia systems: past, present and futuristic development, in *Multimedia Tools and Applications* (2022), pp. 1–66
27. S. Sinha, et al., Impact of dos attack in iot system and identifying the attacker location for interference attacks, in *2021 6th International Conference on Communication and Electronics Systems (ICCES)* (IEEE, 2021), pp. 657–662

28. A.R. Gad, A.A. Nashat, T.M. Barkat, Intrusion detection system using machine learning for vehicular ad hoc networks based on ton-iot dataset. IEEE Access **9**, 142206–142217 (2021)
29. J. Galeano-Brajones, J. Carmona-Murillo, J.F. Valenzuela-Valdés, F. Luna-Valero, Detection and mitigation of dos and ddos attacks in iot-based stateful sdn: an experimental approach. Sensors **20**(3), 816 (2020)
30. A.A. Ghali, R. Ahmad, H. Alhussian, A framework for mitigating ddos and dos attacks in iot environment using hybrid approach. Electronics **10**(11), 1282 (2021)
31. F. Hussain, S.G. Abbas, M. Husnain, U.U. Fayyaz, F. Shahzad, G.A. Shah, Iot dos and ddos attack detection using resnet, in *2020 IEEE 23rd International Multitopic Conference (INMIC)* (IEEE, 2020), pp. 1–6
32. S. El Madani, S. Motahhir, A. El Ghzizal, Internet of vehicles: concept, process, security aspects and solutions, in *Multimedia Tools and Applications* (2022), pp. 1–25
33. C. Lyu, X. Zhang, Z. Liu, C.-H. Chi, Selective authentication based geographic opportunistic routing in wireless sensor networks for internet of things against dos attacks. IEEE Access **7**, 31068–31082 (2019)
34. A.P. Nasution, V. Suryani, A.A. Wardana, Iot object security towards on-off attack using trustworthiness management, in *2020 8th International Conference on Information and Communication Technology (ICoICT)* (IEEE, 2020), pp. 1–6
35. U.U. Rehman, A. Ali, H.S.M. Bilal, M.A. Razzaq, S.-B. Park, S. Lee, A novel mutual trust evaluation method for identification of trusted devices in smart environment, in *2022 16th International Conference on Ubiquitous Information Management and Communication (IMCOM)* (IEEE, 2022), pp. 1–4
36. S. Ayed, L. Chaari, A. Fares, A survey on trust management for wban: investigations and future directions. Sensors **20**(21), 6041 (2020)
37. F. Moradi, A. Sedaghatbaf, S.A. Asadollah, A. Čaušević, M. Sirjani, On-off attack on a blockchain-based iot system, in *2019 24th IEEE International Conference on Emerging Technologies and Factory Automation (ETFA)* (IEEE, 2019), pp. 1768–1773
38. J. Caminha, A. Perkusich, M. Perkusich, A smart middleware to detect on-off trust attacks in the internet of things, in *2018 IEEE International Conference on Consumer Electronics (ICCE)* (IEEE, 2018), pp. 1–2
39. R.R. Sahoo, S. Sarkar, S. Ray, Defense against on-off attack in trust establishment scheme for wireless sensor network, in *2019 2nd International Conference on Signal Processing and Communication (ICSPC)* (IEEE, 2019), pp. 153–160
40. V. Suryani, S. Sulistyo, et al., The detection of on-off attacks for the internet of things objects, in *2018 International Conference on Control, Electronics, Renewable Energy and Communications (ICCEREC)* (IEEE, 2018), pp. 1–5
41. N. Zhou, W. Fang, W. Zhang, X. Lv, J. Huang, A novel trust management scheme for defending against on-off attack based on gaussian distribution, in *2018 IEEE International Conference on Internet of Things (iThings) and IEEE Green Computing and Communications (GreenCom) and IEEE Cyber, Physical and Social Computing (CPSCom) and IEEE Smart Data (SmartData)* (IEEE, 2018), pp. 1806–1811
42. A.H. Farea, K. Küçük, Detections of iot attacks via machine learning-based approaches with cooja. EAI Endorsed Trans. Internet of Things **7**(28), 1–1 (2021)
43. J. Su, S. He, Y. Wu, Features selection and prediction for iot attacks. High-Confid. Comput. **2**(2), 100047 (2022)
44. V. Suryani, S. Sulistyo, W. Widyawan, Simulation of trust-based attacks in internet of things, in *MATEC Web of Conferences*, vol. 154 (EDP Sciences, 2018), p. 03014
45. M. Devi, A. Majumder, Side-channel attack in internet of things: a survey, in *Applications of Internet of Things* (Springer, 2021), pp. 213–222

46. M. Khan, Y. Chen, A randomized switched-mode voltage regulation system for iot edge devices to defend against power analysis based side channel attacks, in *2021 IEEE Intl Conf on Parallel & Distributed Processing with Applications, Big Data & Cloud Computing, Sustainable Computing & Communications, Social Computing & Networking (ISPA/BDCloud/SocialCom/SustainCom)* (IEEE, 2021), pp. 1771–1776
47. M. Conti, E. Losiouk, R. Poovendran, R. Spolaor, *Side-Channel Attacks on Mobile and IoT Devices for Cyber-Physical Systems* (Elsevier, 2022)
48. A. Wiemers, A remark on a success rate model for side-channel attack analysis. J. Cryptogr. Eng. **10**(3), 269–274 (2020)
49. S. Bandara, M.A. Kinsy, Adaptive caches as a defense mechanism against cache side-channel attacks. J. Cryptogr. Eng. **11**(3), 239–255 (2021)
50. D. Jauvart, N. El Mrabet, J.J. Fournier, L. Goubin, Improving side-channel attacks against pairing-based cryptography. J. Cryptogr. Eng. **10**(1), 1–16 (2020)
51. Q. Fang, L. Lin, Y. Zu Wong, H. Zhang, M. Alioto, Side-channel attack counteraction via machine learning-targeted power compensation for post-silicon hw security patching, in *2022 IEEE International Solid-State Circuits Conference (ISSCC)*, vol. 65 (IEEE, 2022), pp. 1–3
52. N. Prates, A. Vergütz, R.I. Macedo, A. Santos, M. Nogueira, A defense mechanism for timing-based side-channel attacks on iot traffic, in *GLOBECOM 2020-2020 IEEE Global Communications Conference* (IEEE, 2020), pp. 1–6
53. R. Abarzúa, C. Valencia, J. López, Survey on performance and security problems of countermeasures for passive side-channel attacks on ecc. J. Cryptogr. Eng. **11**(1), 71–102 (2021)
54. Z. Wu, T. Kinnunen, E.S. Chng, H. Li, E. Ambikairajah, A study on spoofing attack in state-of-the-art speaker verification: the telephone speech case, in *Proceedings of the 2012 Asia Pacific Signal and Information Processing Association Annual Summit and Conference* (IEEE, 2012), pp. 1–5
55. H. Abdulla, H. Al-Raweshidy, W.S. Awad, Arp spoofing detection for iot networks using neural networks, in *Proceedings of the Industrial Revolution & Business Management: 11th Annual PwR Doctoral Symposium (PWRDS)* (2020)
56. S. Sultana, S. Nasrin, F.K. Lipi, M.A. Hossain, Z. Sultana, F. Jannat, Detecting and preventing ip spoofing and local area network denial (land) attack for cloud computing with the modification of hop count filtering (hcf) mechanism, in *2019 International Conference on Computer, Communication, Chemical, Materials and Electronic Engineering (IC4ME2)* (IEEE, 2019), pp. 1–6
57. M.R. Nosouhi, K. Sood, M. Grobler, R. Doss, Towards spoofing resistant next generation iot networks. IEEE Trans. Inf. Forensics Secur. **17**, 1669–1683 (2022)
58. Z. Jiang, K. Zhao, R. Li, J. Zhao, J. Du, Phyalert: identity spoofing attack detection and prevention for a wireless edge network. J. Cloud Comput. **9**(1), 1–13 (2020)
59. S. Rajashree, K. Soman, P.G. Shah, Security with ip address assignment and spoofing for smart iot devices, in *2018 International Conference on Advances in Computing, Communications and Informatics (ICACCI)* (IEEE, 2018), pp. 1914–1918
60. J.S. Meghana, T. Subashri, K. Vimal, A survey on arp cache poisoning and techniques for detection and mitigation, in *2017 Fourth International Conference on Signal Processing, Communication and Networking (ICSCN)* (IEEE, 2017), pp. 1–6
61. K. Archana, N. Harini, Mitigation of spoofing attacks on iot home networks. Int. J. Eng. Adv. Technol. **9**(1S), 240–245 (2019)
62. R. Petrović, D. Simić, S. Stanković, M. Perić, Man-in-the-middle attack based on arp spoofing in iot educational platform, in *2021 15th International Conference on Advanced Technologies, Systems and Services in Telecommunications (TELSIKS)* (IEEE, 2021), pp. 307–310

63. H. Mohammadnia, S.B. Slimane, Iot-netz: practical spoofing attack mitigation approach in sdwn network, in *2020 Seventh International Conference on Software Defined Systems (SDS)* (IEEE, 2020), pp. 5–13
64. V. Rohatgi, S. Goyal, A detailed survey for detection and mitigation techniques against arp spoofing, in *2020 Fourth International Conference on I-SMAC (IoT in Social, Mobile, Analytics and Cloud)(I-SMAC)* (IEEE, 2020), pp. 352–356
65. M. Bellare, K.G. Paterson, P. Rogaway, Security of symmetric encryption against mass surveillance, in *Annual Cryptology Conference* (Springer, 2014), pp. 1–19
66. P. Hodges, D. Stebila, Algorithm substitution attacks: State reset detection and asymmetric modifications, in *IACR Transactions on Symmetric Cryptology* (2021), pp. 389–422
67. L. Jinhui, Y. Yong, W. Fusheng, Y. Cheng, H. Zhang, An algorithm substitution attack on fiat-shamir signatures based on lattice. Wuhan Univ. J. Nat. Sci. **27**(1), 17–25 (2022)
68. M. Armour, B. Poettering, Substitution attacks against message authentication. Cryptology ePrint Archive (2019)
69. S. Berndt, M. Liśkiewicz, Algorithm substitution attacks from a steganographic perspective, in *Proceedings of the 2017 ACM SIGSAC Conference on Computer and Communications Security* (2017), pp. 1649–1660
70. R. Chen, X. Huang, M. Yung, Subvert kem to break dem: practical algorithm-substitution attacks on public-key encryption, in *International Conference on the Theory and Application of Cryptology and Information Security* (Springer, 2020), pp. 98–128
71. Y. An, H.-S. Lee, J. Lee, S. Lim, Key substitution attacks on lattice signature schemes based on sis problem. Secur. Commun. Netw. **2018** (2018)
72. M. Bellare, J. Jaeger, D. Kane, Mass-surveillance without the state: strongly undetectable algorithm-substitution attacks, in *Proceedings of the 22nd ACM SIGSAC Conference on Computer and Communications Security* (2015), pp. 1431–1440
73. B. Alshemali, J. Kalita, Generalization to mitigate synonym substitution attacks, in *Proceedings of Deep Learning Inside Out (DeeLIO): The First Workshop on Knowledge Extraction and Integration for Deep Learning Architectures* (2020), pp. 20–28
74. M. Fischlin, S. Mazaheri, Self-guarding cryptographic protocols against algorithm substitution attacks, in *2018 IEEE 31st Computer Security Foundations Symposium (CSF)* (IEEE, 2018), pp. 76–90
75. Y. Mo, B. Sinopoli, Secure control against replay attacks, in *2009 47th Annual Allerton Conference on Communication, Control, and Computing (Allerton)* (IEEE, 2009), pp. 911–918
76. F. Chen, Y. Huo, J. Zhu, D. Fan, A review on the study on mqtt security challenge, in *2020 IEEE International Conference on Smart Cloud (SmartCloud)* (IEEE, 2020), pp. 128–133
77. Y. Feng, W. Wang, Y. Weng, H. Zhang, A replay-attack resistant authentication scheme for the internet of things, in *2017 IEEE International Conference on Computational Science and Engineering (CSE) and IEEE International Conference on Embedded and Ubiquitous Computing (EUC)*, vol. 1 (IEEE, 2017), pp. 541–547
78. S. Na, D. Hwang, W. Shin, K.-H. Kim, Scenario and countermeasure for replay attack using join request messages in lorawan, in *2017 International Conference on Information Networking (ICOIN)* (IEEE, 2017), pp. 718–720
79. A. Phimphinith, X. Anping, Q. Zhu, Y. Jiang, Y. Shen, An enhanced mutual authentication scheme based on ecdh for iot devices using esp8266, in *2019 IEEE 11th International Conference on Communication Software and Networks (ICCSN)* (IEEE, 2019), pp. 490–496
80. M.S. Wara, Q. Yu, New replay attacks on zigbee devices for internet-of-things (iot) applications, in *2020 IEEE International Conference on Embedded Software and Systems (ICESS)* (IEEE, 2020), pp. 1–6

81. A. Verma, V. Ranga, The impact of copycat attack on rpl based 6lowpan networks in internet of things. Computing **103**(7), 1479–1500 (2021)
82. L. Agilandeeswari, S. Paliwal, A. Chandrakar, M. Prabukumar, A new lightweight conditional privacy preserving authentication and key–agreement protocol in social internet of things for vehicle to smart grid networks. Multimed. Tools Appl. 1–28 (2022)
83. F. Farha, H. Ning, W. Zhang, K.-K.R. Choo, et al., Timestamp scheme to mitigate replay attacks in secure Zigbee networks. IEEE Trans. Mobile Comput. (2020)
84. H. Kurdi, V. Thayananthan, Authentication mechanisms for iot system based on distributed mqtt brokers: review and challenges. Proc. Comput. Sci. **194**, 132–139 (2021)
85. N. Ravanbakhsh, M. Nazari, An efficient improvement remote user mutual authentication and session key agreement scheme for e-health care systems. Multimed. Tools Appl. **77**(1), 55–88 (2018)
86. Z. Banković, J.C. Vallejo, D. Fraga, J.M. Moya, Detecting bad-mouthing attacks on reputation systems using self-organizing maps, in *Computational Intelligence in Security for Information Systems* (Springer, 2011), pp. 9–16
87. M. Bin-Yahya, O. Alhussein, X. Shen, Securing software-defined wsns communication via trust management. IEEE Internet of Things J. (2021)
88. W. Fang, W. Zhang, W. Chen, T. Pan, Y. Ni, Y. Yang, Trust-based attack and defense in wireless sensor networks: a survey. Wirel. Commun. Mobile Comput. **2020** (2020)
89. T. Suh, Y. Cho, An enhanced trust mechanism with consensus-based false information filtering algorithm against bad-mouthing attacks and false-praise attacks in wsns. Electronics **8**(11), 1359 (2019)
90. P. Roy, S. Bhattacharjee, H. Alsheakh, S.K. Das, Resilience against bad mouthing attacks in mobile crowdsensing systems via cyber deception, in *2021 IEEE 22nd International Symposium on a World of Wireless, Mobile and Multimedia Networks (WoWMoM)* (IEEE, 2021), pp. 169–178
91. V.B. Reddy, A. Negi, S. Venkataraman, V.R. Venkataraman, A similarity based trust model to mitigate badmouthing attacks in internet of things (iot), in *2019 IEEE 5th World Forum on Internet of Things (WF-IoT)* (IEEE, 2019), pp. 278–282
92. J. Jiang, Y. Liu, Secure iot routing: Selective forwarding attacks and trust-based defenses in rpl network (2022). arXiv:2201.06937
93. A. Nangue, E. Fute Tagne, E. Tonye, Robust and accurate trust establishment scheme for wireless sensor network. Int. J. Comput. Netw. Inf. Secur. **12**(6) (2020)
94. A. Crescenzi, D. Kelly, L. Azzopardi, Impacts of time constraints and system delays on user experience, in *Proceedings of the 2016 Acm on Conference on Human Information Interaction and Retrieval* (2016), pp. 141–150
95. A. Dhakne, P. Chatur, Detailed survey on attacks in wireless sensor network, in *Proceedings of the International Conference on Data Engineering and Communication Technology* (Springer, 2017), pp. 319–331
96. A. Alnasser, H. Sun, J. Jiang, Recommendation-based trust model for vehicle-to-everything (v2x). IEEE Internet Things J. **7**(1), 440–450 (2019)
97. A.R. Dhakne, P.N. Chatur, Design of hierarchical trust based intrusion detection system for wireless sensor network [htbid]. Int. J. Appl. Eng. Res. **12**(8), 1772–1778 (2017)
98. N. Kandhoul, S.K. Dhurandher, I. Woungang, T_cafe: a trust based security approach for opportunistic iot. IET Commun. **13**(20), 3463–3471 (2019)
99. R.U. Zaman, R. Sultana, Identifying trustworthy nodes in an integrated internet manet to establish a secure communication, in *Proceedings of the Second International Conference on Computational Intelligence and Informatics* (Springer, 2018), pp. 321–328

100. E. Zupancic, D. Trcek, Qade: a novel trust and reputation model for handling false trust values in e-commerce environments with subjectivity consideration. Technol. Econ. Dev. Econ. **23**(1), 81–110 (2017)
101. J.M. Jorquera Valero, P.M. Sánchez Sánchez, A. Lekidis, J. Fernandez Hidalgo, M. Gil Pérez, M.S. Siddiqui, A. Huertas Celdrán, G. Martínez Pérez, Design of a security and trust framework for 5g multi-domain scenarios. J. Netw. Syst. Manag. **30**(1), 1–35 (2022)
102. Y. Wang, R. Chen, J.-H. Cho, J.J. Tsai, Trust-based task assignment with multiobjective optimization in service-oriented ad hoc networks. IEEE Trans. Netw. Serv. Manag. **14**(1), 217–232 (2016)
103. Y. Sun, Y. Zhao, Dynamic adaptive trust management system in wireless sensor networks, in *2019 IEEE 5th International Conference on Computer and Communications (ICCC)* (IEEE, 2019), pp. 629–633
104. M.A. Hassan, Q.-T. Vien, M. Aiash, Software defined networking for wireless sensor networks: a survey. Adv. Wirel. Commun. Netw. **3**(2), 10–22 (2017)
105. P. Agrawal, A. Nakarmi, M.P. Jhanwar, S. Sharma, S. Banerjee, Ballot stuffing and participation privacy in pollsite voting (2022). https://doi.org/10.48550/arxiv.2210.14833. https://arxiv.org/abs/2210.14833
106. N. Anita, M. Vijayalakshmi, Blockchain security attack: a brief survey, in *2019 10th International Conference on Computing, Communication and Networking Technologies (ICCCNT)* (IEEE, 2019), pp. 1–6
107. A. Ugur, Manipulator: a novel collusion attack on trust management systems in social iot, in *Computer Science On-line Conference* (Springer, 2021), pp. 578–592
108. F. Jabeen, Z. Hamid, W. Abdul, S. Ghouzali, A. Khan, S.U.R. Malik, M. Shaukat Khan, S. Nawaz, Anonymity-preserving reputation management system for health sector. PloS one **13**(4), 0195021 (2018)
109. F. Khedim, N. Labraoui, M. Lehsaini, Dishonest recommendation attacks in wireless sensor networks: a survey, in *2015 12th International Symposium on Programming and Systems (ISPS)* (IEEE, 2015), pp. 1–10
110. H. Hu, R. Lu, Z. Zhang, J. Shao, Replace: a reliable trust-based platoon service recommendation scheme in vanet. IEEE Trans. Veh. Technol. **66**(2), 1786–1797 (2016)
111. L.E. Jim, M.A. Gregory, An artificial immune system-based strategy to enhance reputation in manets. J. Telecommun. Digital Econ. **7**(1), 68–82 (2019)
112. A.M. Shabut, K.P. Dahal, S.K. Bista, I.U. Awan, Recommendation based trust model with an effective defence scheme for manets. IEEE Trans. Mob. Comput. **14**(10), 2101–2115 (2014)
113. H.C. Tan, M. Ma, H. Labiod, P.H.J. Chong, J. Zhang, A non-biased trust model for wireless mesh networks. Int. J. Commun. Syst. **30**(9), 3200 (2017)
114. W. Ali, G. Abbas, Z.H. Abbas, Joint sybil attack prevention and energy conservation in wireless sensor networks, in *2019 International Conference on Frontiers of Information Technology (FIT)* (2019), pp. 179–1795. https://doi.org/10.1109/FIT47737.2019.00042
115. D. Gopi, S. Cheng, R. Huck, Comparative analysis of sdn and conventional networks using routing protocols, in *2017 International Conference on Computer, Information and Telecommunication Systems (CITS)* (2017), pp. 108–112. https://doi.org/10.1109/CITS.2017.8035305
116. N. Galloro, M. Polino, M. Carminati, A. Continella, S. Zanero, A systematical and longitudinal study of evasive behaviors in windows malware. Comput. Secur. **113**, 102550 (2022)
117. R. Das, A. Prasad, Survey of blockchain techniques for iot device security, in *Blockchain Technology* (CRC Press, 2022), pp. 57–72
118. J.M. McGinthy, L.J. Wong, A.J. Michaels, Groundwork for neural network-based specific emitter identification authentication for iot. IEEE Internet Things J. **6**(4), 6429–6440 (2019). https://doi.org/10.1109/JIOT.2019.2908759

119. J. Wang, Y. Liu, H. Song, Counter-unmanned aircraft system (s)(c-uas): State of the art, challenges, and future trends. IEEE Aerosp. Electron. Syst. Mag. **36**(3), 4–29 (2021)
120. A. Sargolzaei, A. Abbaspour, C.D. Crane, Control of cooperative unmanned aerial vehicles: review of applications, challenges, and algorithms, in *Optimization, Learning, and Control for Interdependent Complex Networks* (2020), pp. 229–255
121. A.S. Abdalla, V. Marojevic, Communications standards for unmanned aircraft systems: the 3g pp perspective and research drivers. IEEE Commun. Stand. Mag. **5**(1), 70–77 (2021)
122. C.H. Liu, X. Ma, X. Gao, J. Tang, Distributed energy-efficient multi-uav navigation for long-term communication coverage by deep reinforcement learning. IEEE Trans. Mob. Comput. **19**(6), 1274–1285 (2019)
123. J. Wang, M. Liu, J. Sun, G. Gui, H. Gacanin, H. Sari, F. Adachi, Multiple unmanned-aerial-vehicles deployment and user pairing for nonorthogonal multiple access schemes. IEEE Internet Things J. **8**(3), 1883–1895 (2020)
124. P. Garcia-Aunon, A. Barrientos Cruz, Comparison of heuristic algorithms in discrete search and surveillance tasks using aerial swarms. Appl. Sci. **8**(5), 711 (2018)
125. J. Scherer, B. Rinner, Multi-uav surveillance with minimum information idleness and latency constraints. IEEE Robot. Autom. Lett. **5**(3), 4812–4819 (2020)
126. M. Patil, T. Abukhalil, T. Sobh, Hardware architecture review of swarm robotics system: self-reconfigurability, self-reassembly, and self-replication. Int. Sch. Res. Not. **2013** (2013)
127. R. Rumba, A. Nikitenko, The wild west of drones: a review on autonomous-uav traffic-management, in *2020 International Conference on Unmanned Aircraft Systems (ICUAS)* (IEEE, 2020), pp. 1317–1322
128. M. Pino, J.P. Matos-Carvalho, D. Pedro, L.M. Campos, J.C. Seco, Uav cloud platform for precision farming, in *2020 12th International Symposium on Communication Systems, Networks and Digital Signal Processing (CSNDSP)* (IEEE, 2020), pp. 1–6
129. M. Radmanesh, Uav traffic management for national airspace integration. PhD thesis, University of Cincinnati (2016)
130. Y. Choi, Y. Choi, S. Briceno, D.N. Mavris, Energy-constrained multi-uav coverage path planning for an aerial imagery mission using column generation. J. Intell. Robot. Syst. **97**(1), 125–139 (2020)
131. C. Zhan, H. Hu, Z. Liu, Z. Wang, S. Mao, Multi-uav-enabled mobile-edge computing for time-constrained iot applications. IEEE Internet Things J. **8**(20), 15553–15567 (2021)
132. L.D. Nguyen, A. Kortun, T.Q. Duong, An introduction of real-time embedded optimisation programming for uav systems under disaster communication. EAI Endorsed Trans. Ind. Netw. Intell. Syst. **5**(17), 5–5 (2018)
133. J. Tang, S. Lao, Y. Wan, Systematic review of collision-avoidance approaches for unmanned aerial vehicles. IEEE Syst. J. (2021)

Mimonah Al Qathrady received her Ph.D. in Computer Engineering (CE) in 2020 from the University of Florida (UF), Gainesville, USA. She obtained her M.Sc in CE in 2013 from UF, and her Bachelor of Information Systems from King Khalid University, Abha, Saudi Arabia, 2007. Since 2020, she has been working as an assistant professor in Najran University. Her research interests include IoT data analysis and modeling, and their applications in mobility, encounter and infection tracing, and security fields; as well as building data-driven systems incorporating machine and deep learning. She worked on several projects, such as MobiBench, as well as being the lead investigator in a multitude of projects, such as i-Hospital and AI in IoT. She was the coordinator of NOMADS lab at UF, and she is a member of ACM and IEEE.

Sultan Almakdi received the B.S. degree in computer science from King Khalid University, Abha, Saudi Arabia, in 2010, the M.S. degree in computer science from the University of Colorado Denver, Denver, USA, in 2014, and the Ph.D. degree in computer science from the University of Arkansas, Fayettiville, USA, in 2020. He is currently working as an Assistant Professor with the Department of Computer Science and Information Systems, Najran University, Saudi Arabia. His research interests include cloud security, fog security, edge computing security, the IoT security, and computer security. He received a Graduate Certificate in cybersecurity from the University of Arkansas, in 2020.

Mohammed S. Alshehri received the B.S. degree in Computer Science from the King Khalid University, Abha, Saudi Arabia, in 2010, the M.S. degree in Computer Science from the University of Colorado Denver, Denver, USA, in 2014, and the Ph.D. degree in Computer Science with concentration on Information Security from the University of Arkansas, Fayetteville, USA, in 2021. He received a Graduate Certificate in cybersecurity from the University of Arkansas, in 2020. Mohammed's areas of interest are Cybersecurity, Computer Networks, Blockchain, Machine Learning, and Deep Learning.

Samar Alqhtani is currently an associate professor in Najran university, Saudi Arabia. She received her Ph.D. In Information Technology from the University of Newcastle, Australia. Her research interest is in information technology and multimedia, including artificial intelligence, machine learning, deep learning, health informatics, data mining, image processing, computer vision, text processing, information security and IoT. Alqhtani has lectured and developed curricula for courses in computer science and information system. Recently she has led and worked on various projects, including event detection application in social media, medical application, and applying artificial intelligence algorithms to emerging technologies.

Secure Communications for Autonomous Multiple-UAV Media Production

Ioannis Mademlis, Paraskevi Nousi, Damien Lavaux, Thomas Aubourg, Cedric Le Barz, and Ioannis Pitas

Abstract

Equipping Unmanned Aerial Vehicles (UAVs/drones) with professional cameras has rapidly transformed the media production landscape in recent years. However, their creative potential in aerial cinematography applications can only be fully exploited by enhancing their cognitive autonomy and deploying them in a collaborative, multi-drone fleet setting. Thus, networking, security and data streaming issues arise naturally. In this Chapter, we assume a stand-alone UAV fleet coordinated in real-time by a central on-ground compute station for live outdoor event media coverage, with high-definition, low-latency video streaming from many moving sources. This is the most general and technically difficult filming scenario: on top of security concerns, fluctuations in wireless signal

I. Mademlis (✉) · P. Nousi · I. Pitas
Aristotle University of Thessaloniki, Aristotle University of Thessaloniki Campus, Thessaloniki, Greece
e-mail: imademlis@csd.auth.gr

P. Nousi
e-mail: paranous@csd.auth.gr

I. Pitas
e-mail: pitas@csd.auth.gr

D. Lavaux · T. Aubourg · C. Le Barz
Advanced Studies Department THERESIS, Thales, Palaiseau, France
e-mail: damien.lavaux@thalesgroup.com

T. Aubourg
e-mail: thomas.aubourg@thalesgroup.com

C. Le Barz
e-mail: cedric.lebarz@thalesgroup.com

M. Abdelkader and A. Koubaa (eds.), *Unmanned Aerial Vehicles Applications: Challenges and Trends*, Synthesis Lectures on Intelligent Technologies,
https://doi.org/10.1007/978-3-031-32037-8_11

power inevitably make stable wireless communications a real challenge with current technology. Motivated by these difficulties, we designed and evaluated a novel multiple-UAV platform for live outdoor media production, featuring a communications architecture able to handle and overcome the relevant communication issues. Both 4G/LTE and WiFi are utilized to make this infrastructure easy to deploy, secure and robust, as indicated by the included empirical evaluation. Notably, this is an innovative, prototype platform: the first one specifically designed for handling difficult professional filming scenarios with multiple autonomous UAVs.

Keywords
Autonomous drones • Unmanned aerial vehicle • Media production • UAV cinematography • UAV communications • Wireless network security • 4G • LTE • WiFi • Data streaming

1 Introduction

Unmanned Aerial Vehicles (UAVs, or "drones") have revolutionized the media production landscape during the past decade, by allowing flexible and easy aerial cinematographic shot acquisition, access to narrow, or difficult-to-reach spaces and the possibility to easily obtain impressive footage (e.g., dynamic panoramas or novel, multiview and 360° shots). Their ability to fly and/or hover, as well as their small size, high agility and low cost, have made them an indispensable tool in the film/TV/media industry.

The cumbersome logistics of fully manual, professional UAV filming (a trained pilot and a separate camera operator are required per drone, acting with precise coordination during filming) and technological progress are slowly leading the market towards increased adoption of cognitive autonomy functionalities. Commercial cinematography UAVs already provide several autonomous capabilities for both safe flight and filming, such as obstacle detection and avoidance, automated landing, physical target tracking or orbiting (for low-speed, manually preset targets), as well as with automatic central composition framing: that is, constantly rotating the camera so that the preselected target always remains properly centred on-frame.

Clearly, the foreseeable future holds the promise of fully autonomous UAVs that only require minimal, high-level supervision by a human operator. This can be achieved via Artificial Intelligence (AI) algorithms, which run either solely on-board the vehicle, or are split between on-board execution and remote execution on a powerful ground compute station. The AI methods most relevant to autonomous UAVs are computer vision/machine learning algorithms, which conduct semantic visual analysis of each captured video frame (e.g., for object detection on images), and task/motion planning methods, which assign tasks to the UAV and plan/adapt its flight trajectories.

However, filming with a single UAV, which is still the norm in professional media production, severely limits the creative potential. Each target can only be captured from a specific view angle and with a specific framing shot type at any given time, while there can only be a single target at each time instance. Finally, the need to move the UAV from one point to another in order to shoot from a different angle, aim at a different target, or return to the recharge platform, results in "dead" periods that prevent smooth and unobstructed shooting.

A fleet of multiple, cooperating UAVs that are deployed simultaneously for a shooting mission can easily solve the above issues. However, the burden of manual multi-UAV operation makes cognitive autonomy a necessity, achievable through AI and robotics technologies which can greatly enhance the attractiveness of UAV fleets in media production. But that's not the end of the story, since additional issues arise due to the need for communication and coordination between the UAVs. In addition, if the purpose of the shooting mission is live broadcast, an on-ground compute station must be able to handle video streams transmitted simultaneously by all members of the fleet. Depending on the platform's architectural design, this central station may also act as a coordinator between them.

As expected, security is of the utmost importance in a UAV-based media production environment. The consequences of a compromised platform can include serious legal implications and privacy violations. The main risks involved are data hijacking, drone hacking and/or capture during flight by a malicious third party, with the aim of studying it and convert it (e.g., for carrying explosives) or crash it directly on people/infrastructures. In addition, in UAV-based live broadcast, the air-to-ground communication specifications must always ensure a minimum bitrate, sufficient for acceptable real-time video broadcast quality; therefore data streaming is also not trivial.

Overall, communication challenges are especially prominent among the issues an autonomous multiple-UAV platform for media production/cinematography applications must be able to handle. This Chapter assumes an autonomous UAV fleet coordinated by a centralized on-ground compute station for live coverage of outdoor events. This is the most general and technically difficult filming scenario, where wireless signal strength fluctuations are inevitably making radio communications a real challenge with current technology.

Such difficulties have motivated us to design a novel, autonomous multiple-UAV platform specifically targeting live outdoor media production, i.e., the so-called MULTIDRONE platform. This Chapter details said innovative, prototype platform, showcasing that it is indeed able to overcome the relevant communication challenges. MULTIDRONE features a robust and secure architecture, providing resilient and secure communications for UAV control and payload data transport. Both 4G/LTE and WiFi are employed to make this infrastructure easily deployable, secure and resilient, as indicated by the included empirical evaluation. Notably, MULTIDRONE is the first platform specifically designed for handling difficult multiple-UAV professional filming scenarios. However, although the emphasis is on semi-autonomous live filming in extended outdoor environments, the determined challenges and solutions apply equally to a vast range of different applications, with only minor modifications.

Note that this Chapter focuses on the relevant networking, security and data streaming aspects. For the cinematography, artificial intelligence, robotics and software architecture

Table 1 Previous literature stemming from MULTIDRONE

Artificial intelligence	Software architecture	Cinematography	Robotics
[29]	[21]	[17]	[33]
[26]	[20]	[18]	[31]
[27]	[23]	[19]	[5]
[37]	[16]	[14]	[2]
[28]		[15]	[1]
[12]		[13]	[32]
[10]		[3]	[4]
[11]			[7]
[36]			
[35]			

aspects, previous literature can be consulted; it is summarized in Table 1. The remainder of this Chapter is organized in the following manner. Section 2 presents and discusses the various networking, streaming and security issues arising in semi-autonomous multiple-UAV live filming in extended outdoor environments. Then, Sect. 3 briefly presents the overall MULTIDRONE platform architecture, so as to put into perspective the analysis of its communications subsystems and modules. The following Sect. 4 details exactly the MULTIDRONE communications architecture, derived from a set of choices aiming at overcoming the challenges mentioned in Sect. 2. Next, Sect. 5 details the conducted, partial experimental evaluation of the MULTIDRONE communications platform, before the Chapter is concluded with potential future research and innovation directions in Sect. 6.

2 Communications for Autonomous Multiple-UAV Filming

2.1 Autonomous Multiple-UAV Filming Architectures

Autonomous multiple-UAV filming typically requires the following types of cognitive software components to be present in the platform:

- **perception** modules; e.g., semantic visual analysis (e.g., object detection/tracking), obstacle/collision detection, localization and mapping (SLAM), etc.
- **planning** modules; e.g., mission/task planning, path planning, etc.
- **control** modules; e.g., UAV trajectory control, UAV formation control, camera/gimbal control, etc.

These SW components can be distributed among the various platform nodes, i.e., the UAVs and, optionally, a central on-ground compute station. Typically, only a subset of them interacts with UAV sensors and motors, but they must all intercommunicate at a high frequency, either via memory (when the components reside on the same node), or wireless networking (in UAV-to-UAV, UAV-to-ground or ground-to-UAV communication exchanges). Additionally, in cinematography applications, high-resolution video stream(s) may also have to be transmitted over-the-air with low latency, regardless of the cognitive SW components.

Three different scenarios of multiple-UAV cinematography applications can be defined:

- off-line shooting with full post-production editing (i.e., for TV programmes or movies);
- filming of live events for deferred broadcast and, thus, with potential post-production modifications (i.e., for deferred TV programmes);
- full live event shooting (i.e., for live TV programmes) with limited post-production effects.

Obviously, full live event shooting is both the most general and the most challenging scenario, therefore it is emphasized by the presented communications architecture. Below, communication issues are first introduced from a high-level perspective.

2.2 Overview of Communications Challenges

In order to successfully deploy fleets of UAVs, especially for live event media coverage applications, having a strong communications infrastructure is crucial. It is difficult to stream high-resolution video (particularly 4K) down to a ground station with Quality-of-Service (QoS) guarantees and simultaneously execute real-time algorithms, even in single-UAV missions. While professional cameras and open-source software can handle on-the-fly video acquisition, compression, synchronization, and transmission, the scarcity of commercial media production-quality camera models with Camera Serial Interface (CSI) connectivity, which guarantees reliable high-speed, low-power data transfer between a camera and a computer, gives rise to practical issues. Using dedicated hardware is not a viable solution due to additional energy consumption, cost, and weight, which are critical factors in UAVs. A trade-off must be made between broadcast video resolution, hardware cost, and degree of vehicle intelligence.

In non-live coverage scenarios, such as filming for deferred broadcast or shooting a scripted sequence, video can simply be stored onboard and retrieved later. However, communications are necessary for all other cases, including non-live single-UAV filming where a subset of the AI algorithms are being executed remotely on a ground compute station. Private QoS-guaranteeing 4G/LTE infrastructure is suitable for outdoor event filming due to the high mobility of the UAVs and long distances that need to be covered. Traditional WiFi is a suboptimal alternative with higher latency and smaller range. Public LTE networks are unreliable because UAV communications cannot be prioritized over telephony.

Live broadcasting is the most challenging scenario, even with private LTE infrastructure, leading to a fall back on FullHD resolution for video streaming.

When using a fleet of multiple cooperating UAVs, more problems arise. Available bandwidth may not be enough to support live FullHD video streaming from all drones simultaneously, resulting in a limit on the number of drones. Direct coordination between drones may be required for distributed variants of algorithms or redundancy/fault tolerance, requiring an intra-fleet Flying Ad Hoc Network (FANET, WiFi mesh) that enables ad hoc routing and accounts for dynamic network topology with high node mobility. However, FANETs are not yet a mature technology, and either custom, optimized WiFi extensions, or falling back to LTE infrastructure is necessary for actual deployment, at the cost of increased latency.

Regarding security, simple ways for an attacker to intervene are radio communication or GPS signal jamming, GPS spoofing, UAV autopilot firmware hacking and/or communication hijacking. Additionally, a Man-in-the-Middle attack can allow an unauthorized person to pose as the ground station and take command of the UAV. Weak security in communications can also allow obtaining the video captured by the drone, or its intended flight path.

Moreover, during a shooting session, a subset of the generated data is stored on-board, while another subset is being on-the-fly transmitted over the air. Depending on the application, video footage may be both stored and broadcasted, or only stored on-board for offline processing at a later time. In contrast, real-time transmission of telemetry and control data is necessary for ensuring a secure and safe flight, making them vulnerable to potential security threats. Thus:

- Video footage must be protected during storage, using authentication mechanisms to restrict its accessibility to the copyright owners.
- Telemetry and control commands need to be transmitted at the highest priority and be protected against misuse at the same time.

2.3 Data Streaming Challenges

To stream outdoor events in real-time, there are several cinematography factors that need to be considered, such as tracking targets of interest, adjusting zoom, composition and focus, and using multiple UAVs to get different perspectives. However, these UAVs must also fly autonomously and adhere to safety regulations (e.g., avoid to fly over human crowds, or exceed a specific altitude).

When designing a communications architecture for real-time live video streaming, several factors must be considered, including weak wireless communication, the large size of video data, and the need to use video compression to reduce this size (1 min of raw 720p 8-bit video at 30 Frames Per Second is 1.55 GBs). H264 and H265 coding are good options, but they introduce delays and quality loss. Using Real-time Transport Protocol (RTP) with User Datagram Protocol (UDP) is a good choice for the network protocol stack, as it prefers timeliness over reliability.

If multiple UAVs are involved in the streaming, clock synchronization becomes important. Network Time Protocol (NTP) can be used to ensure that all devices use a single clock. However, on-board analysis of the captured video frames introduces delays, so metadata, including the results of AI-enabled semantic visual analysis and UAV telemetry, must be transmitted along with the respective video frame, in a manner immune to any compression-induced corruption. This is the only way for any software running at the receiving end, e.g., on a ground compute station, to properly correspond metadata with the respective visual content. The best approach to achieve this is to insert such metadata as an RTP header extension.

GStreamer, an open-source multimedia framework with multiple programming language bindings, can be used for low-level handling of these issues.

3 The MULTIDRONE Architecture

This Section briefly describes an innovative prototype platform that attempts to handle the issues and challenges outlined in Sect. 2. It has been developed in the context of the EU-funded research and development project MULTIDRONE,[1] which aimed exactly at designing and implementing a semi-autonomous multiple-UAV platform for live outdoor filming, featuring a small fleet of drones and a ground compute station. The purpose here is not to fully describe the overall MULTIDRONE platform, which has been detailed in [21], but to provide the background necessary for discussing the MULTIDRONE communications architecture in Sect. 4.

The presented MULTIDRONE architecture includes a fleet of cooperating, camera-equipped UAVs and a central *Ground Station* (GS). GS is employed for preplanning the shooting mission by the Director's team (using an appropriate GUI named "Dashboard") and for dynamic, autonomous mission replanning and execution monitoring. Additionally, it is used for on-line semantic environment mapping concerning human crowds [12], since such areas constitute no-flight zones due to regulations and safety issues. Finally, the *Supervision Station* (SS) is included on-ground, i.e., a GUI permitting a human operator to constantly monitor the status of the UAVs, so as to cancel the mission in case any security issues arise during its execution.

The UAVs are responsible for collectively executing the mission, i.e., mainly filming and physically following prespecified moving targets (e.g., athletes), in an adaptive manner, so as to capture the desired cinematographic shot types. Additionally, they gather visual data to facilitate semantic mapping and on-map localization of targets. Each UAV carries a PixHawk/PX4 Autopilot [22] (i.e., a popular low-level flight trajectory control system), an Intel NUC computer (with a powerful CPU), an NVIDIA Jetson Tegra X2 embedded computing board (containing a CPU and a powerful GP-GPU), two cameras (a "navigation cam-

[1] https://multidrone.eu/.

era", or NavCam, and a "cinematographic camera", or CinCam), a gimbal and a lightweight LiDAR. Multiple functionalities are implemented by on-board software components, such as autonomous UAV localization and control, gimbal/camera control and obstacle/collision avoidance.

The Jetson TX2 board is employed for on-board, on-the-fly semantic visual analysis of the captured video frames [26]. In general, a wide range of human-centered visual analysis algorithms could be employed, although the emphasis of MULTIDRONE is on on-board object detection and tracking [35, 36] and on on-ground human crowd detection [37]. The Intel NUC computer is employed for executing on-board localization and control SW modules [31]. Overall, the majority of the platform's software components are implemented using the popular Robot Operating System (ROS) middleware [30], which provides the abstractions of *topics* (following a publisher/subscriber model) and *services*, to permit easy inter-process communication across devices.

Backup human pilots on stand-by are supported for safety reasons; the platform allows them to fully take manual control of the UAVs in case of emergency or serious error. The complete MULTIDRONE architecture is presented in detail in [21], while Fig. 1 summarizes it at a high-level.

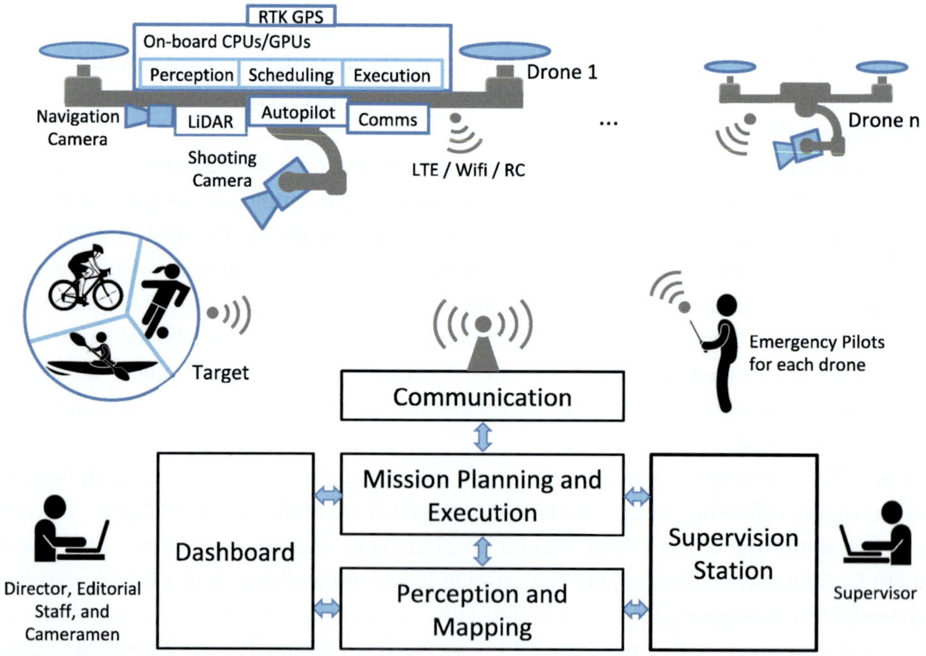

Fig. 1 High-level diagram of the MULTIDRONE architecture: on-board (top) and Ground Station (bottom)

4 MULTIDRONE Communications Module

The majority of the communication exchanges between the UAVs and the GS, including real-time live video streaming, are assured by an LTE system [24], composed of an LTE user equipment on-board and an LTE base station on-ground. Inter-UAV communications are assured by an auxiliary WiFi mesh (WiFi) [9]. This acts as: (i) a redundancy feature, in case a UAV loses 4G/LTE access due to signal fluctuations, and (ii) a medium of inter-UAV coordination for on-the-fly UAV formation control algorithms (e.g., to facilitate capturing multiview cinematic shots of moving targets). Each UAV carries on-board a dedicated *Communication Module* (CM) that is responsible for:

- Acting as a default IP communication gateway/router to the ground and to other UAVs.
- Scheduling IP flows depending on applications' precedence and assigned IP QoS.
- Traffic shaping/admission control when congestion occurs.
- Authentication, encryption and other security-related mechanisms.

The infrastructure is fully independent of any public LTE operator and provides a high level of real-time performance and cybersecurity: it offers plenty of available network bandwidth, long range of transmission, fine-grained QoS management and enhanced streams protection, with Virtual Private Network (VPN) tunneling and strong encryption.

The network topology provided for MULTIDRONE is designed to be as transparent as possible for the various software modules. LTE and WiFi subnetworks are routed to simplify the communication between each node, but certain constraints have been taken into account to ensure an efficient system behavior.

- All communications with the LTE system must be IP-based.
- IP addressing of on-board UAV computers and the GS conforms to the one detailed by Figs. 3 and 4.

CM can be considered as a default IP router for the rest of the system. As such, it exposes an Ethernet interface to the computers on-board the UAV and implements a full IP protocol stack. Since it is fully independent from the other modules in the architecture, it has its own hardware and operating system (Linux OpenWRT). In addition, a separate *Video Streamer* (VS) module is necessary for video transmission and interacts heavily with CM. For each UAV, two video streams are generated: one by the NavCam (H.264 compressed, 4:2:0 chroma subsampling, 640×480 resolution), another one by the CinCam (H.264 compressed, 4:2:2 chroma subsampling, 1920×1080 resolution, @25fps).

A Blackmagic Micro Cinema Camera with a motorised Panasonic x3 lens was selected as the CinCam, supporting FullHD resolution. On the other hand, the NavCam does not require FullHD, since its main purpose is simply to provide the SS with good situational awareness. Compression takes place on-board the NVIDIA Jetson TX2 platform, which offers hardware-accelerated image/video compression. Video streams are then transmitted

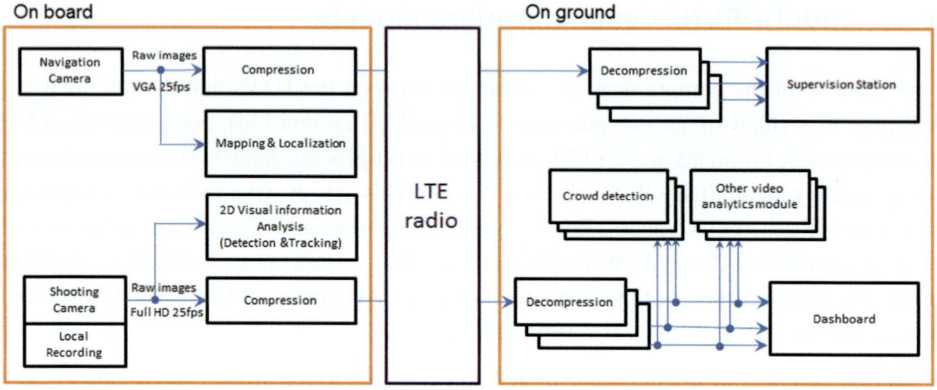

Fig. 2 Data flow for video streaming from the UAVs to the ground

through the LTE radio network using RTP. The RTP Control Protocol (RTCP) is also used for on-ground synchronization of video streams coming from different UAVs. The RTP packets hold a 32-bit RTP timestamp. Several consecutive RTP packets may have equal timestamps if they are (logically) generated at once, e.g., they belong to the same video frame. The Sender Report packet holds the correspondence between the RTP timestamp and the absolute 64-bit timestamp (system hour), that is broadcasted through the LTE network thanks to NTP.

Figure 2 depicts the data flow for all video streams. It is assumed that 3 UAVs are connected to the GS. The CinCam video streams are transmitted to the Dashboard through the radio network. In parallel, the streams are also resized so that they can be processed on-board by the perception modules. VS can process either the images coming from the CinCam, or from the NavCam, depending on the situation. On-ground, these streams will be uncompressed to be displayed on the Dashboard and, also, resized to be processed by any video analysis modules.

Regarding system scalability, increasing the number of UAVs only implies updating the configuration and hardware of the LTE modules. Of course, the required bandwidth should be manageable by the communication base station. Apart from that, there is no other major impact on the overall communication architecture.

Redundant RF communications are also provided for safety, through additional links. An example would be in case of LTE streaming failure. The navigation stream is sent to the backup pilot via RF in manual mode, thus an analog signal is required. RF can also handle the commands to control the gimbal and the camera from a transmitter. The Pixhawk may receive at the same time commands coming from the RF receiver and from the on-board computer, which received it from the Dashboard through the LTE.

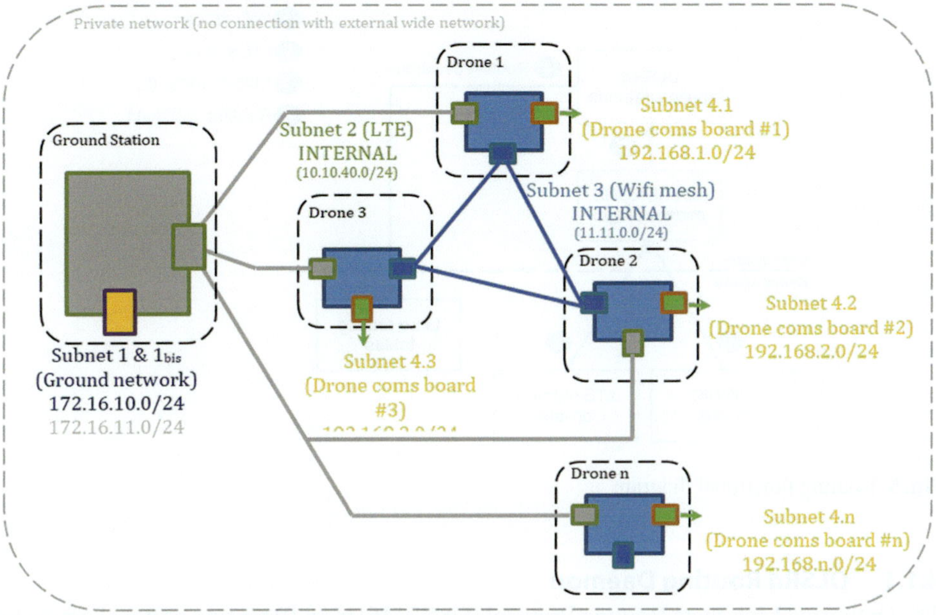

Fig. 3 Communication network addressing (complete)

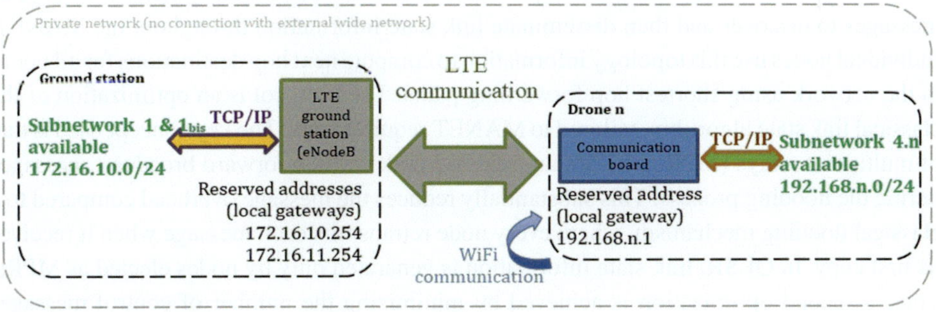

Fig. 4 Communication network addressing (user point-of-view)

4.1 MULTIDRONE Routing and Link Monitoring

Multiple IP subnetworks are being used in MULTIDRONE, requiring IP routing between them to achieve end-to-end IP communications. While manual IP routing could be used, such a solution would be completely static and thus would not allow for dynamically and automatically changing the data transfer path. In order to introduce the required dynamics, the routing configuration is setup as depicted in Fig. 5.

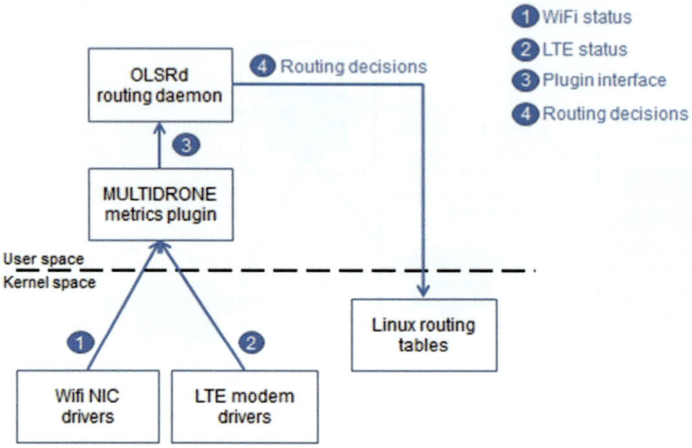

Fig. 5 Routing functional diagram

4.1.1 OLSRd Routing Daemon

The Optimized Link State Routing Protocol (OLSR) [6] is an IP routing protocol optimized for Mobile Ad Hoc Networks (MANETs). It is standardised at IETF under RFC 3626. OLSR is a proactive link-state routing protocol, which uses hello and topology control (TC) messages to discover and then disseminate link state information throughout the MANET. Individual nodes use this topology information to compute next hop destinations for all nodes in the network using shortest hop forwarding paths. The protocol is an optimization of the classical link state algorithm, tailored to MANET requirements. The key concept used is that of multipoint relays (MPRs). MPRs are selected nodes which forward broadcast messages during the flooding process. This substantially reduces the message overhead compared to a classical flooding mechanism, where every node retransmits each message when it receives its first copy. In OLSR, link state information is generated only by nodes elected as MPRs. Thus, a second optimization is achieved by minimizing the number of control messages flooded in the network. As a third optimization, an MPR node may choose to report only links between itself and its MPR selectors. Hence, contrary to the classic link state algorithm, partial link state information is distributed in the network and used for route calculation. OLSR provides optimal routes (in terms of number of hops). The OLSR Daemon (OLSRd) used in MULTIDRONE is the reference implementation of the IETF RFC 3626 from the OLSR.org association.

4.1.2 MULTIDRONE Metrics Plugin

The MULTIDRONE Metrics plugin has been developed in order to monitor link status (availability, performances) and report the results to the OLSRd routing daemon for route computation. The MULTIDRONE Metrics plugin comes as an OLSRd plugin. OLSRd sup-

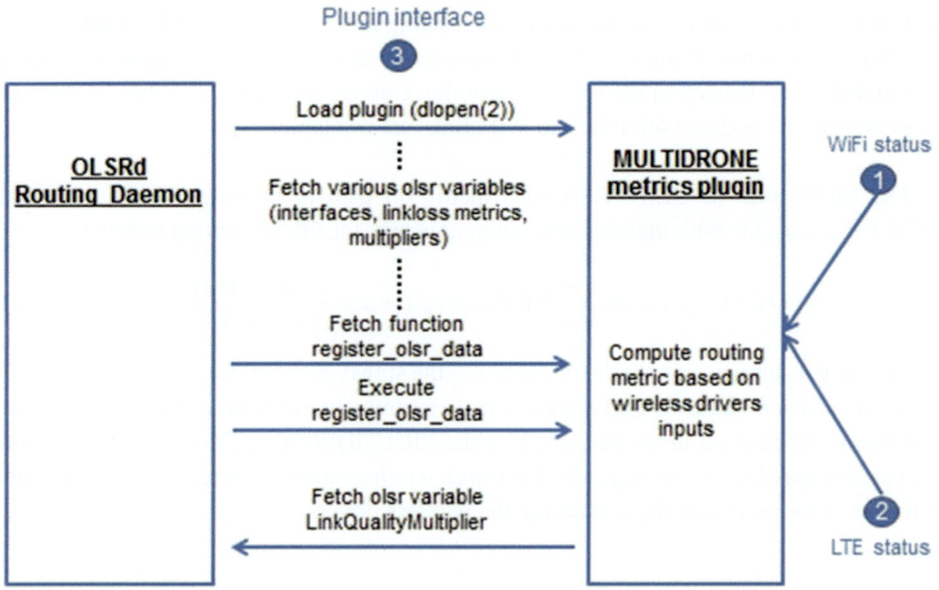

Fig. 6 The MULTIDRONE metrics plugin interface with OLSRd

ports loading of dynamically linked libraries (called plugins) for generation and processing of private package-types and any other custom functionality. The plugin design was chosen for, among others, the following reasons:

- No need to change any code in the OLSR daemon to add our MULTIDRONE custom metric or functionality.
- Plugins can be written in any language that can be compiled as a dynamic library.
- The plugin interface will always be backwards compatible.

Its interface with OLSRd is depicted in Fig. 6.

4.1.3 Metric Computation

Computation of the routing metric injected into OLSR is based on the wireless link status/info provided from the LTE and WiFi drivers. The metric value is influenced by:

- Signal strengths ($SStrengthI_{LTE}$ and $SStrenghI_{WiFi}$)—By selecting paths with strong signal, paths which can support high-data rates with small error rates are typically preferred.
- Link status: down or up.

- LTE bias: a bias (α) is introduced into the routing metric computation in order to avoid changing path too frequently. Indeed, frequent metric changes will introduce routing instability and latency in the routing algorithm convergence time. This bias is intended to favour LTE in case both LTE and WiFi links are available.

The legacy sum of OLSR ETX [25] metric on path is evaluated by OLSRd itself ($\sum ETX_{WiFiPath}$). With these values and parameters, the OLSR routing metric is:

$$OLSR_{LQ_{WiFi}} = \sum ETX_{WiFiPath} * \alpha * \frac{SStrengthI_{LTE}}{SStrenghI_{WiFi}}. \quad (1)$$

We use non-intrusive methods for obtaining the signal strength, for both LTE and WiFi, and estimate bandwidths. We do not generate measurement overhead that could potentially hurt the active payload flows generated by the MULTIDRONE platform. These methods are based on passive monitoring, which allows a wireless node to intercept the transmission activities of its own radios by consulting the wireless drivers.

4.2 MULTIDRONE Security Functions

MULTIDRONE's CM is not only in charge of routing IP traffic, but also of enforcing security functions for secure communications. Indeed, even if 4G LTE used in MULTIDRONE uses symmetric-key cryptography to (1) authenticate the subscriber (phone), and (2) encrypt data sent over the 4G wireless link (128 bit encryption), potential threats that can be exploited by a malicious attacker still exist. Even if this risk is considered minimal, the criticality of UAV control commands and the value of cinematography content require additional security measures that are described in this Subsection.

4.2.1 IPsec Deployment

In order to add end-to-end authentication and encryption on top of LTE, we chose the IPsec protocol [8] in a site-to-site configuration, in which each IPsec gateway is embedded into the MULTIDRONE CMs. IP flows are encrypted using AES-256 encryption on top of LTE's own encryption (based on AES 128 bit encryption). The designed IPsec security topology is based on site-to-site tunnelling used with the on-board and the ground CMs as tunnel end-points, thus efficiently encrypting all traffic generated by MULTIDRONE applications and expected to be transported by the LTE network.

The greatest advantage of IPsec is its transparency to MULTIDRONE applications. Since IPsec operates at OSI Layer 3 (IP), it has no significant impact on the higher network layers and no impact at all on the application layer. It is indifferent as to whether application traffic is being transported using TCP or UDP protocols. Consequently, IPsec is equally appropriate for securing real-time traffic (such as UAV video streams) and for traditional data applications using the ROS middleware.

IPsec is a suite of related protocols for cryptographically securing communications at the IP layer. IPsec also provides methods for the manual and automatic negotiation of *Security Associations* (SAs) and key distribution, with all the necessary attributes gathered in a *Domain of Interpretation* (DOI). The IPsec DOI is a document containing definitions for all the security parameters required for the successful negotiation of a VPN tunnel; essentially, all the attributes required for SA and Internet Key Exchange (IKE) negotiations. Additionally, since IPsec is deployed between the ground and the on-board CMs, it is not required for other MULTIDRONE computers to be IPsec capable.

4.2.2 Internet Key Exchange (IKEv2)

In order to establish the SAs required by IPsec between the ground and the on-board CMs, it is required to deploy a keying daemon. For this purpose, StrongSwan [34] is deployed on all CMs. StrongSwan is a keying daemon capable of using the *Internet Key Exchange* protocols (IKEv1 and IKEv2). IKE provides strong authentication of CMs and derives unique cryptographic session keys. Besides authentication and key material, IKE also provides the means to exchange configuration information (e.g., virtual IP addresses) and to negotiate IPsec SAs, which are often called CHILD-SAs. IPsec SAs define which network traffic is to be secured and how it has to be encrypted and authenticated. A CHILD-SA consists of two components:

- The actual IPsec SAs (there are two, one in each direction) describing the algorithms and keys used to encrypt and authenticate the traffic.
- The policies (there are at least two) that define which network traffic shall use such an SA.

The policies work both ways, that is, only traffic matching an inbound policy is allowed after decryption. Policies are derived from the *Traffic Selectors* (TS) negotiated via IKE when establishing a CHILD-SA. Unprotected traffic that the kernel receives and for which there is a matching inbound IPsec policy are dropped for security reasons. The actual IPsec traffic is not handled by strongSwan, but instead by the network and the IPsec stack of the operating system kernel.

Generally, IPsec processing and routing are two different topics. For the purpose of MULTIDRONE, IPsec is bumped into the stack and the original routing decision for the unprotected packet also applies to the protected packet. In this sense, IPsec was configured to be deployed in a policy-based manner, instead of route-based which is a less flexible mode.

Authentication is required to ensure that the peer with an IKE-SA is really who it claims to be. StrongSwan provides several methods to do this:

- Public Key Authentication
- Pre-Shared-Key (PSK)
- Extensible Authentication Protocol (EAP)
- eXtended Authentication (XAuth).

In the context of MULTIDRONE, Pre-Shared Key (PSK) was used for its ease of deployment, although it comes at the "cost" of requiring strong secrecy to be secure.

4.2.3 Advanced Encryption Standard (AES-256) and Cryptographic Hardware Acceleration

The purpose of cryptographic hardware acceleration is to offload the computation-intensive tasks of IP packet encryption/decryption and compression/decompression. Acceleration is achieved by executing any arithmetic calculations required by the AES-256 algorithm on dedicated hardware, instead of having the CPU in charge of it. A Cryptographic Hardware Accelerator can be:

- integrated into the System-on-Chip (SoC) as a separate, special-purpose processor.
- integrated in a co-processor on the circuit board.
- contained on a chip on an extension circuit board, that can be connected to the mainboard via some bus, e.g., PCI.

The i.MX6 Processors used in the MULTIDRONE on-board CM offer hardware encryption through Freescale's Cryptographic Accelerator and Assurance Module (CAAM, also known as SEC4). It offers the following support:

- Security Control.
- Advanced High Assurance Boot (A-HAB) System (HAB with embedded enhancements).
- SHA-256, 2048-bit RSA key.

For AES-256 encryption, the MULTIDRONE on-board CM is set-up to make use of the hardware cryptographic accelerators on-board the i.MX6 processors. The latter leverage a CAAM driver to make use of the above features, via the Linux CryptoAPI. The driver itself is integrated with the Crypto API kernel service, in which the algorithms supported by CAAM can replace the native software implementations. The Cryptodev module is implemented as an out-of-kernel module and, therefore, must be compiled against the i.MX6 kernel.

Table 2 MULTIDRONE QoS specifications

Service type	Protocol	Precedence	QoS QCI	QoS Priority	Max Latency (ms)	Max Packet loss
Telemetry	MAVLink	High	5	1	100	10^{-6}
Control and signalling	ROS	High	2	4	150	10^{-10}
NavCam video stream	RTP/RTCP/UDP	Medium	3	3	50	10^{-10}
NTP	NTP	Medium	7	7	100	10^{-3}
CinCam video stream	RTP/RTCP/UDP	Low effort	Best effort	Best effort	Best effort	Best effort

4.3 MULTIDRONE Quality-of-Service

In order to exploit the full capacity of the wireless links provided by the CMs, QoS is deployed. It is based on the Diffserv model and uses DSCP marking and different LTE radio bearers in order to classify, mark, and then enforce QoS on each IP flow depending on its Class-of-Service (COS). Table 2 provides the classification and mapping of each data stream (according to the protocols used) to a dedicated LTE QoS bearer.

The CinCam video stream has its precedence set to Low and a best-effort method, due to the higher criticality of the other streams. Without an optimal control/command and safety services (such as ROS) the UAVs are not able to fly. The NavCam video stream needs the QoS with the lowest latency to allow the backup pilot to have the best control of the vehicle if needed (only one QoS bearer allows a latency with a maximum of 50 ms).

The different ROS messages cannot be separated in different QoS classes without a specific DSCP marking. The approach chosen for MULTIDRONE was to aggregate all the ROS messages into the same QoS class, thanks to the UDP/TCP port management. Alternatively, one could employ ROS proxies with a dedicated ROS node per UAV and a dedicated ROS node on-ground that subscribes to all ROS topics, using TCP sockets to transport the messages over-the-air.

5 MULTIDRONE Communications Evaluation

The MULTIDRONE communications infrastructure was evaluated with regard to performance compared to the attenuation, i.e., understanding the impact of distance between the LTE ground station and the UAVs on QoS and the IPsec tunnel.

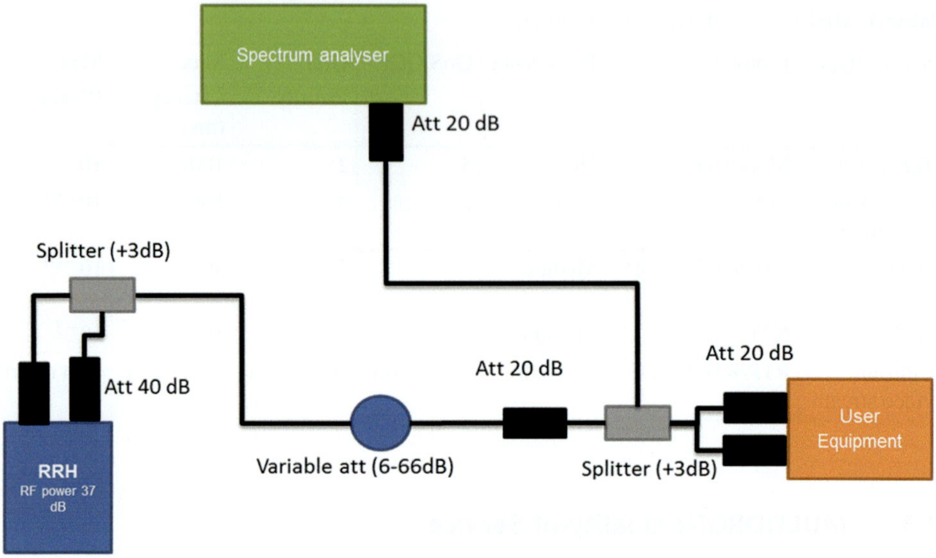

Fig. 7 LTE testbed setup

5.1 LTE Evaluation

All tests were performed with an RF cabled test bed (no over-the-air communication) to keep the environment under control by minimizing interference. The testbed, which is depicted in Fig. 7, was composed of:

- 1 Base Band Unit server
- 1 Remote Radio Head
- 1 User Equipment (UE).

Figure 8 depicts the results of the attenuation tests. **UE-ID** is the S1 eNodeB UE identity (unique among all cells), **CQI** is the Channel Quality Indicator (between 0, i.e., bad, and 15, i.e., very good), **mcs** is the Average Modulation and Coding Scheme, **Retx** is the number of transport blocks retransmissions, **txok** is the number of successfully transmitted transport blocks, while **Brate** is the average bitrate (in bits per second).

The conclusion is that a maximum throughput of around 15 Mb/s in downlink and 30 Mb/s in uplink are available until 108 dB of attenuation. It is possible to extrapolate from this attenuation result an approximate distance. To calculate it (assuming free environment, Line-of-Sight condition, no RF loss), the following equation is useful:

$$a = 32.45 + 20log(f) + 20log(d), \tag{2}$$

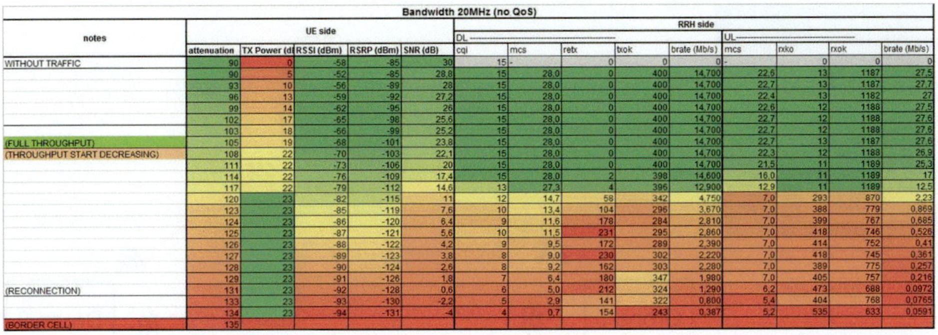

Bandwidth 20MHz (no QoS)														
notes	UE side					RRH side								
						DL					UL			
	attenuation	TX Power (d	RSSI (dBm)	RSRP (dBm)	SNR (dB)	cqi	mcs	retx	txok	brate (Mb/s)	mcs	rxko	rxok	brate (Mb/s)
WITHOUT TRAFFIC	90	0	-58	-85	30	15	-	0	0	0	-	0	0	0
	90	5	-52	-85	28,8	15	28,0	0	400	14,700	22,6	13	1187	27,5
	93	10	-56	-89	28	15	28,0	0	400	14,700	22,7	13	1187	27,7
	96	13	-59	-92	27,2	15	28,0	0	400	14,700	22,4	13	1182	27
	99	14	-62	-95	26	15	28,0	0	400	14,700	22,6	12	1188	27,5
	102	17	-65	-98	25,6	15	28,0	0	400	14,700	22,7	12	1188	27,6
	103	18	-66	-99	25,2	15	28,0	0	400	14,700	22,7	12	1188	27,6
(FULL THROUGHPUT)	105	19	-68	-101	23,8	15	28,0	0	400	14,700	22,7	13	1187	27,6
(THROUGHPUT START DECREASING)	108	22	-70	-103	22,1	15	28,0	0	400	14,700	22,3	12	1188	26,9
	111	22	-73	-106	20	15	28,0	0	400	14,700	21,5	11	1189	25,3
	114	22	-76	-109	17,4	15	28,0	2	398	14,600	16,0	11	1189	17
	117	22	-79	-112	14,6	13	27,3	4	396	12,900	12,9	11	1189	12,5
	120	23	-82	-115	11	12	14,7	58	342	4,750	7,0	293	870	2,23
	123	23	-85	-119	7,6	10	13,4	104	296	3,670	7,0	388	779	0,869
	124	23	-86	-120	6,4	9	11,6	178	284	2,810	7,0	399	767	0,685
	125	23	-87	-121	5,6	10	11,5	231	295	2,860	7,0	418	746	0,526
	126	23	-88	-122	4,2	9	9,5	172	289	2,390	7,0	414	752	0,41
	127	23	-89	-123	3,8	8	9,0	230	302	2,220	7,0	418	745	0,361
	128	23	-90	-124	2,6	8	9,2	162	303	2,280	7,0	389	775	0,257
	129	23	-91	-126	1,8	7	6,4	180	347	1,980	7,0	405	757	0,216
(RECONNECTION)	131	23	-92	-128	0,6	6	5,0	212	324	1,290	6,2	473	688	0,0972
	133	23	-93	-130	-2,2	5	2,9	141	322	0,800	5,4	404	768	0,0765
	134	23	-94	-131	-4	4	0,7	154	243	0,387	5,2	535	633	0,0591
(BORDER CELL)	135													

Fig. 8 LTE attenuation tests

where a is attenuation in dB, f is center frequency in MHz and d is distance in km. Then, to approximately calculate the distance according to the attenuation and the frequency:

$$d = 10\left[\frac{a - 32.45}{20} - log(f)\right] \tag{3}$$

Thus, according to the tests results and Eqs. (2) and (3), it is possible to achieve approximately 2.55 km with a full throughput, using a center frequency at 2350 MHz.

5.2 WiFi Mesh Evaluation

The main goal of these tests is to compare and measure the capacity and performances of the WiFi mesh (IEEE 802.11s protocol) with multiple UEs. Due to the number of UEs, it was not possible to test with RF cable. All these tests are done "over-the-air" in real radio conditions. Compared to wired tests, interference and RF multipaths may occur. The testbed is depicted in Fig. 9, while evaluation results are depicted in Figs. 10, 11 and 12.

5.3 Overall Evaluation

More generally, the overall prototype MULTIDRONE platform evaluation requirements demanded the ability to handle a fleet of three UAVs, with each one transmitting live through LTE a FullHD video stream (H.264 @ 30 Frames Per Second, from the cinematographic camera, at a maximum latency of 100 ms) plus a standard-definition video stream (640×480 pixels, H.264 @ 30 Frames Per Second, from the navigation camera, at a maximum latency of 50 ms). Thus, the system should be demonstrated as able to robustly handle a required

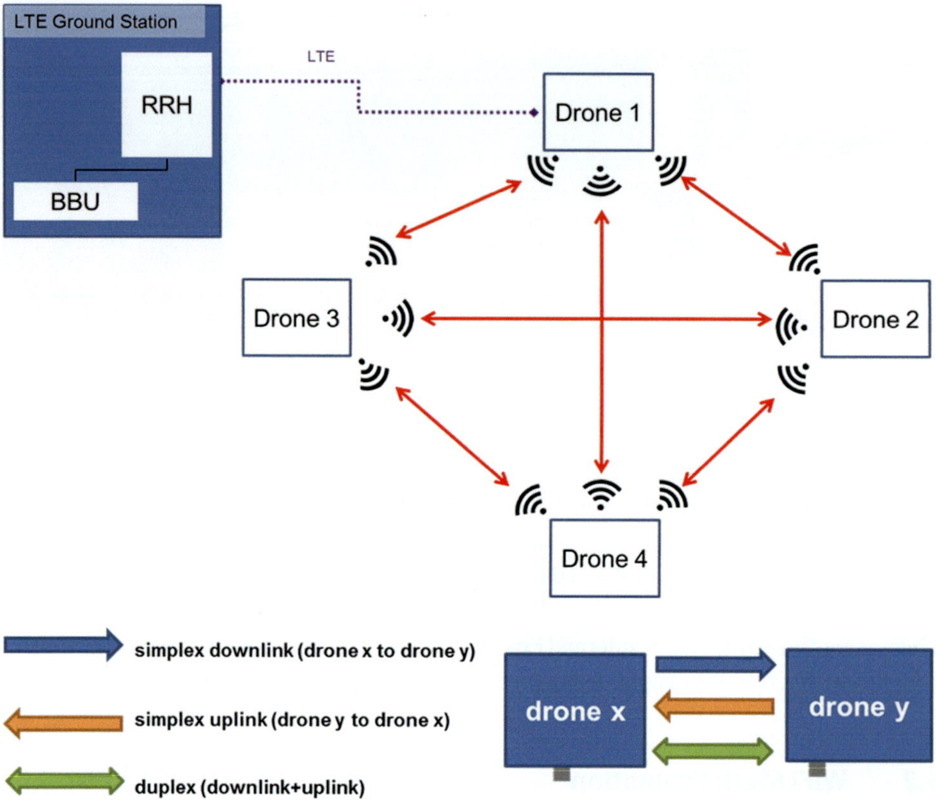

Fig. 9 WiFi mesh testbed setup

UE + direction		Average throughput (60 sec duration tests)		signal strength		comment
UE	direction	DL (Mb/s)	UL (Mb/s)	left UE (dBm)	right UE (dBm)	Throughput can't exceed 95,7Mb/s because of PoE switch limitation
1-2	simplex	95.7	95.7	-41	-44	
1-2	duplex	52.4	50.6	-41	-44	
2-3	simplex	95.7	95.6	-47	-37	
2-3	duplex	47.4	48.9	-47	-37	
3-4	simplex	95.7	80.2	-34	-48	
3-4	duplex	54.1	34.4	-34	-48	
1-4	simplex	95.6	95.6	-46	-52	
1-4	duplex	53.1	50.0	-46	-52	
4-2	simplex	95.6	95.6	-45	-43	
4-2	duplex	56.4	53.7	-45	-43	
3-1	simplex	72.5	78.5	-41	-51	
3-1	duplex	33.2	38.5	-41	-51	

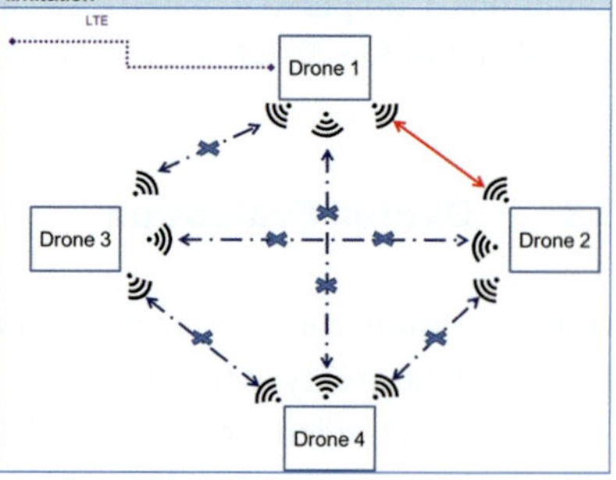

Fig. 10 WiFi mesh evaluation

UE + direction		Average throughput (60 sec duration tests)		comment
UE	direction	DL (Mb/s)	UL (Mb/s)	Throughput can't exceed 95,7Mb/s because of PoE switch limitation
1-2-3	simplex	48.5	43.6	
1-2-3	duplex	13.8	10.9	
2-3-4	simplex	45.4	42.9	
2-3-4	duplex	15.6	13.9	
3-4-1	simplex	45.0	33.2	
3-4-1	duplex	9.21	10.1	

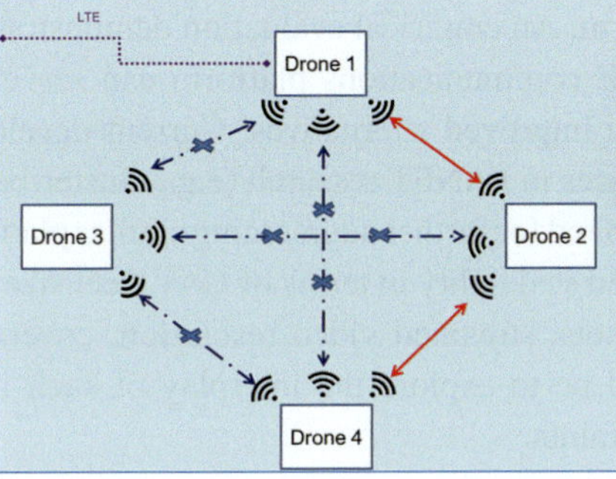

average signal strength during data exchange

	1	2	3	4
1		-36	-47	-43
2	-41		-49	-45
3	-40	-36		-40
4	-44	-47	-50	

Fig. 11 WiFi mesh evaluation

UE + direction		Average throughput (60 sec duration tests)		comment
UE	direction	DL (Mb/s)	UL (Mb/s)	Throughput can't exceed 95,7Mb/s because of PoE switch limitation
1-2-3-4	simplex	25.5	30.2	
1-2-3-4	duplex	9.26	10.2	

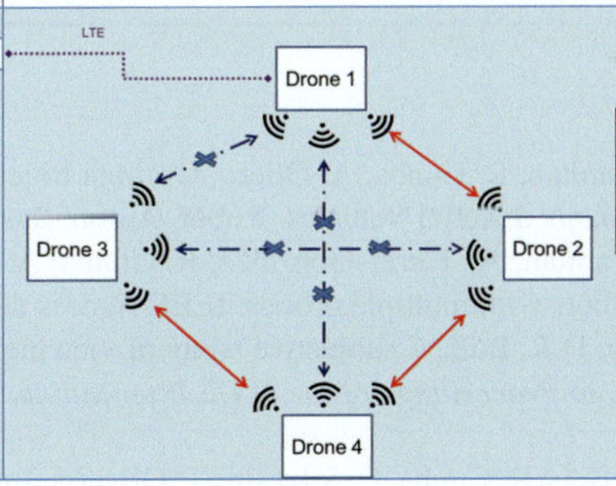

average signal strength during data exchange

	1	2	3	4
1		-36	-47	-43
2	-41		-49	-45
3	-40	-36		-40
4	-44	-47	-50	

Fig. 12 WiFi mesh evaluation

bandwidth of about 10 Mbps per UAV. Final evaluation of the overall MULTIDRONE platform indeed confirmed that the system met these goals, achieving a combined uplink throughput of 31 Mbps.

6 Conclusions

Communication issues (networking, security, and data flow) are especially challenging when using autonomous UAV fleets for media production/cinematography applications. The presented innovative, domain-specific, secure and robust communications architecture handles them, while emphasizing the more general and difficult filming scenario, i.e., live outdoor coverage. The designed communications infrastructure (utilizing both 4G/LTE and a WiFi mesh) for a partially autonomous UAV fleet coordinated by a central ground station was described in detail and the rationale behind the choices made was explained in the context of the presented platform. An empirical evaluation demonstrates the latter's effectiveness.

The MULTIDRONE communications platform can serve as a baseline prototype for facilitating research on improved alternatives. Current developments, such as the advent of 5G and recent advances in FANET research (e.g., cluster-based and bio-inspired routing protocols) may be exploited for further augmenting similar platforms with improved Quality-of-Service and enhanced scalability in terms of UAV fleet size, UAV formation range when capturing multiview shots, streamed video resolution, coverage area, etc. An interesting research avenue would be to explore the interplay of such improvements with aesthetic cinematographic constraints.

Acknowledgements The research leading to these results has received funding from the European Union's European Union Horizon 2020 research and innovation programme under grant agreement No 731667 (MULTIDRONE). This publication reflects only the author's views. The European Union is not liable for any use that may be made of the information contained therein.

References

1. A. Alcantara, J. Capitan, R. Cunha, A. Ollero, Optimal trajectory planning for cinematography with multiple Unmanned Aerial Vehicles. Robot. Auton. Syst. **140**, 103778 (2021)
2. A. Alcantara, J. Capitan, A. Torres-Gonzalez, R. Cunha, A. Ollero, Autonomous execution of cinematographic shots with multiple drones. IEEE Access 201300–201316 (2020)
3. S. Boyle, F. Zhang, D.R. Bull, A subjective study of viewing experience for drone videos using simulated content, in *Proceedings of the IEEE International Conference on Image Processing (ICIP)* (2019)
4. D. Cabecinhas, S. Brás, R. Cunha, C. Silvestre, P. Oliveira, Integrated visual servoing solution to quadrotor stabilization and attitude estimation using a pan and tilt camera. IEEE Trans. Control Syst. Technol. **99**, 1–16 (2017)
5. L.E. Caraballo, A. Montes-Romero, J.M. Diaz-Bañez, J. Capitan, A. Torres-Gonzalez, A. Ollero, Autonomous planning for multiple aerial cinematographers, in *Proceedings of the International Conference on Intelligent Robots and Systems (IROS)* (2020)
6. T. Clausen, P. Jacquet, C. Adjih, A. Laouiti, P. Minet, P. Muhlethaler, A. Qayyum, L. Viennot, Optimized link state routing protocol (OLSR). Technical report (2003)
7. R. Cunha, M. Malaca, V. Sampaio, B. Guerreiro, P. Nousi, I. Mademlis, A. Tefas, I. Pitas, Gimbal control for vision-based target tracking, in *Proceedings of the European Signal Processing Conference (EUSIPCO), Satellite Workshops* (2019)
8. C.R. Davis, *IPSec: Securing VPNs* (McGraw-Hill Professional, 2001)
9. A. Guillen-Perez, R. Sanchez-Iborra, M.D. Cano, J.C. Sanchez-Aarnoutse, J. Garcia-Haro, WiFi networks on drones, in *Proceedings of ITU Kaleidoscope: ICTs for a Sustainable World (ITU WT)* (IEEE, 2016)

10. E. Kakaletsis, I. Mademlis, N. Nikolaidis, I. Pitas, Bayesian fusion of multiview human crowd detections for autonomous UAV fleet safety, in *Proceedings of the European Signal Processing Conference (EUSIPCO)* (IEEE, 2020)
11. E. Kakaletsis, I. Mademlis, N. Nikolaidis, I. Pitas, Multiview vision-based human crowd localization for UAV fleet flight safety. Signal Process.: Image Commun. **99**, 116484 (2021)
12. E. Kakaletsis, C. Symeonidis, M. Tzelepi, I. Mademlis, A. Tefas, N. Nikolaidis, I. Pitas, Computer vision for autonomous UAV flight safety: an overview and a vision-based safe landing pipeline example. ACM Comput. Surv. (CSUR) **54**(9), 1–37 (2021)
13. I. Karakostas, I. Mademlis, N. Nikolaidis, I. Pitas, UAV cinematography constraints imposed by visual target tracking, in *Proceedings of the IEEE International Conference on Image Processing (ICIP)* (2018)
14. I. Karakostas, I. Mademlis, N. Nikolaidis, I. Pitas, Shot type constraints in UAV cinematography for target tracking applications. Inf. Sci. **506**, 273–294 (2019)
15. I. Karakostas, I. Mademlis, N. Nikolaidis, I. Pitas, Shot type feasibility in autonomous UAV cinematography, in *IEEE International Conference on Acoustics, Speech and Signal Processing (ICASSP)* (2019)
16. I. Mademlis, V. Mygdalis, N. Nikolaidis, M. Montagnuolo, F. Negro, A. Messina, I. Pitas, High-level multiple-UAV cinematography tools for covering outdoor events. IEEE Trans. Broadcast. **65**(3), 627–635 (2019)
17. I. Mademlis, V. Mygdalis, N. Nikolaidis, I. Pitas, Challenges in autonomous UAV cinematography: an overview, in *Proceedings of the IEEE International Conference on Multimedia and Expo (ICME)* (2018)
18. I. Mademlis, N. Nikolaidis, A. Tefas, I. Pitas, T. Wagner, A. Messina, Autonomous unmanned aerial vehicles filming in dynamic unstructured outdoor environments. IEEE Signal Process. Mag. **36**(1), 147–153 (2018)
19. I. Mademlis, N. Nikolaidis, A. Tefas, I. Pitas, T. Wagner, A. Messina, Autonomous UAV cinematography: a tutorial and a formalized shot type taxonomy. ACM Comput. Surv. **52**(5), 105:1–105:33 (2019)
20. I. Mademlis, A. Torres-Gonzalez, J. Capitan, R. Cunha, B. Guerreiro, A. Messina, F. Negro, C. Le Barz, T. Goncalves, A. Tefas, N. Nikolaidis, I. Pitas, A multiple-UAV software architecture for autonomous media production, in *Proceedings of the European Signal Processing Conference (EUSIPCO), Satellite Workshops* (2019)
21. I. Mademlis, A. Torres-González, J. Capitán, M. Montagnuolo, A. Messina, F. Negro, C. Le Barz, T. Gonçalves, R. Cunha, B. Guerreiro, F. Zhang, S., Boyle, G. Guerout, A. Tefas, N. Nikolaidis, D. Bull, A. Ollero, I. Pitas, A multiple-UAV architecture for autonomous media production. Multimedia Tools Appl. 1–30 (2022)
22. L. Meier, P. Tanskanen, F. Fraundorfer, M. Pollefeys, Pixhawk: A system for autonomous flight using onboard computer vision, in *Proceedings of the IEEE International Conference on Robotics and Automation (ICRA)* (2011)
23. A. Montes-Romero, A. Torres-González, J. Capitán, M. Montagnuolo, S. Metta, F. Negro, A. Messina, A. Ollero, Director tools for autonomous media production with a team of drones. Appl. Sci. **10**(4) (2020)
24. S.D. Muruganathan, X. Lin, H.L. Maattanen, Z. Zou, W. Hapsari, S. Yasukawa, An overview of 3GPP release-15 study on enhanced LTE support for connected drones (2018). arXiv preprint arXiv:1805.00826
25. X. Ni, K.C. Lan, R. Malaney, On the performance of expected transmission count (ETX) for wireless mesh networks, in *Proceedings of the 3rd International Conference on Performance Evaluation Methodologies and Tools* (ICST, 2008)

26. P. Nousi, I. Mademlis, I. Karakostas, A. Tefas, I. Pitas, Embedded UAV real-time visual object detection and tracking, in *Proceedings of the IEEE International Conference on Real-time Computing and Robotics (RCAR)* (2019)
27. P. Nousi, D. Triantafyllidou, A. Tefas, I. Pitas, Re-identification framework for long-term visual object tracking based on object detection and classification. Signal Process.: Image Commun. **88**, 115969 (2020)
28. C. Papaioannidis, I. Mademlis, I. Pitas, Autonomous UAV safety by visual human crowd detection using multi-task deep neural networks, in *Proceedings of the IEEE International Conference on Robotics and Automation (ICRA)* (2021)
29. F. Patrona, P. Nousi, I. Mademlis, A. Tefas, I. Pitas, Visual object detection for autonomous UAV cinematography, in *Proceedings of the Northern Lights Deep Learning Workshop* (2020)
30. M. Quigley, K. Conley, B. Gerkey, J. Faust, T. Foote, J. Leibs, R. Wheeler, A.Y. Ng, ROS: an open-source robot operating system, in *Proceedings of the IEEE International Conference on Robotics and Automation (ICRA) Workshop on Open Source Software* (2009)
31. F. Real, A. Torres-González, P. Ramón-Soria, J. Capitán, A. Ollero, UAL: an abstraction layer for unmanned aerial vehicles, in *Proceedings of the International Symposium on Aerial Robotics* (2018)
32. F. Real, A. Torres-González, P. Ramón Soria, J. Capitán, A. Ollero, Unmanned aerial vehicle abstraction layer: an abstraction layer to operate unmanned aerial vehicles. Int. J. Adv. Rob. Syst. **17**(4), 1–13 (2020)
33. B. Sabetghadam, A. Alcantara, J. Capitan, R. Cunha, A. Ollero, A. Pascoal, Optimal trajectory planning for autonomous drone cinematography, in *Proceedings of the European Conference on Mobile Robots* (2019)
34. A. Steffen, StrongSwan: the opensource IPsec-based VPN solution (2005), https://www.strongswan.org
35. C. Symeonidis, I. Mademlis, N. Nikolaidis, I. Pitas, Improving neural Non-Maximum Suppression for object detection by exploiting interest-point detectors, in *Proceedings of IEEE International Workshop on Machine Learning for Signal Processing (MLSP)* (2019)
36. D. Triantafyllidou, P. Nousi, A. Tefas, Fast deep convolutional face detection in the wild exploiting hard sample mining. Big Data Research **11**, 65–76 (2018)
37. Tzelepi, M., Tefas, A.: Graph-embedded convolutional neural networks in human crowd detection for drone flight safety. IEEE Transactions on Emerging Topics in Computational Intelligence (2019)

Ioannis Mademlis Dr. Ioannis Mademlis is a computer scientist, specialized in artificial intelligence. He received a M.Sc. degree in intelligent/cognitive systems (2014) and a Ph.D. in machine learning and computer vision (2018) from the Aristotle University of Thessaloniki, Greece (AUTH). Presently, he is employed as a postdoctoral research associate at AUTH. He has participated in 5 European Union-funded R&D projects, having co-authored more than 50 publications in academic journals and international conferences. He is an IEEE Senior Member and a committee member of the International AI Doctoral Academy (IAIDA). His current research interests include machine learning, computer vision, autonomous robotics and intelligent cinematography.

Paraskevi Nousi Paraskevi Nousi obtained her B.Sc. and Ph.D. degree in Informatics from the Aristotle University of Thessaloniki in 2014 and 2021 respectively. She is currently a post-doctoral researcher in the Artificial Intelligence and Information Analysis Laboratory in the Department of

Informatics at the Aristotle University of Thessaloniki. Her research interests include deep learning for computer vision, robotics, timeseries forecasting and gravitational waves analysis.

Damien Lavaux Damien Lavaux is an advanced studies engineer at Thales Communications & Security, France. He received his engineering degree from ESME-Sudria Paris engineering school, specializing in telecommunications networks in 2006. He has been involved in collaborative R&D projects since 2007. His current research topics cover wireless, mobile and ad hoc networks, from the software and IP protocol perspective, with a focus on mission-critical communications. He has participated in multiple different EC-funded R&D projects, having coordinated the awarded FP-7 HIT-GATE European R&D project and been technical manager of PPDR-TC.

Thomas Aubourg Thomas Aubourg received a M.Eng. on computer systems and networks from ESME Sudria in 2015. Since then, he has been working as an engineer with Thales Group in France, specializing in telecommunication networks and aeronautics.

Cedric Le Barz Cédric Le Barz, Ph.D., is a computer vision research engineer. In 2000, he first joined the imaging group of Philips Semiconductor Company, where he participated in R&D of algorithms for an ASIC dedicated to digital still cameras. He then joined Thales Company, where he initially worked in the Signal Processing team and acquired an expertise in image and video codecs (JPEG2000 & H.264), streaming over wireless networks and video analytics. In 2009, he joined the "Vision Lab" laboratory, a Common CEA/Thales endeavour focused on video analytics. In 2015 he received a Ph.D. on computer vision for UAVs, in collaboration with the French Aerospace Lab (ONERA) and the machine learning team of Paris Sorbonne University. He is now a research engineer with active interests in video streaming/processing and AI for autonomous systems. He has participated in, or coordinated several national or European R&D projects related to computer vision.

Ioannis Pitas Ioannis Pitas (IEEE Fellow, IEEE Distinguished Lecturer, EURASIP Fellow) received the Diploma and PhD degree in Electrical Engineering, both from the Aristotle University of Thessaloniki (AUTH), Greece. Since 1994, he has been a Professor at the Department of Informatics of AUTH and Director of the Artificial Intelligence and Information Analysis (AIIA) lab. He served as a Visiting Professor at several Universities. His current interests are in the areas of computer vision, machine learning, autonomous systems, intelligent digital media, image/video processing, human-centred interfaces, affective computing, 3D imaging and biomedical imaging. He has published over 906 papers, contributed in 47 books in his areas of interest and edited or (co-)authored another 11 books. He has also been member of the program committee of many scientific conferences and workshops. In the past he served as Associate Editor or co-Editor of 9 international journals and General or Technical Chair of 4 international conferences. He participated in 70 R&D projects, primarily funded by the European Union and is/was principal investigator/researcher in 42 such projects. He has 31600+ citations to his work and h-index 87+ (Google Scholar). Prof. Pitas leads the International AI Doctoral Academy (IAIDA) of the European H2020 R&D project AI4Media https://ai4media.eu/. He coordinates the HE project "TEMA" (g.a.n. 101093003).

Unmanned Aerial Vehicles Communications Security Challenges: A Survey

Moez Krichen and Alaeddine Mihoub

Abstract

It is well-established that UAVs are advantageous and effective for certain sorts of operations. The advancement of technology has made feasible what appeared unrealistic a few years ago, and the pace of research in this field augurs new UAV applications in a future that might not be that far after all. Additionally, the current state of UAV development and the outlook for this industry have positive repercussions, notably for industrial applications and numerous other connected industries. When human aid is either unavailable or problematic, UAVs can operate in emergency situations. These could help save lives in natural disasters, nuclear mishaps, and humanitarian crises, among other scenarios. Stakeholders, regulators, and institutions must carefully assess certain risks, as with any technology, in order to prevent, limit, and avoid the negative effects of certain implementations of this technology. Today's UAVs also pose the dangers of cybersecurity incidents and criminal usage. This study focuses on the security issues associated with the usage of unmanned aerial vehicles. Specifically, we provide an overview of the potential dangers, assaults, and possible responses pertaining to UAV communications. Software Defined

M. Krichen (✉)
FCSIT, Al-Baha University, Al-Baha 65528, Saudi Arabia
e-mail: mkreishan@bu.edu.sa

ReDCAD Laboratory, University of Sfax, Sfax 3038, Tunisia

A. Mihoub
Department of Management Information Systems and Production Management, College of Business and Economics, Qassim University, P.O. Box 6640, Buraidah 51452, Saudi Arabia
e-mail: a.mihoub@qu.edu.sa

M. Abdelkader and A. Koubaa (eds.), *Unmanned Aerial Vehicles Applications: Challenges and Trends*, Synthesis Lectures on Intelligent Technologies,
https://doi.org/10.1007/978-3-031-32037-8_12

Networks, Machine Learning Based techniques, Fog Computing, and Blockchain Technology constitute the majority of the countermeasures under consideration.

Keywords
UAV • Communications • Security • Threats • Attacks • Countermeasures • Survey

1 Introduction

An Unmanned Aerial Vehicle (UAV) [1, 2] is an unmanned aircraft that may either fly independently or be operated from the ground. UAVs may vary in size from a few centimeters for tiny versions to several meters for specialist ones. Long endurance UAVs have flight times ranging from a few minutes to more than forty hours.

The notion of the UAV originated during World War I. The UAV was initially a military objective [3–5]. Its growth followed the tempo of the twentieth century's major wars. Particularly for reconnaissance and surveillance, UAVs are more cost-effective while avoiding risking the lives of pilots and mobilizing land soldiers. Their prevalence among militaries and police services has substantially increased last years. As a result of their capacity to include cameras, infrared cameras, and environmental sensors, UAVs have spawned new applications in a variety of private sector disciplines such as medicine [6], farming [7] and environment [8], etc. Several transport corporations (UPS, UPS, Allship, and La Poste) and the e-commerce behemoth Amazon are developing also delivery UAV solutions [9].

Due to the introduction of tiny, inexpensive, and user-friendly entertainment UAVs in the 2010s, the leisure UAV industry has undergone a substantial rise. There are some lesser-known but more specialized uses for UAVs that are not corporate nor leisure or militaristic. When they contain an installed camera [10], they may be used to observe the wildlife or the infrastructure in inaccessible places, such as old sites, and wildfires. Additionally, they may have sophisticated instruments for geological studies.

UAVs are not all safe. They pose several cyber threats [11]. The only moderately secure aircraft is the military UAV. As with any other linked device, all others may be compromised or abused. They may endanger our privacy, cybersecurity, and physical security [12]. Non-military UAVs are comparatively simple to control. A few years ago, a security analyst created a gadget (named Icarus) that enabled him to tweak the UAV's transmission frequency [13]. The lines of communication would switch every 11 milliseconds, but while Icarus was waiting for a channel, he could break the aircraft's encryption and seize control of it. If the researchers are successful, you can bet that the hackers will be as well.

In addition, small Raspberry Pi CPUs are small enough to be carried by recreational UAVs [14, 15]. It is possible to program them to recognize Wi-Fi signals. Ethical scientists utilized these UAVs to evaluate the security of infrastructure (such as power plants) that cannot be directly accessible over the Internet. Moreover, UAVs represent a variety of threats to our physical security [16], including inadvertent harm, targeted assaults, and miscalculations. An uncontrolled UAV might also create mishaps. It might be a genuine user or a hacker who loses control, or it could be a software or hardware failure within the device. UAVs that drop from the sky and strike a person may inflict severe injuries, and much more so if the UAV is huge and heavy [17].

In this study, we are concerned with security issues associated with UAV usage. In particular, we want to identify the most common forms of security assaults that may be launched against UAVs, as well as the many countermeasures that can be implemented to combat these risks and attacks. The countermeasures under consideration rely mainly on Blockchain [18, 19], Machine Learning (ML) [20, 21], Fog Computing [22] and Software Defined Networks (SDN) [23, 24].

Our primary purpose in this study is to provide interested researchers and practitioners (who might not be experts in the field) with an easy-to-read paper that introduces them to the fundamental concepts linked to UAVs in general and the underlying communication security concerns that they face. Our contributions in this chapter may be summarized as follows:

1. We provide a comprehensive overview of UAVs, emphasizing their general architecture, their various subtypes and control mechanisms, their employment in various fields, and their primary advantages and disadvantages.
2. We list the main security gaps that exist for UAVs as well as the many sorts of assaults and security threats that could endanger them.
3. We identify the primary defenses that may be used to address the various security communication threats.

The structure of our article is as follows. In Sect. 2, we review some introductory information on UAV Technology. In Sect. 3, we highlight the main security flaws of UAVs. Section 4 lists the main potential assaults against UAVs. Section 5 provides related defense measures. Finally, Sect. 6 concludes the article and and makes suggestions for potential future expansions.

2 Preliminaries about UAVs

In this part, we provide an overview of the UAVs [25, 26], including its architecture, communication types, UAV kinds, and various control techniques.

2.1 Architecture

As seen in Fig. 1, the fundamental components of any UAV architecture are the Flight Controller, the Data link and Ground Control Station.

Fig. 1 UAV architecture

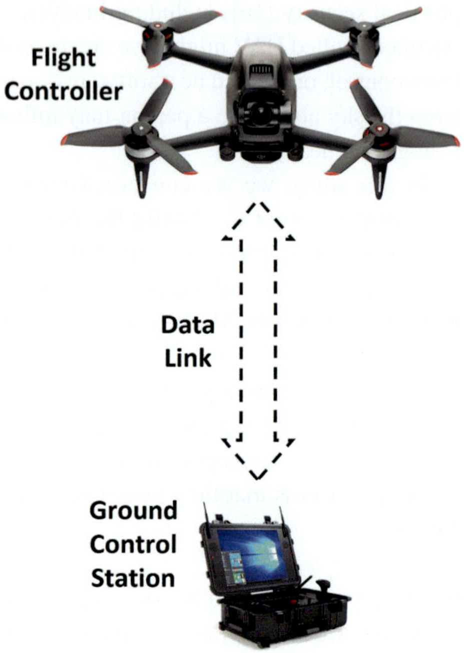

- Flight Controller: It is basically the UAV's primary processing module.
- Data Link: It is used to transmit and regulate flow of data between the UAV and the ground-based control station.
- Ground Control Station: It provides operators with the remote control and monitoring capabilities required for UAV operations.

As illustrated in Fig. 2, There are four kinds of UAV communications: UAV-To-Ground Station, UAV-To-Network, UAV-To-Satellite, and UAV-To-UAV.

- UAV-To-Ground Station: This kind of connectivity employs well-established and defined protocols, such as Wi-Fi and Bluetooth. Nevertheless, the vast majority of these communications are accessible publicly and unsecured, making them vulnerable to a variety of assaults.
- UAV-To-Network: This form of communication enables users to choose a network depending on the required degree of security. Additionally, cellular communication may be incorporated. When these wireless connectivity networks are implemented, it is crucial to ensure their security.
- UAV-To-Satellite: This kind of communication is used by the GPS network to transmit real-time positions. This permits any UAV that has strayed outside the control line to be

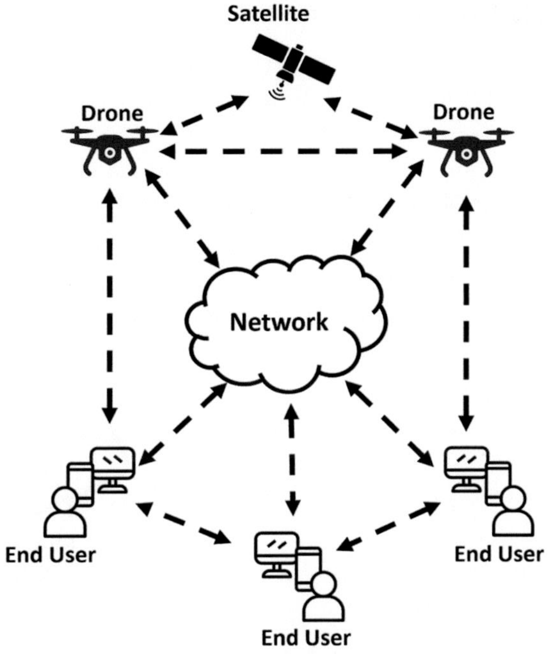

Fig. 2 Possible types of communications between UAVs and end-users

called back to its initial station. It is believed that satellite communications are secure and trustworthy. However, they are expensive and need extensive maintenance [27, 28].
- UAV-To-UAV: This method of communication is not yet standardized. It may be considered a P2P (peer-to-peer) interaction, which makes it susceptible to numerous P2P attacks.

2.2 Types of UAVs and Control Methods

As shown in Fig. 3, UAVs can be categorized in three main types which are: Hybrid-Wing UAV, Fixed-Wing UAV and, Multi-Rotor UAVs.

- Multi-Rotor UAV [29]: It employs the concept of vertical landings and takeoffs. It is operated with a high degree of precision and exactness. However, its movement is restricted and it uses a significant amount of energy.
- Fixed-Wing UAV [30]: It can glide and move at high speeds while delivering enormous freight loads. Its primary advantage is its energy efficiency. Due to its Horizontal Takeoff and Landing capabilities, a runway is required for these operations. Another drawback is that it cannot hover over fixed locations.

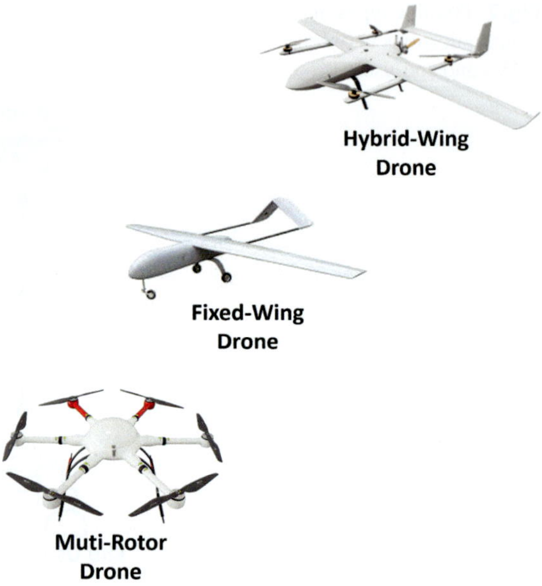

Fig. 3 Different types of UAVs

– Hybrid-Wing UAV [31]: Recently, UAV including both static and movable wings has joined the market. Using four rotors, this kind of UAV can swiftly reach its destination by floating through the air and hovering.

As illustrated in Fig. 4, there are three sorts of controlling UAVs which are pilot control, supervised control and autonomous Control.

– Autonomous Control: In this scenario, the UAV is capable of making all choices required to effectively accomplish the mission without human intervention [32, 33].

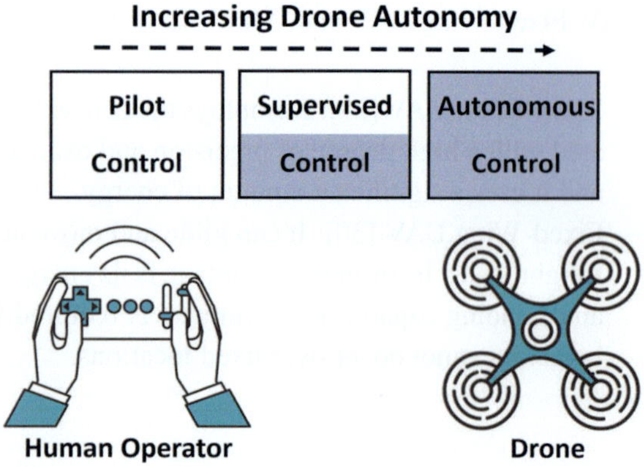

Fig. 4 UAV autonomy spectrum

- Supervised Control: This kind of control enables the UAV to start and finish a mission independently, while allowing for human intervention if required.
- Pilot Control: In this scenario, all decisions are made by a human distant operator.

2.3 Use of UAVs in Different Fields

UAVs can have several types of applications in various fields such as:

- Make War: Unfortunately, we cannot be exhaustive without recalling the primary use of UAVs. These military UAVs that are used in contemporary warfare are used to eliminate strategic targets without endangering the pilots. A use that is still disputed in the army because it puts a cold distance between the operator and his target, between the control screen and the eliminated target. The United States are the largest users of combat UAVs, which entered their armies in the 1990s. It was for example in 1994 that the Predators made their first test flights and were launched the following year in the Balkan. In general, the UAV has a triple use in contemporary armies: surveillance (inside and outside), reconnaissance of the terrain for ground and air troops and elimination.
- Have fun and learn to fly: Far from military concerns, the UAV has become what we know today under this name thanks to the so-called leisure UAV. Aeromodelling did not wait for the 2000s and the appearance of smartphones to create planes and other remote-controlled helicopters, but it is with a new generation of UAVs that use has passed from experienced professionals to the general public.
- Precision Agriculture: At the beginning of the year 2016, the agricultural UAV had a place of choice in the aisles of the innovation pole. Yes, these little flying machines will become valuable allies for farmers who want to monitor their fields. And for traditional agriculture or precision agriculture, the UAV can do a lot of things:
  - Identify and recognize soil heterogeneities;
  - Allow zones to be created on a plot to vary the density of seedlings;
  - Use zoning to homogenize a parcel in order to increase its potential;
  - Establish precisely the needs of the plantations;
  - Quickly know the damage on a plot and its cause.
- Help the police: Police UAVs will hunt bandit UAVs in the near future. But before reality meets fiction, there are these UAVs that some European gendarmerie want to use for reconnaissance or surveillance of hard-to-reach areas. For example in Paris, the police prefer the use of UAVs to helicopters, which are more expensive and more difficult to handle for reconnaissance in urban areas. The mini-UAV, which is easier to use and more responsive, can carry out covert missions of surveillance and observation over brief periods of time and distance. The use of UAVs is also being studied for the surveillance of road networks. The mobile radar could become an aerial mobile radar in the next few years.

- Broadcast Wifi: If Google has rather relied on balloons and Elon Musk on satellites, Facebook has imagined a UAV with the wingspan of a Boeing to accomplish this task. This machine named Aquila can stay for several months at very high altitude thanks to its solar panels and can communicate live with other UAVs to cover an area permanently. On the Internet broadcasting side, it is equipped with an optical laser capable of transferring data at 10 GB per second.
- Delivering Parcels: The UAV is a godsend for all companies that have a delivery activity: economical, mobile, fast and autonomous, the delivery UAV only has a size and weight disadvantage compared to a motorized equivalent. Thus, we see Amazon and other postal services positioning themselves on the research and development of UAV delivery. Smaller companies like PopChef believe it will be the next revolution in convenience delivery. UAVs can also be used to make deliveries in inaccessible areas.
- Journalism: UAV journalism also makes it possible to reach disaster areas or conflict zones to report without endangering reporters. It also allows to have new general plans (for example of a demonstration) without using expensive helicopters. A practice that is tending to spread: in the United States, there is already a course in UAV journalism at the University of Missouri or a professional association of UAV journalists. In practice, the journalists of the Daily Dot, for example, used a Phantom to document the explosion in Harlem.

2.4 Main Advantages of the Use of UAVs

Among the main advantages of using UAVs we mention:

- Their capacity for traveling far: It takes time to square or rake a vast area on foot. You can swiftly arrive on the scene for at least to delineate the search area and coordinate the teams thanks to a UAV that can fly at a respectable altitude and more than 50 km/h, sparing precious time. Consider the area that several UAVs operating at once could cover.
- They can be deployed quickly. The size of the search region can be swiftly reduced through the rapid deployment of search resources to high likelihood areas. Naturally, moving quickly is essential since as the hours go by, the odds of discovering people decrease. In contrast to a helicopter's rather leisurely deployment, UAVs may be deployed swiftly because to their size, weight, and battery capabilities.
- Staff Safety Enabling rescuers to search steep and rugged areas without endangering them is essential. In this context, UAVs represent an effective means of reducing the risks threatening human beings in areas deemed dangerous.

- Precision UAVs can be programmed and steered precisely to specific areas because they use the Global Positioning System (GPS). This is especially beneficial in a number of circumstances. UAVs are used in precision agriculture, for instance, to identify weed infestations, spray insecticide and fertilizer, and check on the condition of the crops. UAV precision allows farmers to save money and time.

2.5 Main Limitations of the Use of UAVs

Despite their obvious benefits, UAVs have certain limitations:

- Compared to human aircraft, UAVs fly at a lesser speed. Moreover, the UAV's flight time is constrained by its battery life; to increase that time, additional batteries may need to be installed to the UAV. The battery life is impacted by cold weather.
- It could be rather expensive to purchase a UAV with all the functions you require. The federal regulation stipulates that UAVs for various purposes must be equipped with particular hardware, software, and camera functions, which could be expensive. The expense of the UAVs is increased by the need for special training for people using them.
- Through a remote control, UAVs survey a location. It is simple to take pictures or record activities in a location because to the high-power zoom lenses, see-through imaging, and its night vision. People prefer to employ UAVs for investigations since they may fly undetectedly to a scene and monitor activity there.
- UAV use is impacted by local weather fluctuations. Since the majority of UAVs are built to fly at 30 mph, you cannot use them where the wind speed is less than 20 mph. UAVs can sustain technological damage and have communication issues with their controllers when flying in the rain or snow. They are only able to fly a few meters above the ground.
- UAVs are remotely controlled vehicles, hence a data link to the base control is necessary. Hackers can gain access to your control system by intercepting the data link network. Your opponents can interfere with local surveillance or even track your remote location once they have access to link over their local mobile telephone network.

3 Key Weaknesses in UAV Security

We highlight the following vulnerabilities of UAVs that might lead to security issues [34]:

- Both the Flight Controller and the Ground Control Station of a UAV are susceptible to cyber and physical assaults. Threats and vulnerabilities at the hardware level include issues like hardware breakdowns, physical UAV accidents, and hardware trojans [35].
- UAVs are technological devices that rely heavily on sensors [36]. As a consequence, their functionality is dependent on sensor readings. Due to the fact that these sensors process

sensitive data, a malicious operator might exploit them to compromise flight operations. For example, civil GPS broadcasts are unprotected and vulnerable to unauthorized access [37]. As a consequence, an attacker may exploit this issue by replicating a GPS signal in order to trick the operator.
- Malware that infects a UAV [38] may cause the loss of critical data and the system's control. Theoretically, an adversary with access to the UAV's navigation stack may shut it down, causing a denial-of-service attack and interrupting the flying mission. Thus, This spyware's incorporation into UAVs may compromise their security and privacy.
- Communication is a vital part for flight management and data transmission. WiFi is used by the majority of UAVs to connect with the ground-based control station [39, 40]. The complexity and dynamic nature of the wireless communications may be a source of potential vulnerabilities and grave threats.

4 Different forms of Attacks and Threats

As shown in Fig. 5, we list the following potential forms of attacks [41]: GPS Spoofing Attack [42], Data Interception Attack [43], Denial of Service Attack [44], Malware Infection Attack [45], Man In The Middle Attack [46], Wormhole Attack [47] and Jamming Attack [48].

- Denial of Service Attack [49]: It involves sending the UAV an overwhelming amount of requests that it cannot handle. The UAV enters a mode that prevents authorized users

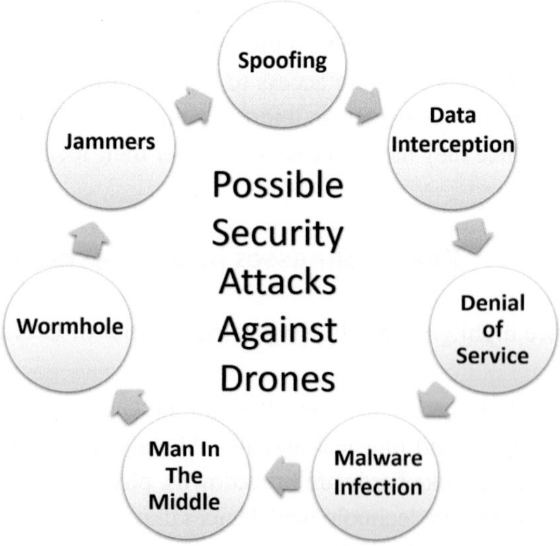

Fig. 5 Possible security attacks against UAVs

from gaining access. If a UAV is attacked this way, it will lose communication with its operator and might be unable to finish the mission.
- Man In The Middle Attack [50]: It happens when an attacker inserts himself into a conversation between the system and the user in order to eavesdrop or impersonate one of the parties, making it seem as if a regular information flow is occurring.
- Wormhole Attack [51]: is among the most prevalent assaults in which an adversary node captures packets from a specific network site and transmits them to a distant malicious node.
- Jamming Attack [52]: is a sort of Denial of Service attack wherein one node prevents other nodes from communicating by occupying the communication channel.
- Data Interception Attack [53]: is performed with packet sniffing program, which analyzes data packets when they transit over a network. The collected information is transmitted to the hacker.
- Malware Infection Attack [54]: is a sort of cyberattack in which malicious software seizes control of the UAV and uses it for illegal purposes. Malicious software encompasses a broad variety of attacks, such as malware, ransomware, command and control, and others.
- GPS Spoofing Attack [55]: is a kind of attack in which a close radio transmitter is deployed to disrupt genuine GPS signals. The hacker might deliver either no data or inaccurate coordinates.

5 Principal Countermeasures for Improving the Security of UAV Communications

In this paragraph, we will present four main technologies that are being utilized and researched extensively to improve the security and reliability of UAV communications (Fig. 6).

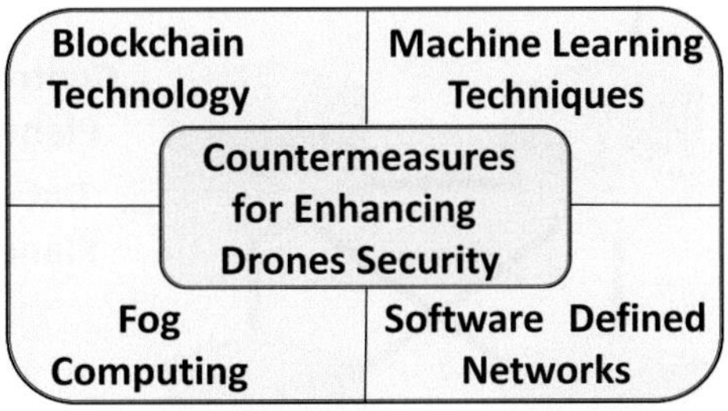

Fig. 6 Different types of countermeasures for enhancing UAVs security

5.1 Software Defined Languages for Improving Communications Security of UAVs

A Software Defined Network (SDN) [56, 57] is a network management method that enables dynamic, efficient network architecture in order to boost network performance. SDN was developed to address the reality that the static design of older networks is decentralized and complicated, but modern networks demand more flexibility and troubleshooting simplicity. By separating the transmission of network packets from the routing process, SDN aims to centralize network intelligence inside a single network component (a.k.a. controller machine). The difference between software-defined networks and traditional networks is illustrated in Fig. 7.

The following are the key benefits of using SDNs in general:

- Optimization: When using SDN, a new method of hardware device optimization is implemented. With the aid of SDN controllers, any gear, old or new, may be given a defined role. As a result, it does away with the limitation of hardware that serves a particular purpose only.
- Scalability: The degree of scalability that SDN provides is an additional advantage. The network's infrastructure can be altered immediately. All of this may be done without the need to buy or set up resources.
- Cost: SDN generally results in cost reductions because all network administration is automated and centralized. By leveraging SDN and improving virtualization and server

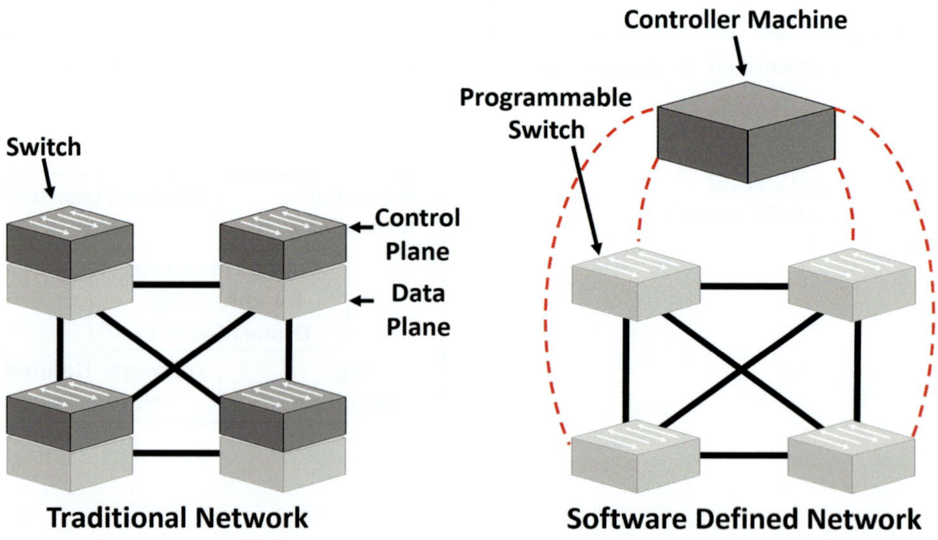

Fig. 7 Traditional network versus software defined network

utilization, users reduce wasteful costs. Aside from this, multitasking made possible by SDN decreases network operations. As a result, there is no longer a need for pricey hardware.

- Centralization: The network can be managed centrally thanks to SDN. From a single location, all networking devices may be managed and monitored. It removes the impediment to managing the infrastructure that was brought about by conventional methods. SDN makes it feasible to administer each system independently, even if that is necessary.
- Security: A controller included in SDN offers network-wide security. This controller makes sure the network is using the appropriate security information and guidelines. Additionally, SDN has a single management system. The security and features will be under the control of a single body. Due to the usage of a single central point, this type of technique is well renowned for being extremely safe. Security threats could be readily blocked by the administrator, preventing infection.

In the following, we enumerate some applications of the SDN-based Technology for improving UAV communications security:

- In [58], a lightweight approach based on the usage of SDN controller to prevent malfunctioning devices was described. The job of the SDN controller is to authenticate network devices wanting to enter the network.
- In order to create scalable networks that are immune to DoS attacks, the NetFence SDN-based protocol was suggested in [59].
- The Middlebox-Guard introduced in [60], represent a security model (based on software-defined network) that manages dataflows to maintain network efficiency and decrease network latency.
- In [61], a strategy for detecting and mitigating Distributed DoS attacks against UAVs was given. This approach is also designed to quickly identifying the device that initiated the attack.

Although the usage of SDNs may be helpful for enhancing the security of UAV communications, they still have several restrictions:

- Configuration: A SDN network configuration is not an easy task because it entails significant costs. Specifically, configuring each of them separately will not allow for the implementation of SDN protocols and controller. Reconfiguring the network as a whole is necessary.
- Maintenance: In order for networking to function, maintenance is a crucial element. An SDN is missing in its maintenance component. Especially when a network is expanding, it makes managing the actual devices practically impossible.

– Latency: A network requires room for each device that is linked to it. The amount of virtualized resources has an impact on how rapidly networks and devices can communicate. If greater speed is desired, more virtualized resources can be added. At the moment, virtualizing resources may result in a significant amount of latency.

5.2 Fog Computing for Improving Communications Security of UAVs

Fog computing [62], also known as distributed computing, is a platform that distributes computing resources across logical locations between a data source and the cloud, including data and programs. The goal is to deliver essential analytic services to the network's edge, which will improve performance by bringing resources closer to where they are needed, ultimately improving the performance and efficiency of the entire network. Fog computing can be utilized for network security as well because it can partition bandwidth usage and offer various firewalls for increased security. The general scheme of a Fog Computing Architecture is represented by Fig. 8.

The following are the general key benefits of using fog computing:

– Latency: The reduction in delay that results from processing some data locally is a significant advantage. The closest data source to the user in terms of distance can process the data. Instant responses may result from this, especially for services that require quick responses.

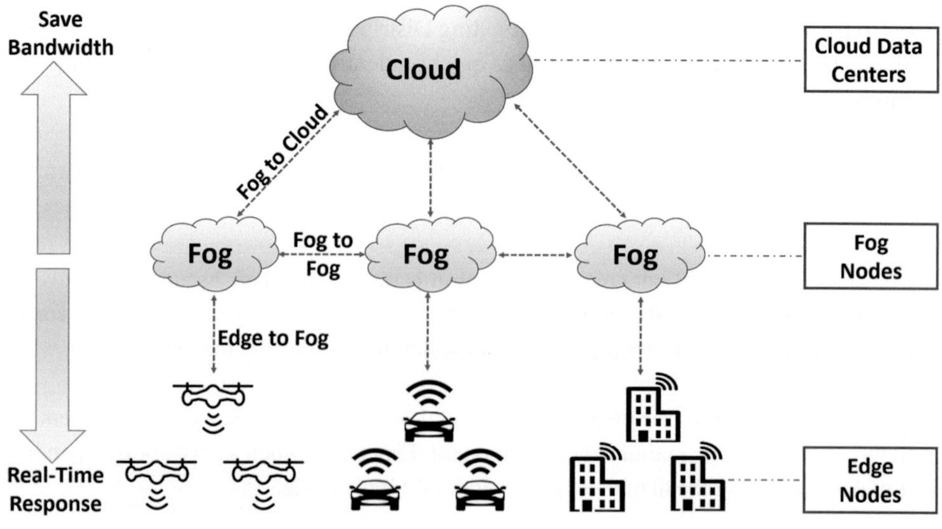

Fig. 8 Fog computing general scheme

- Bandwidth: Depending on the resources, the bandwidth needed to transfer data can be expensive. There are comparatively few bandwidth needs since the chosen data can be processed locally rather than being sent to the cloud. The reduction in bandwidth will be especially helpful as the number of IoT devices rises.
- Productivity: Customers can make use of fog applications to modify the machine's behavior to suit their needs. The correct tools allow developers to create these fog applications with ease. It can be deployed whenever they want when development is complete.
- Privacy: You may manage how much information is private by using fog computing. Instead of transferring the user's sensitive data to a centralized cloud infrastructure, it can be evaluated locally. The IT team would be able to monitor and manage the concerned device in this manner. Additionally, the cloud can be used to examine any subset of data that has to be delivered there.
- Security: The ability to connect many devices to the same network is a feature of fog computing. As a result, rather than being performed in a single location, the operations are carried out at numerous end points in a complicated distributed system. As a result, it is simpler to spot possible risks before they affect the entire network.

In the following, we list some examples that illustrate the way fog computing may be employed for improving communication security of UAVs:

- As stated in [63], heavy security and cryptographic tasks (such as key generation and hashing) are outsourced to more powerful nodes in the fog to ensure data confidentiality.
- In [64], a system for intrusion detection and intrusion prevention that can handle man-in-the-middle assaults at the fog layer was introduced.
- In [65], a fog-to-cloud computing solution based on proven army anti-UAV technology for detecting GPS jamming assaults was presented.
- In [66], a fog computing solution that optimizes latency in UAV connections was suggested.

5.3 Machine Learning Approaches for Improving Communications Security of UAVs

Machine Learning [67–70] is a sub-field of artificial intelligence (AI) that enables computers to learn and improve themselves without being explicitly programmed. Machine learning deals with the development of computer programs that can read data, learn without human intervention, and adapt their behavior appropriately. Figure 9 illustrates the difference between classic programming and machine learning techniques.

The key benefits of using machine learning techniques are summed up in the following sentences:

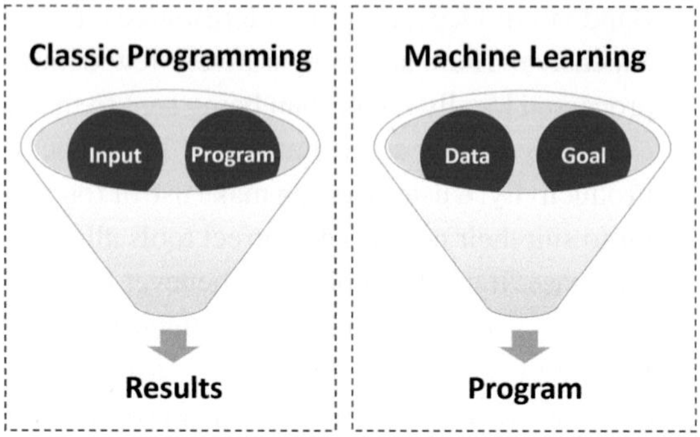

Fig. 9 Classic programming versus machine learning

- Used in a variety of fields: In many spheres of life, including education, medicine, engineering, etc., machine learning is used ranging from a very modest application to very large and complex structured machines that aid in data analysis and prediction. It not only turns into a healthcare provider but also offers prospective clients more individualized services.
- Automatic: With machine learning, a computer performs all data interpretation and analysis. For the prediction or interpretation of the results, no human interaction is necessary. The entire machine learning process begins with machine learning and anticipating the algorithm or program that will produce the best outcome.
- Find patterns and trends: A machine can learn more as it accumulates data, and as it accumulates data, it also learns patterns and trends. For instance, on social networking sites like Facebook, users surf and browse a variety of data, and their interests are recorded and understood through patterns that are then displayed to them in order to maintain their interest in the same app. Machine learning assists in recognizing trends and patterns in this way.
- Range of development: Machine learning improves itself and becomes more accurate and effective at work in the same way that humans do as they gain experience. This resulted in smarter decisions.

The study presented in [71] describes the key UAV security concerns that have been investigated by researchers, as well as the machine learning techniques that have been used to solve them. In the examples below, we demonstrate how machine learning can be used to enhance the communication security of UAVs:

- In [72], machine learning methods were used in order to lower latency and improves the consistency of data that must be sent to the cloud.
- In [73, 74], Support Vector Machines (SVM) as well as Artificial Neural Networks were used to identify Denial of Service Attacks.

- For adversary identification and avoiding adversarial assaults, a machine learning approach that relies on Convolution Neural Networks (CNN) and Recurrent Neural Networks (RNN) was applied [75].
- In radar detection, machine learning methods may be used to tackle a range of recognition and tracking challenges that plague traditional radar detection techniques [76].
- Recent research [77, 78] has also examined the use of machine learning techniques to avoid privacy breaches in UAV communications.

In this context, the main limitations of the use of Machine Learning techniques are:

- Time-consuming and requiring additional resources: The machine may occasionally take a long time to learn because efficacy and efficiency may only be attained via experience, which again takes time. Additionally, more resources are needed, like possibly more computers.
- The need for data is greater: For better forecasting or decision-making, a computer needs to be fed with more data since the more data it receives, the more accurate and effective it becomes. But occasionally, it might not be achievable. Additionally, the information must be accurate and neutral. Data requirements can be challenging at times. The amount of storage space needed to store the data increases as more data is needed for interpretation. It takes a lot of storage space to handle or keep data for further decision-making because more data means the computer has more information or material to learn from.
- Greater likelihood of error or malfunction: Even if machine learning is thought to be more accurate, it is very open to attack. For instance, the machine might be given a collection of biased or flawed programs. When the same program is used to create multiple forecasts or predictions, a chain of errors may develop that, while recognizable, may take some time to identify the source of. There is a danger of inaccurate interpretation since even minor data tampering or bias might start a long chain of errors. Sometimes data that is error-free can nonetheless be interpreted incorrectly by a machine because the data it was given may not have met all of its requirements.

5.4 Blockchain for Improving Communications Security of UAVs

Blockchain [79, 80] is a data storage method that makes it difficult or impossible to modify, damage, or falsify the data. Blockchain are digital records of transactions that are duplicated and shared throughout the devices and computers that constitute the whole blockchain network. Each block on the chain includes a collection of transactions, and if new transactions are added to the blockchain, new records of them are recorded to the local ledgers of all participants. A Distributed Ledger can be seen as a decentralized database that is administered by a wide group of individuals. Figure 10 illustrates the difference between classic systems and blockchain-based systems.

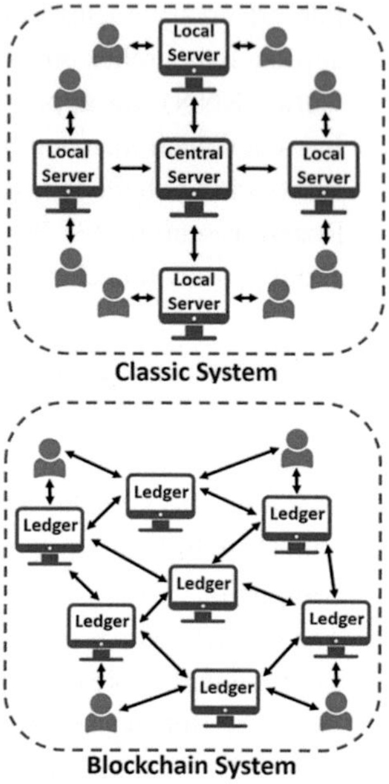

Fig. 10 Classic system versus blockchain System

The main advantages of the use of Blockchain Technology are:

- Transparency: One of the major problems in the current industry is transparency. Organizations have attempted to impose more rules and regulations in an effort to increase openness. However, centralization prevents any system from being completely transparent. With blockchain, a company may create a fully decentralized network with no requirement for a centralized authority, increasing the transparency of the system.
- Costs: Businesses currently spend a lot of money trying to manage their current system better. They aim to save costs so they can invest the savings in creating something new or streamlining existing procedures. Organizations can significantly reduce the costs related to third-party providers by embracing blockchain. Blockchain eliminates the requirement for any vendor charges because it lacks an inherited centralized player. Additionally, authenticating a transaction requires less engagement, which further reduces the requirement for spending money or effort on simple tasks.
- Efficacy and quickness: Conventional paper-based processes are time-consuming, prone to human error, and frequently require third-party mediation. By using blockchain to automate these processes, transactions may be completed more quickly and efficiently. The blockchain might combine transaction and documentation data, eliminating the need for paper-based exchange. There is no need to reconcile several ledgers, therefore clearing and settlement can go much more swiftly.

- Automation: Smart contracts allow for the automation of transactions, which will increase productivity and speed up the process even more. Upon the fulfillment of pre-specified requirements, the next stage of a transaction or process is automatically started. Smart contracts require less human involvement and rely less on third parties to verify that a contract's terms have been followed. For instance, once the customer has provided the necessary proof when filing an insurance claim, the claim may be instantly resolved and paid.
- Security: The security used by blockchain technology is greater than that of prior systems. The consensus strategy must be utilized to obtain consensus on all recorded transactions. In addition, each transaction is totally encrypted and linked to the preceding transaction using a hashing technique. Each node holds a copy of each transaction that has ever occurred on the network, hence enhancing its security. So long as other nodes are refusing the request to write transactions to the network, a malicious actor will be unable to alter the stored data.

By lowering security concerns, blockchain algorithms have the ability to establish new norms within the UAV industry. Even if a UAV's data slips into the hands of unscrupulous people, the encoded data it holds will remain secure. In particular, blockchain technology may improve the safety of UAV communications on the following fronts:

- Using blockchain technology, the current position of UAVs may be updated. Since the data of the blockchain will be broadcasted, neighboring uavs will be able to know locations of other UAVs, enabling them to escape and eliminate collisions totally. Real-time updates are also possible for information on restricted places. Due of this, UAVs will be prohibited to access prohibited areas.
- A blockchain identification system will monitor the flying of authorized UAVs and generate reports for further investigation by authorities. This can be completely accomplished without compromising the owner's privacy. Information classed as confidential will be kept indeed entirely private.
- The distributed ledger of Blockchain may be used to provide UAVs with encrypted data. The data sent by these devices will be encrypted and hence safe. Therefore, UAVs may be used for sensitive missions and operations.

However, blockchain technology, like many other techniques, still faces numerous and important challenges such as:

- As opposed to its equivalent centralized systems, blockchains are not as scalable. If you have ever used the Bitcoin network, you are aware that the speed at which transactions are executed depends on how congested the network is. This issue is connected to blockchain network scalability problems. Simply put, the likelihood of a network slowing down increases as more users or nodes join it.

- The Proof-of-Work consensus algorithm used by blockchain technology required miners to put in a lot of labor. The miners are rewarded for figuring out challenging mathematical puzzles. These challenging mathematical puzzles are less than optimal for use in the actual world because of their high energy requirements. The miners must answer the puzzles each time a new transaction is added to the ledger, which consumes a lot of energy.
- Interoperability is another drawback of blockchain technology. There are several kinds of blockchain networks that operate differently and attempt to address the DLT issue in their own particular ways, as was noted in the previous point. As a result, there are problems with interoperability if various chains can't efficiently communicate. When it comes to conventional systems and systems utilizing blockchain technology, the interoperability problem still exists.
- Blockchain technology is a very new invention. This indicates that the technology is new and needs time to develop. If you consider the various consortiums, you will see that many actors are each attempting to use their particular way to address the decentralized issue. For instance, we have Enterprise Ethereum, Corda, Hyperledger, Ripple, and more! Overall, there is still a long way to go until blockchain technology reaches maturity and organizations are less reluctant to adopt it. Maturity is another issue that blockchain must address, making it one of its drawbacks as with any other new technology.

6 Conclusion

UAV technology is still in its infancy. In the future decades, their capabilities will grow. Society and political institutions must be cognizant of the dangers they pose. Developed for good causes, these devices may be unfortunately abused. China has created solar-powered UAVs capable of continuous flight, for instance. When a UAV is equipped with a new, ultra-powerful camera and face recognition technology, it may fly indefinitely until it discovers pre-registered human targets. It is possible to locate and automatically eliminate a target with small weapons without the need of military-grade and sophisticated UAVs. Despite its appearance like science fiction, it has already become a reality. Moreover, UAVs are not effectively regulated, which represent a major concern. Yet no one has asserted the right to govern them so they do not constitute a grave danger to civilization. For the technical side and in order to acquire more robust findings, adopting formal approaches based on the usage of advanced mathematical models will be an intriguing strategy for future UAV security research. In the future, we intend to conduct a more in-depth investigation on the subject matter of this chapter. We are also eager to study additional intriguing issues about the environmental, social, and economic effects of using UAVs in the ensuing several decades while taking into account the unpredictable political environment that is currently dominating a significant portion of the world.

References

1. B. Aydin, Public acceptance of drones: knowledge, attitudes, and practice. Technol. Soc. **59**, 101180 (2019)
2. N.S. Labib, M.R. Brust, G. Danoy, P. Bouvry, The rise of drones in internet of things: a survey on the evolution, prospects and challenges of unmanned aerial vehicles. IEEE Access **9** (2021)
3. C. Candelmo, Drones at war: the military use of unmanned aerial vehicles and international law, in *Use and Misuse of New Technologies* (Springer, 2019), pp. 93–112
4. C. Paucar, L. Morales, K. Pinto, M. Sánchez, R. Rodríguez, M. Gutierrez, L. Palacios, Use of drones for surveillance and reconnaissance of military areas," in *International Conference of Research Applied to Defense and Security* (Springer, 2018), pp. 119–132
5. G. Choudhary, V. Sharma, I. You, Sustainable and secure trajectories for the military internet of drones (iod) through an efficient medium access control (mac) protocol. Comput. Electr. Eng. **74**, 59–73 (2019)
6. B. Hiebert, E. Nouvet, V. Jeyabalan, L. Donelle, The application of drones in healthcare and health-related services in north America: a scoping review. Drones **4**(3), 30 (2020)
7. Y. Al-Mulla, A. Al-Ruehelli, Use of drones and satellite images to assess the health of date palm trees, in *IGARSS 2020-2020 IEEE International Geoscience and Remote Sensing Symposium* (IEEE, 2020)
8. Ö. Yildirim, K. Diepold, R.A. Vural, Decision process of autonomous drones for environmental monitoring, in *IEEE International Symposium on Innovations in Intelligent Systems and Applications*, vol. 2019 (IEEE, 2019), pp. 1–6
9. C.D. Burzichelli, Delivery drones: will amazon air see the national airspace. Rutgers Comput Tech. LJ **42**, 162 (2016)
10. C.W. Chen, Drones as internet of video things front-end sensors: challenges and opportunities. Discov. Internet Things **1**(1), 1–12 (2021)
11. M. Krichen, W.Y.H. Adoni, A. Mihoub, M.Y. Alzahrani, T. Nahhal, Security challenges for drone communications: Possible threats, attacks and countermeasures, in *2nd International Conference of Smart Systems and Emerging Technologies (SMARTTECH)*, vol. 2022 (IEEE, 2022), pp. 184–189
12. M. Yahuza, M.Y.I. Idris, I.B. Ahmedy, A.W.A. Wahab, T. Nandy, N.M. Noor, A. Bala, Internet of drones security and privacy issues: taxonomy and open challenges. IEEE Access **9** (2021)
13. J. O'Malley, Pirates of the skies. Eng. Technol. **12**(3), 32–35 (2017)
14. S. Benhadhria, M. Mansouri, A. Benkhlifa, I. Gharbi, N. Jlili, Vagadrone: intelligent and fully automatic drone based on raspberry pi and android. Appl. Sci. **11**(7), 3153 (2021)
15. O. Westerlund, R. Asif, Drone hacking with raspberry-pi 3 and wifi pineapple: security and privacy threats for the internet-of-things, in *2019 1st International Conference on Unmanned Vehicle Systems-Oman (UVS)* (IEEE, 2019), pp. 1–10
16. J.A. Johnson, M.R. Svach, L.H. Brown, Drone and other hobbyist aircraft injuries seen in us emergency departments, 2010–2017. Am. J. Prev. Med. **57**(6), 826–829 (2019)
17. C.H. Koh, K. Low, L. Li, Y. Zhao, C. Deng, S.K. Tan, Y. Chen, B.C. Yeap, X. Li, Weight threshold estimation of falling uavs (unmanned aerial vehicles) based on impact energy. Transp. Res. Part C: Emerg. Technol. **93**, 228–255 (2018)
18. B. Bera, D. Chattaraj, A.K. Das, Designing secure blockchain-based access control scheme in iot-enabled internet of drones deployment. Comput. Commun. **153**, 229–249 (2020)
19. R. Jabbar, N. Fetais, M. Krichen, K. Barkaoui, Blockchain technology for healthcare: Enhancing shared electronic health record interoperability and integrity, in *2020 IEEE International Conference on Informatics, IoT, and Enabling Technologies (ICIoT)* (IEEE, 2020)

20. R. Gupta, A. Kumari, S. Tanwar, Fusion of blockchain and artificial intelligence for secure drone networking underlying 5g communications. Trans. Emerg. Telecommun. Technol. **32**(1), e4176 (2021)
21. H. Fatemidokht, M.K. Rafsanjani, B.B. Gupta, C.-H. Hsu, Efficient and secure routing protocol based on artificial intelligence algorithms with uav-assisted for vehicular ad hoc networks in intelligent transportation systems. IEEE Trans. Intell. Transp. Syst. (2021)
22. M. Al-Khafajiy, T. Baker, A. Hussien, A. Cotgrave, Uav and fog computing for ioe-based systems: a case study on environment disasters prediction and recovery plans, in *Unmanned Aerial Vehicles in Smart Cities* (Springer, 2020), pp. 133–152
23. C. Pan, J. Yi, C. Yin, J. Yu, X. Li, Joint 3d uav placement and resource allocation in software-defined cellular networks with wireless backhaul. IEEE Access **7**, 104 279–104 293 (2019)
24. A. Alioua, S.-M. Senouci, S. Moussaoui, H. Sedjelmaci, M.-A. Messous, Efficient data processing in software-defined uav-assisted vehicular networks: a sequential game approach. Wireless Pers. Commun. **101**(4), 2255–2286 (2018)
25. J.-P. Yaacoub, H. Noura, O. Salman, A. Chehab, Security analysis of drones systems: attacks, limitations, and recommendations. Internet Things **11**, 100218 (2020)
26. Z. Zhang, R. Zhang, Z. Jihong, G. Tong, C. Fei, W. Zhang, Integrated batteries layout and structural topology optimization for a solar-powered drone. Chinese J. Aeronaut. **34**(7), 114–123 (2021)
27. S. Bhatnagar, L. Gill, S. Regan, S. Waldren, B. Ghosh, A nested drone-satellite approach to monitoring the ecological conditions of wetlands. ISPRS J. Photogramm. Remote. Sens. **174**, 151–165 (2021)
28. Y. Hu, M. Chen, W. Saad, Joint access and backhaul resource management in satellite-drone networks: a competitive market approach. IEEE Trans. Wireless Commun. **19**(6), 3908–3923 (2020)
29. A. Bahabry, X. Wan, H. Ghazzai, G. Vesonder, Y. Massoud, Collision-free navigation and efficient scheduling for fleet of multi-rotor drones in smart city, in *IEEE 62nd International Midwest Symposium on Circuits and Systems (MWSCAS)*, vol. 2019 (IEEE, 2019), pp. 552–555
30. T. Elijah, R.S. Jamisola, Z. Tjiparuro, M. Namoshe, A review on control and maneuvering of cooperative fixed-wing drones. Int. J. Dynam. Control **9**(3), 1332–1349 (2021)
31. D. Todeschini, L. Fagiano, C. Micheli, A. Cattano, Control of vertical take off, dynamic flight and landing of hybrid drones for airborne wind energy systems, in *American control conference (ACC)*, vol. 2019 (IEEE, 2019), pp. 2177–2182
32. G.A. Venkatesh, P. Sumanth, K. Jansi, Fully autonomous uav, in *2017 International Conference on Technical Advancements in Computers and Communications (ICTACC)* (IEEE, 2017), pp. 41–44
33. A. Suleiman, Z. Zhang, L. Carlone, S. Karaman, V. Sze, Navion: a fully integrated energy-efficient visual-inertial odometry accelerator for autonomous navigation of nano drones, in *IEEE Symposium on VLSI Circuits*, vol. 2018 (IEEE, 2018), pp. 133–134
34. Y. Mekdad, A. Aris, L. Babun, A.E. Fergougui, M. Conti, R. Lazzeretti, A.S. Uluagac, A survey on security and privacy issues of uavs (2021). arXiv preprint arXiv:2109.14442
35. K. Xiao, D. Forte, Y. Jin, R. Karri, S. Bhunia, M. Tehranipoor, Hardware trojans: lessons learned after one decade of research. ACM Trans. Design Autom. Electron. Syst. (TODAES) **22**(1), 1–23 (2016)
36. P. Sniatala et al., Drone as a sensors' platform. Modern Technol. Enabl. Safe Secure UAV Oper. Urban Airspace **59**, 115 (2021)
37. X.-C. Zheng, H.-M. Sun, Hijacking unmanned aerial vehicle by exploiting civil gps vulnerabilities using software-defined radio. Sens. Mater. **32**(8), 2729–2743 (2020)

38. W. Niu, X. Zhang, X. Zhang, X. Du, X. Huang, M. Guizani et al., Malware on internet of uavs detection combining string matching and fourier transformation. IEEE Internet Things J. (2020)
39. J. Gordon, V. Kraj, J.H. Hwang, A. Raja, A security assessment for consumer wifi drones, in *2019 IEEE International Conference on Industrial Internet (ICII)* (IEEE, 2019), pp. 1–5
40. A. Guillen-Perez, R. Sanchez-Iborra, M.-D. Cano, J.C. Sanchez-Aarnoutse, J. Garcia-Haro, Wifi networks on drones, in *2016 ITU Kaleidoscope: ICTs for a Sustainable World (ITU WT)* (IEEE, 2016)
41. O.B. Fredj, O. Cheikhrouhou, M. Krichen, H. Hamam, A. Derhab, An owasp top ten driven survey on web application protection methods, in *International Conference on Risks and Security of Internet and Systems* (Springer, Cham, 2020), pp. 235–252
42. G. Panice, S. Luongo, G. Gigante, D. Pascarella, C. Di Benedetto, A. Vozella, A. Pescapè, A svm-based detection approach for gps spoofing attacks to uav, in *2017 23rd International Conference on Automation and Computing (ICAC)* (IEEE, 2017), pp. 1–11
43. Y. Zhao, K. Xu, R. Mijumbi, M. Shen, Design and evaluation of a policy-based security routing and switching system for data interception attacks, in *International Conference on Big Data Computing and Communications* (Springer, 2015), pp. 179–192
44. F.E. Salamh, U. Karabiyik, M. Rogers, F. Al-Hazemi, Drone disrupted denial of service attack (3dos): Towards an incident response and forensic analysis of remotely piloted aerial systems (rpass), in *15th International Wireless Communications and Mobile Computing Conference*, vol. 2019 (IEEE, 2019), pp. 704–710
45. S. Kumari, R.K. Upadhyay, *Exploring the behavior of malware propagation on mobile wireless sensor networks: stability and control analysis*. Math. Comput. Simul. (2021)
46. F. Ahmad, F. Kurugollu, A. Adnane, R. Hussain, F. Hussain, Marine: man-in-the-middle attack resistant trust model in connected vehicles. IEEE Internet Things J. **7**(4) (2020)
47. M. Tahboush, M. Agoyi, A hybrid wormhole attack detection in mobile ad-hoc network (manet). IEEE Access **9** (2021)
48. N. Giuri, Jamming techniques for protection against drone attack, Ph.D. dissertation. Politecnico di Torino (2020)
49. J. Chen, Z. Feng, J.-Y. Wen, B. Liu, L. Sha, A container-based dos attack-resilient control framework for real-time uav systems, in *Design, Automation and Test in Europe Conference and Exhibition (DATE)*, vol. 2019 (IEEE, 2019), 1222–1227
50. B. Bhushan, G. Sahoo, A.K. Rai, Man-in-the-middle attack in wireless and computer networking-a review, in *2017 3rd International Conference on Advances in Computing, Communication and Automation* (IEEE, 2017), pp. 1–6
51. A. Bhawsar, Y. Pandey, U. Singh, Detection and prevention of wormhole attack using the trust-based routing system, in *2020 International Conference on Electronics and Sustainable Communication Systems (ICESC)* (IEEE, 2020)
52. Z. Li, W. Qiao, Y. Lu, H. Lei, Optimal controller placement in mec-aided software-defined uav networks against jamming attack, in *Proceedings of the 2020 2nd International Conference on Robotics, Intelligent Control and Artificial Intelligence* (2020), pp. 74–79
53. S. Gogineni, S. Umar, A. Hemanth, A study of data interception in wireless sensor networks. Int. J. Sci., Eng. Comput. Technol. **3**(7), 239 (2013)
54. R.P. Ojha, P.K. Srivastava, G. Sanyal, N. Gupta, Improved model for the stability analysis of wireless sensor network against malware attacks. Wireless Pers. Commun. **116**(3), 2525–2548 (2021)
55. M. Majidi, A. Erfanian, H. Khaloozadeh, Prediction-discrepancy based on innovative particle filter for estimating uav true position in the presence of the gps spoofing attacks. IET Radar, Sonar Navigat. **14**(6) (2020)

56. A. Shaghaghi, M.A. Kaafar, R. Buyya, S. Jha, Software-defined network (sdn) data plane security: issues, solutions, and future directions, in *Handbook of Computer Networks and Cyber Security* (2020)
57. Y. Duan, W. Li, X. Fu, Y. Luo, L. Yang, A methodology for reliability of wsn based on software defined network in adaptive industrial environment. IEEE/CAA J. Autom. Sin. **5**(1) (2017)
58. F. Olivier, G. Carlos, N. Florent, New security architecture for iot network. Proc. Comput. Sci. **52**, 1028–1033 (2015)
59. X. Liu, X. Yang, Y. Xia, Netfence: preventing internet denial of service from inside out. ACM SIGCOMM Comput. Commun. Rev. **40**(4), 255–266 (2010)
60. Y. Liu, Y. Kuang, Y. Xiao, G. Xu, Sdn-based data transfer security for internet of things. IEEE Internet Things J. **5**(1), 257–268 (2017)
61. D. Yin, L. Zhang, K. Yang, A ddos attack detection and mitigation with software-defined internet of things framework. IEEE Access **6**, 24 694–24 705 (2018)
62. Opportunities and challenges, A. Kumari, S. Tanwar, S. Tyagi, N. Kumar, Fog computing for healthcare 4.0 environment. Comput. Electr. Eng. **72**, 1–13 (2018)
63. N. Abbas, M. Asim, N. Tariq, T. Baker, S. Abbas, A mechanism for securing iot-enabled applications at the fog layer. J. Sens. Actuator Netw. **8**(1), 16 (2019)
64. F. Aliyu, T. Sheltami, E.M. Shakshuki, A detection and prevention technique for man in the middle attack in fog computing. Proc. Comput. Sci. **141**, 24–31 (2018)
65. G. Ding, Q. Wu, L. Zhang, Y. Lin, T.A. Tsiftsis, Y.-D. Yao, An amateur drone surveillance system based on the cognitive internet of things. IEEE Commun. Mag. **56**(1) (2018)
66. X. Hou, Z. Ren, J. Wang, S. Zheng, W. Cheng, H. Zhang, Distributed fog computing for latency and reliability guaranteed swarm of drones. IEEE Access **8**, 7117–7130 (2020)
67. A. Mihoub, O.B. Fredj, O. Cheikhrouhou, A. Derhab, M. Krichen, Denial of service attack detection and mitigation for internet of things using looking-back-enabled machine learning techniques. Comput. Electr. Eng. **98**, 107716 (2022)
68. O. Ben Fredj, A. Mihoub, M. Krichen, O. Cheikhrouhou, A. Derhab, Cybersecurity attack prediction: a deep learning approach, in *13th International Conference on Security of Information and Networks* (2020), pp. 1–6
69. A. Mihoub, H. Snoun, M. Krichen, R.B.H. Salah, M. Kahia, Predicting covid-19 spread level using socio-economic indicators and machine learning techniques, in *First International Conference of Smart Systems and Emerging Technologies (SMARTTECH)*, vol. 2020 (IEEE, 2020), pp. 128–133
70. S. Mian Qaisar, N. Alyamani, A. Waqar, M. Krichen, Machine learning with adaptive rate processing for power quality disturbances identification. SN Comput. Sci. **3**(1), 1–6 (2022)
71. P.S. Bithas, E.T. Michailidis, N. Nomikos, D. Vouyioukas, A.G. Kanatas, A survey on machine-learning techniques for uav-based communications. Sensors **19**(23), 5170 (2019)
72. B. Wang, D. Liu, Y. Peng, X. Peng, Multivariate regression-based fault detection and recovery of uav flight data. IEEE Trans. Instrum. Meas. **69**(6), 3527–3537 (2019)
73. S. Seufert, D. O'Brien, Machine learning for automatic defence against distributed denial of service attacks, in *2007 IEEE International Conference on Communications* (IEEE, 2007), pp. 1217–1222
74. S. Umarani, D. Sharmila, Predicting application layer ddos attacks using machine learning algorithms. Int. J. Comput. Syst. Eng. **8**(10), 1912–1917 (2015)
75. P. Fraga-Lamas, L. Ramos, V. Mondéjar-Guerra, T.M. Fernández-Caramés, A review on iot deep learning uav systems for autonomous obstacle detection and collision avoidance. Remote Sens. **11**(18), 2144 (2019)
76. B. Taha, A. Shoufan, Machine learning-based drone detection and classification: state-of-the-art in research. IEEE Access **7**, 138 669–138 682 (2019)

77. C. Lin, D. He, N. Kumar, K.-K.R. Choo, A. Vinel, X. Huang, Security and privacy for the internet of drones: challenges and solutions. IEEE Commun. Mag. **56**(1), 64–69 (2018)
78. A. Kurniawan, M. Kyas, A privacy-preserving sensor aggregation model based deep learning in large scale internet of things applications, in *IEEE 17th World Symposium on Applied Machine Intelligence and Informatics (SAMI)*, vol. 2019 (IEEE, 2019), pp. 391–396
79. A.A. Monrat, O. Schelén, K. Andersson, A survey of blockchain from the perspectives of applications, challenges, and opportunities. IEEE Access **7**, 117 134–117 151 (2019)
80. M. Nofer, P. Gomber, O. Hinz, D. Schiereck, Blockchain. Business and information. Syst. Eng. **59**(3), 183–187 (2017)

Moez Krichen obtained his HDR (Ability to Conduct Researches) in Computer Science from the University of Sfax (Sfax, Tunisia) in 2018. He obtained his PhD in Computer Science in 2007. He is currently an Associate Professor at the University of Al-Baha (KSA) and a member of the Research Laboratory on Development and Control of Distributed Applications—REDCAD (Tunisia). His main Research Interest is Model-Based Testing Methodologies for Real-Time and Distributed Systems. Moreover, he works on applying Formal Methods to several Modern Technologies like Smart Cities, Smart Vehicles, Healthcare, etc.

Alaeddine Mihoub received his Ph.D. degree in Computer Science from Grenoble-Alpes University (France, 2015). Before, he received his MSc degree in Computer Science from Paris Descartes University (France, 2012) and his engineering diploma in Statistics and Data Analysis from ESSAIT (Tunisia, 2011). Actually, Dr. Alaeddine Mihoub is an Assistant Professor of Computer Science in the College of Business and Economics at the Qassim University (Saudi Arabia). Previously, he was a Research and Data Scientist at Maskott (France) and earlier, a Postdoctoral Researcher at Orange Labs (Meylan, France). He is the co-author of many innovative patents as well as several publications in well-ranked international journals and conferences. His-research interests include Social Signal Processing, Statistical and Probabilistic Models for Interactive Systems, and more largely Artificial Intelligence, Machine Learning, and Data Science

Legal and Regulatory Challenges

Challenges and Trends in the Aviation Industry: Integrating UAVs in Non-segregated Airspace

Sofia Michaelides-Mateou

Abstract

Unmanned Aerial Vehicles (UAVs), referred to as drones, are an integral part of the aviation industry. According to a study conducted by the Single European Sky Air Traffic Management Research (SESAR), it is estimated that the drone industry will be worth € 10 billion annually by 2035. The rapid growth in the use of drones undoubtedly provides many commercial benefits and unique opportunities for future use, but, at the same time, it also raises many challenges particularly with regards to using UAVs for civilian purposes. Safely integrating UAVs into non-segregated airspace while safeguarding the rights of all airspace users requires careful planning, the cooperation of all stakeholders as well as a very robust and effective legal framework. Unmanned aircraft are subject to civil regulation under the International Civil Aviation Organization (ICAO) and under regional and national aviation authorities. The chapter will provide an overview of the legal and regulatory framework with regards to integrating UAVs in nonsegregated airspace by providing a discussion and critical analysis of regulations by the ICAO on an international level and the European Aviation safety Agency (EASA) and the American Federal Aviation Administration (FAA) on a regional level before going on to highlight and discuss some of the legal complexities and challenges of UAVs operating in non-segregated airspace.

S. Michaelides-Mateou (✉)
College of Business Administration—Aviation Management, Prince Sultan University, P. O. Box 66833, Riyadh 11586, Kingdom of Saudi Arabia
e-mail: smateou@psu.edu.sa

M. Abdelkader and A. Koubaa (eds.), *Unmanned Aerial Vehicles Applications: Challenges and Trends*, Synthesis Lectures on Intelligent Technologies,
https://doi.org/10.1007/978-3-031-32037-8_13

1 Introduction

The aviation industry, comprised of a complex network of airlines, airports and air traffic management organizations, provides interconnectivity and accessibility across the globe. It is also a major contributor to global economic development providing $961.3 billion to global gross domestic product (GDP) and supporting a total of 87.7 million jobs worldwide by transporting over four billion passengers a year and approximately 61 million tonnes of cargo [1].

The aviation industry is essential for global business, it facilitates international tourism, advances mobility, and also fosters the advancement of cutting edge technology and innovations such as inter alia, artificial intelligence, biometrics, robotics, block chain, alternative fuels and unmanned aircraft or drones as they are commonly referred to.

Drones are said to date back to the mid-1800 [7] and have been used mainly by the military, however, there has been a great increase in using drones for different purposes in various sectors. The size, types and designs of drones vary considerably and this has led to an almost limitless expansion for the use of drones which can range from inter alia, law enforcement, recreational, commercial and other non-commercial purposes including border control, surveillance, security, infrastructure, media, telecommunication, aerial photography, agriculture, mining, humanitarian aid and disaster relief [38]. Other examples of the use of drones include homeland security, traffic control, aerial imaging, precision agriculture, farming management, infrastructure, pipeline and power inspection, resource planning, weather monitoring, scientific research, excavation, risk mitigation and even food and package delivery. Technological advancements such as inter alia, advanced motion and cameras, temperature sensors in cameras, 3D mapping software and extended range are boosting the growth in the commercial market and increasing the ever-expanding use of drones, and hence, their demand. For example, in 2020 Parrot SA launched a unique drone which combines data security with easy-to-use features specifically built for use by first responders, firefighters, search-and-rescue teams, security agencies, surveying, and inspection professionals [2].

Novel and recent uses for drones include: [29]

Multiple unmanned aircraft, equipped with innovative pods, were used by farmers in Israel to dispense pollen from the air as a means to compensate for the decrease in the bee population and the labour shortage caused by the COVID-19 pandemic;

An unmanned aircraft was used by Japan Airlines to transport 20 kg of fresh fish from Kamigoto airport to Saiki city. The drone, an 80 kg unmanned aircraft, operated a 35 km beyond visual line-of-sight flight to connect remote islands;

The State of New York created a 50-mile unmanned aircraft systems corridor to connect the cities of Rome and Syracuse, in New York, to be used for government agencies and first responders to operate the UAS beyond visual line-of-sight for emergency response and deliver medical and test-kits between hospitals and labs;

During the Covid-19 pandemic, a cargo unmanned aircraft was used in a pilot project to transport samples for COVID-19 testing. The flight was operated beyond visual line of sight between two hospitals in Warsaw and was managed by the Polish Air Navigation Services Agency;

In February 2022, three hospitals in Poland launched the first regular long-haul drone delivery service aimed at shortening the delivery time of medical samples. The vertical take-off (VTOL) drones, flew along an approximate 60-km route connecting three towns, Warsaw, Pułtusk and Sochaczew, 100 m above the ground at an airspeed of 72 km/h [42].

An unmanned aircraft or unmanned aerial vehicle (UAV) is defined by ICAO as a pilotless aircraft which is flown without a pilot-in-command on-board and is either remotely and fully controlled from another place, or programmed and fully autonomous [30]. The term was later replaced by unmanned aircraft systems (UAS), namely, an aircraft and its associated elements which are operated with no pilot on board (ICAO 2011). A remotely piloted aircraft system (RPAS) is a subset of a UAS, and is defined as a set of configurable elements comprised of a remotely piloted aircraft, its associated remote pilot station(s), the essential command and control links, and any other system elements as may be required at any point during flight operations [31]. The term UAV, UAS, RPAS and drone are used interchangeably in this chapter for simplicity and easy reading.

Estimates of the expected growth and value of the drone industry vary. A European Drones Outlook Study conducted in 2016 by the Single European Sky Air Traffic Management Research (SESAR) anticipated that the drone industry would be worth € 10 billion annually by the year 2035, with more than 400,000 active commercial and government drones expected in Europe alone by 2050 [47]. More recent studies on the Unmanned Aerial Vehicle Market published by Market and Market in 2021 state that the overall UAV market, including the original equipment manufacturer and the aftermarket (parts produced by different companies) was estimated to be USD 27.4 billion in 2021, with a projection of USD 58.4 billion by 2026 at compound annual growth rate of 16.4% from 2021 to 2026 [36]. Another study valued the global commercial drone market at $2.72 billion in 2020 with a projection to reach $21.69 billion in 2030, registering a compound annual growth rate of 23.7% [2]. Despite the different estimates, there is no doubt that UAVs are an integral part of the aviation industry and that the drone industry will impact the mode of transportation as well as many other industries.

The growth of the drone industry undoubtedly presents considerable opportunities, encourages innovation and competition and generates employment, but, at the same time it presents many complexities and challenges. The expanded uses of drones have emerged largely as a result of legal amendments and an improved regulatory framework moving towards fully integrating UAVs into unsegregated airspace. The unprecedented challenge for the aviation industry and the main challenge for using UAVs for civilian purposes is their safe integration into non-segregated airspace.

Safely integrating UAS into the unsegregated airspace systems presents many safety and security challenges, legal complexities, operational issues, and technological considerations which need to be considered. Deficiencies in the current regulatory and legal framework must be identified and addressed before establishing a safe, secure, sustainable and reliable airspace and seamlessly introduce UAVs into non-segregated civil airspace. In addition to providing a safe and secure environment for UAV operations, relevant laws and regulations need to be flexible enough to keep up with innovation, encourage economic competitiveness and allow for growth in the industry.

Safely integrating UAVs into non-segregated airspace while safeguarding the rights of all airspace users requires careful planning, the cooperation of all stakeholders as well as a very solid, robust and effective legal and regulatory framework. It is essential that safety, technical and operational requirements are dealt with on the entire legal spectrum, namely, internationally, regionally (for example, the European Union), and nationally, and that that the right balance between safeguarding the rights of all stakeholders and adequately addressing the many complex societal concerns and legal complexities is achieved.

The next section will set the scene by outlining the international regulatory framework established by the body that oversees and regulates the development of international aviation, namely, the International Civil Aviation Organization (ICAO) which lays down the foundation for the integration of remotely piloted aircraft in non-segregated airspace.

The second section provides a discussion of how, on a regional level, the European Union and the Unites States of America are amending or establishing relevant regulations, policies and procedures for the safe and secure integration of unmanned aircraft and the progress made with regards to the traffic management of unmanned aircraft, before presenting some recent developments and outlining some major concerns.

In the third and last section, some of the safety and security challenges emerging from integrating unmanned aircraft operations **into non-segregated airspace** will be discussed before briefly outlining additional legal complexities with regards to privacy, nuisance and liability issues in the event of an accident involving an unmanned aircraft.

2 Legal Framework

2.1 International—ICAO

The 1944 International Civil Aviation Convention known as the Chicago Convention established the International Civil Aviation Organization (ICAO) which is responsible for overseeing and regulating the development of international aviation. It is noteworthy to point out that even though drone technology started in the twentieth century, Article 8 of the 1944 Chicago Convention refers to a pilotless aircraft as follows "Pilotless aircraft—No aircraft capable of being flown without a pilot shall be flown without a pilot over the territory of a contracting State without special authorization by that State and in accordance with the terms of such authorization. Each contracting State undertakes to ensure that the flight of such aircraft without a pilot in regions open to civil aircraft shall be so controlled as to obviate danger to civil aircraft," [4].

The aim of ICAO with regards to UASs is to develop an international regulatory framework through Standards and Recommended Practices (SARPs), Procedures for Air Navigation Services (PANS) and guidance material, for the operation of UAS in a safe, harmonized and seamless manner. To this end, ICAO has established a number of procedures and guidance material, some of which are outlined below.

The 'Unmanned Aircraft Systems Study Group' was stablished in 2007 aiming to provide the basis for the global harmonisation of terms, strategies, concepts and principles needed to support the integration of remotely piloted aircraft in non-segregated airspace. The study group was instrumental in establishing the foundation for an international UAS framework and contains material for all UAS stakeholders, including regulators, manufacturers, operators, pilots and air navigation service providers. The Remotely Piloted Aircraft Systems Manual (RPASM) developed was aimed at providing guidance to member states on technical and operational issues applicable to the integration of RPA in non-segregated airspace. The final meeting of the Study Group was in June 2014, [3] when a dedicated ICAO Panel composed of experts, nominated by States and international organizations, was established.

In 2014, the Air Navigation Commission meeting of its 196th Session, agreed to the establishment of the Remotely Piloted Aircraft Systems Panel to replace the study group and to continue the work of the study group on unmanned aircraft systems. The Panel was specifically mandated to:

(a) serve as the focal point and coordinator of all ICAO RPAS related work, with the aim of ensuring global interoperability and harmonization;
(b) develop an RPAS regulatory concept and associated guidance material to support and guide the regulatory process;
(c) review ICAO SARPs, propose amendments and coordinate the development of RPAS SARPs with other ICAO expert groups;
(d) assess impacts of proposed provisions on existing manned aviation; and
(e) coordinate, as needed, to support development of a common position on bandwidth and frequency spectrum requirements for command and control of RPAS for the International Telecommunications Union World Radio Conference negotiations' [3].

The framework on unmanned aircraft systems includes the ICAO Model UAS Regulations titled Parts 101, 102 and 149, and the advisory circulars (ACs) which provide member states with a guide to implement or supplement their existing regulations. Member states are urged to notify ICAO of any developments and enhancements in their regulations.

Part 101 provides, inter alia, that all unmanned aircraft should be registered. Unmanned aircraft weighing 25 kg or less and operating in standard operating conditions do not require any additional operational review, however, those weighing over 15 kg must be inspected and approved. Part 102 addresses all unmanned aircraft weighing more than 25 kg or those weighing 25 kg or less but do not adhere to Part 101 requirements. The provisions permit on-going operations or one-time operations through certification and also provide for a quicker review when manufacturers declare a type or model of unmanned aircraft as being sufficiently tested for a specific operational category or having received an approval through an Approved Aviation Organization. Part 149 stipulates that an Approved Aviation Organization can serve as a designee authorized by the Civil Aviation Authority (CAA) to perform specific tasks. Once the organization has been certified, authorized tasks including remote pilot licensing, inspection and approval of the unmanned aircraft may then be processed quicker as well as reduce the workload for the CAA inspectors. The CAA can however adapt the authorizations as needed and directly oversee the authorization [32].

The ICAO advisory circulars with regards to the Model UAS Regulations, include inter alia: [32]

- AC 101-1 providing guidance associated with rule 101 regarding unmanned aircraft system operations in the Open Category;
- AC 102-1 providing guidance associated with rule 102 regarding the specific category, UAS authorizations or a UAS operator certificate and addresses requirements for manufacturers;
- AC 102-37 providing guidance for the carriage of dangerous goods transported by unmanned aircraft including the risks and responsibilities for safe carriage and information regarding packing and marking; (ICAO Model UAS Regulations, 2020).
- Canada AC 922-001, RPAS Safety Assurance aiming to assist states to set and develop standards for the manufacturer's compliance;
- AC 922-001 providing an example of performance-based criteria specifying to what standard the manufacturer must comply with for operations in controlled airspace, over people, or near people and also provides a classification of injury severity;
- Humanitarian Guidance which applies to unmanned aircraft operations providing assistance during emergency response and humanitarian deliveries.

ICAO has also provided information and assistance regarding unmanned traffic management. The 'Unmanned Aircraft Systems Traffic Management (UTM)-A Common Framework with Core Principles for Global Harmonization', currently in its third edition, provides member states with a framework of a typical UTM system and guidance aimed at harmonizing UTM activities across all member states. The UTM system will assist civil aviation authorities and air navigation service providers to provide real-time information regarding airspace constraints as well as the intentions of other aircraft which will be made available to UAS operators and their remote pilots directly or through a UTM service provider. The responsibility will then be on the operator of the UAS to safely manage its operations within those constraints and without receiving positive air traffic control services from the air navigation service providers [34].

The framework emphasises the importance of aircraft operating in the UTM being separated from each other and other hazards such as buildings, terrain and bad weather and also highlights this as being a major challenge. Adequate separation management must be achieved with proper guidance and responsibilities as well as a number of other tools and procedures which include inter alia:

- a detect and avoid capability for identifying, detecting and avoiding conflicting aircraft and any other hazards;
- methodologies providing enhanced detectability of unmanned aircraft by manned aviation;
- assessing existing and future separation standards between unmanned and manned aircraft and developing additional UA separation standards within the UTM system which include a number of safety margins;
- assessing the surveillance capability to support the integration of novel aircraft;
- determining means of compliance or a system to approve the UTM systems; assigning responsibility for conflict management and separation provision;
- achieving data quality regarding accuracy, resolution, integrity, timeliness, completeness, traceability and format of the system, all of which need to be compatible and interoperable; and
- forecasting and disseminating micro-weather information [34].

Additionally, the ICAO UAS Toolkit provides UAS best practices, lessons learned as well as regulations for member states to consider and adopt. A very important and comprehensive publication is the 'Remotely Piloted Aircraft System (RPAs) Concept of Operations (CONOPS) For International Instrument Flight Rules Operations' which aims to describe the operational environment of manned and unmanned aircraft in order to provide a common understanding of RPAS, their special characteristics and challenges, and how RPAS can ultimately be integrated into the airspace for international instrument flight rules operations. It provides a general framework for all stakeholders, including regulators, operators, airspace users, manufacturers, air navigation service providers and aerodrome operators to consider [33].

Although comprehensive in nature, it has a number of drawbacks. It is limited in scope as it only applies to certificate RPAS who operate internationally within controlled airspace under instrument flight rules in non-segregated airspace and at aerodromes form 2031 onwards. It is also limited in application to aircraft and the kind of RPAS operations as fully autonomous aircraft operations, visual line-of sight, very low altitude airspace operations and very high-altitude operations (e.g. above a flight level of 60 000 feet). In addition, the carriage of persons and domestic operations are not included (Table 1).

Table 1 Below illustrates the broad regulatory framework regarding unmanned aircraft systems, which leads to a number of concerns discussed below [32]

ICAO	ICAO Model UAS Regulations titled Parts 101, 102 and 149, and the advisory circulars (ACs) include:
Part 101	All unmanned aircraft should be registered Unmanned Aircraft Systems (UAS) [25 kg or less] Operating in compliance with [Part 101] Rules
Part 102	Unmanned aircraft weighing more than 25 kg or those weighing 25 kg or less but do not adhere to Part 101 requirements
Part 149	Civil Aviation Authority (CAA) can authorize and designate an Approved Aviation Organization to perform specific tasks
AC 101-1	Guidance associated with rule 101—in the Open Category
AC 102-1	Guidance associated with rule 102—in the Specific category UAS authorizations or a UAS operator certificate Requirements for manufacturers
AC 102-37	Unmanned Aircraft Systems (UAS) Carrying Dangerous Goods
Canada AC 922-01	Remotely Piloted Aircraft Systems Safety Assurance
AC 922-001	Performance-based criteria specifying for manufacturers
UTM	The 'Unmanned Aircraft Systems Traffic Management—A Common Framework with Core Principles for Global Harmonization'—provides states considering the implementation of a UTM system with a framework of a typical UTM system
ICAO UAS Toolkit	UAS best practices, lessons learned, regulations for member states to consider and adopt
CONOPS	Concept of Operations For International Instrument Flight Rules Operations

2.1.1 Concerns

The framework for unmanned aircraft traffic management (UTM) provided by ICAO highlights the many gaps which need to be focused on and tackled in a suitable and fitting manner for the integration of unmanned vehicles into non-segregated airspace [34], including:

- Airspace classification—The current airspace classification may not effectively support visual line-of-sight or beyond visual line-of-sight operations and this may require modifications of current classes of airspace or creating new classes of airspace;
- Airspace access—Policies, rules and priorities to support equitable access to airspace need to be developed
- Rules of the Air—Current Rules of the Air specifying for example, flight rules, right-of-way, altitude above people and obstructions, distance from obstacles and types of flight rules, are incompatible with unmanned aircraft operations within UTM systems.

- Operational procedures—Procedures specific to the UTM system such as normal, contingency and emergency scenarios need to be harmonized with ATM systems;
- Liability—Liability and insurance implications for unmanned service providers need to be determined;
- Certification—Certification of the UTM system interacting with an ATM system, for unmanned aircraft which conform to the airworthiness requirements must be scaled to the appropriate risk-based level;
- Data standards—Appropriate data standards and protocols to support UTM safety-related services and the exchange of data between UTM and ATM systems and between multiple UTM systems are needed;
- Positional references—Common altitude, navigation and temporal references for manned and unmanned operations are needed and gaps need to be resolved;
- Interface between UTM and ATM—Procedures and adequate tools must be developed to ensure the sharing of information, the interoperability of the two systems, and to identify roles, responsibilities and limitations;
- Data recording—Data-recording policies and capabilities, such as the ATC data retention and aircraft flight recorder requirements, are needed to support accident/incident reporting and investigative requirements;
- Communications—Remote pilot interfaces, capabilities and performance requirements for communications with the UTM system must be developed, including the ability to interface and communicate with ATC and the pilots of manned aircraft;
- Alerting systems—The safety and integrity of the UTM system, failure-alerting and failure management need to be addressed;
- Contingency management protocols—Operating protocols that account for contingencies both of the UTM system(s) providing multiple services and of the aircraft operating within the UTM system need to be developed;

The timeframe before any of the gaps highlighted are adequately addressed and the necessary procedures are adopted and implemented can be lengthy. In addition, implementation is dependent on member states amending or establishing and finally adopting relevant policies, procedures and regulations. Technological advancements, research and testing, increased needs and usage and market growth, amongst many other factors, will surely test and illustrate the effectiveness and applicability of such guidance material.

An overview of the some of the policies, procedures and regulations adopted and amended on a regional level is provided by discussing the progress made by the European Union and the United States of America.

2.2 Regional

2.2.1 EASA and Europe

In Europe, the body responsible for civil aviation safety is the European Union Aviation Safety Agency (EASA), which was established in 2002. It is an agency of the European Union with responsibility for civil aviation safety and is responsible for certification, regulation and standardisation by setting rules, standard guidelines and standards and also performs investigation and monitoring. Its headquarters are in in Cologne, Germany, and it also has a branch office in Brussels and representation offices in Beijing, China; Montreal, Canada; Singapore and Washington D.C. in the United States [8−11].

The legal framework for the safe operations of civil drones in Europe is provided by EU Regulations 2019/947 and Regulation 2019/945, which are applicable from 31 December 2020 and contain the rules and procedures for the operation of unmanned aircraft, unmanned aircraft systems and third-country operators of unmanned aircraft systems. The Regulations are formulated based on risk, and take into account the weight, specifications and the intended conduct of the drone without differentiating between recreational or commercial use of the drone. Three categories of civil drone operations, namely, the open, the specific and the certified category are specified. Lower-risk civil drone operations fall in the open category without a need to obtain operational authorisation before a flight and safety is ensured so long as the operator is compliant with the specified requirements for its intended operation. Riskier civil drone operations fall in the 'specific category' and require prior operational authorisation from the national competent authority to ensure safety. In order to get the authorisation, the drone operator is required to conduct a risk assessment which will determine the requirements for the safe operation of the civil drone. Drone operation involving a considerably high safety risk is placed in the 'certified' category with stricter requirements including the certification of the drone operator and licensing of the remote pilot(s) to ensure safety. The categorisations focus on safety risks with regards to mid-air collision with manned aircraft, harm to people, and damage to property, specifically critical and sensitive infrastructure, while also recognising the risks to security and privacy.

Amendments and new means of compliance and guidance material to Regulation 2019/947 were provided by EASA in 2021, after taking into account stakeholders' feedback, in the form of the Executive Director Decision which was specifically aimed at maintaining a high level of safety for the operation of drones in the open and specific categories. This includes inter alia, the establishment of geographical zones; revising the forms for the application and issue of operational authorisations in the specific category; a detailed description of the process to allow operations in different EASA member states; defining a list of training objectives for remote pilots operating in the specific category, and here, the drone operator is requested to select the most appropriate for its operation; as well as a new predefined risk assessment which are dedicated to beyond visual line of sight operations for linear inspections of infrastructures such as power lines or rail tracks (ED

Decision 2022) [11]. The amended and new means of compliance and guidance material are expected to maintain safety, provide better clarity with regards to the interpretation and implementation of regulations and increase harmonisation of drone operations within the European Union.

2.2.2 Unmanned Aircraft Traffic Management—U-Space

An air traffic management framework to enable the safe, secure, and efficient management and integration of unmanned aircraft is required, and the European Commission's U-space initiative aims to provide a policy framework to enable the safe integration of drones in non-segregated airspace and manage drone traffic with European Union airspace. The U-space is a set of services to be used in airspace where heavier traffic is expected such as urban areas, and was developed in order to support commercial drone operations specifically in light of technological advancements, automation and increased complexity. It will be more automated than the traditional air traffic control and will have less human interaction and more capacity to handle flights simultaneously. The U-space system will support increased volume of complex drone operations in challenging environments including air mobility applications in urban environments and close to the vicinity of airports.

U-space provides the framework to enable safe integration and management of unmanned aircraft traffic and can be defined as 'a set of specific services and procedures designed to ensure safe and efficient access to airspace for a large number of drones that embody high levels of digitalisation and automation. It is an unmanned aircraft traffic management solution which will allow the scaling up of the volume of drone operations that are complex, in environments that are challenging. [13]. The purpose of U-space is to achieve automated unmanned aircraft systems by managing and integrating in a seamless manner a large series of operations, which may be simultaneous, into the current air traffic management system. In order to achieve and implement this system, EU member states are required to firstly define and designate U-space airspaces for safe, secure and efficient operations.

In 2017 the SESAR Joint Undertaking drafted the U-space blueprint, resulting in the following four U-space mandatory services which are required to be established in every U-space airspace [17]:

1. Network Identification Service which provides the identity of the operators of unmanned aircraft systems and the location and trajectory of drones during operations.
2. Geo-Awareness Service which provides information on operational conditions, airspace limitations or existing time restrictions;
3. Flight Clearance Service to ensure free-of-conflict operations with other unmanned aircraft systems operating in the same volume of airspace; and
4. Traffic Information Service to alert operators of air traffic that may be found near the aircraft.

In addition to these four mandatory U-space services, two additional ones may be established as mandatory for certain U-space airspaces, if and when deemed necessary by the relevant state, after having conducted a safety evaluation, namely:

5. Weather Information Service to support the flight planning and execution phases and to enhance the performance of other U-space services; and
6. Compliance Monitoring Service to warn of non-compliance with the granted flight clearance and inform operators of any deviations.

Central to developing and effectively implementing an air traffic management system for the safe integration and management of unmanned aircraft is a solid legal and regulatory framework. To do so, states must first define and designate U-space airspaces.

In February 2021, the U-space regulatory package was approved by the European Aviation Safety Agency Committee. In April 2021 the European Commission adopted and published the framework regulating U-space within the EU comprising of three implementing regulations which are applicable from 26 January 2023.

This regulatory package consists, on the one hand, of the Commission Implementing Regulation [14] 2021/664 of 22 April 2021 on a regulatory framework for U-space, which regulates the technical and operational requirements for the U-space system. In addition, two implementing regulations were amended to incorporate requirements and obligations for air navigation service providers and manned aviation in U-space airspaces and which complement the regulatory regime on U-space, namely:

- Implementing Regulation [16] 2021/665 amending Implementing Regulation (EU) 2017/373, establishing common requirements for air traffic management and air navigation service providers to establish the specific coordination procedures and communication facilities between air traffic system units, U-space service providers and operators of unmanned aircraft systems, and
- Implementing Regulation [15] 2021/666 amending Regulation (EU) No. 923/2012 (laying down the rules of the air (SERA Regulation)), establishing the common rules for effectively making the presence of manned aircraft operating in U-space airspace electronically conspicuous.

The 2021 U-space Regulation establishes and harmonises requirements for both manned and unmanned aircraft to operate safely in the U-space airspace, to prevent collisions between aircraft and to mitigate air and ground risks. EASA is continuously working to develop acceptable means of compliance and guidance material to assist member states in implementing the U-space regulation package.

Additionally, agencies and organisations such as ICAO, EASA and Eurocontrol provide assistance by providing inter alia, specialists and expert knowledge, trainings, webinars, conferences, disseminating best practices, validation centres, assessments of airspace and

Table 2 Below is a snapshot of a number of the EU policies and regulations

EU Regulations 2019/947 & 2019/947	Regulations—formulated based on risk Weight, specifications and the intended conduct of the drone are considered No differentiating between recreational or commercial use of the drone 3 categories: open, specific and certified category
EU 2019/947	Rules and procedure for unmanned aircraft, inter alia: geographical zones; revising application/authorization in specific category; training objectives for specific category; a new predefined risk assessment for beyond visual line of sight operations for linear inspections of infrastructures
EU 2019/945	Unmanned aircraft systems and on third-country operators
UTM	Unmanned Aircraft Traffic Management European Commission's U-space initiative
SESAR 2017 U-space blueprint	Single European Sky ATM Research 4 U-space Mandatory Services required to be established in every U-space airspace; Network Identification; Geo-Awareness; Flight Clearance; Traffic Information Service 2 additional ones if deemed necessary by the relevant state, after conducted a safety evaluation; Weather Information and Compliance Monitoring Service
EU 2021/664	Framework for the U-space—Comprising 3 implementing Regulations
EU 2021/665	Amending Implementing Regulation (EU) 2017/373—requirements for air traffic management/air navigation services and other air traffic management network functions in the U-space airspace designated in controlled airspace
EU 2021/666	Amending Regulation (EU) No 923/2012 regarding requirements for manned aviation operating in U-space airspace

other assistance to align policies and procedures with international and national laws and regulations (Table 2).

2.2.3 Some Recent Development

In June 2022, EASA published a proposed new regulatory framework aimed at achieving a safe and secure integration of certified UAS and vertical take-off landing (VTOL)-capable aircraft operations in EU airspace, to provide a safe environment for operators to operate the VTOL-capable aircraft in the single European sky, and to further support and advance urban air mobility. The proposed regulations deal with the technical aspects of airworthiness, air operations, flight crew licensing and rules of the air. The rules complement the existing regulatory framework and include the certification of vertical take-off landing (VTOL) aircraft which will offer air taxi and similar services as well as guidance material on the design of vertiports. They are the first comprehensive proposals for such regulations to be issued globally. VTOL aircraft are a great step forward in achieving new air mobility for people and cargo and ultimately creating 'smart, green and digital' cities [8–9, 11].

The specific objectives of the Notice of Proposed Amendment (NPA), which is open to public consultation until 30 September 2022, are to:

- ensure a high and uniform level of safety for UAS subject to certification and operated in the 'specific' category and for operations with manned VTOL-capable aircraft;
- enable operators to safely operate manned VTOL-capable aircraft in the single European sky;
- create the conditions for the safe operation of UAS and of manned VTOL-capable aircraft in the U-space airspace;
- promote innovation and development in the field of innovative air mobility while establishing an efficient, proportionate, and well-designed regulatory framework, free of burdensome rules that could hinder the UAS market development;
- harmonise the regulatory framework across the EU Member States by enhancing clarity, filling the gaps, and removing the inconsistencies that are inherent to fragmented regulatory systems;
- foster an operation-centric, proportionate, as well as risk- and performance-based regulatory framework, considering important aspects such as privacy, personal data protection, security, and safety' [12].

In addition to the proposed new regulatory framework which is undoubtedly a major step towards achieving a comprehensive regulatory framework that incorporates and fosters future innovation, research and testing are also continuously advancing.

Extensive research and progress have been made towards achieving the safe and secure integrating of UAS in non-segregated airspace. In June 2022, the Belgian authorities approved a flight by the partners of a SESAR's SAFIR-Med project to operate a beyond visual line of sight flight to deliver urgent medical deliveries. The beyond visual line-of-sight flight, which enables drones to cover greater distances as it is coordinated through a command and control centre system interfacing with a U-space services provider, was operated over a populated area between two hospitals in Antwerp in compliance with EASA's specific operations risk assessment and unmanned aircraft system regulations. The Belgian civil aviation authorities validated the project's processes and technologies in three areas, namely, combined mitigations aimed at reducing risks to other airspace users as well as individuals on the ground, risk mitigation and a robust ground organisation, emergency response plan. Further tests in additional cities are planned [46].

2.2.4 Concerns

EU member states are now required to implement the U-space regulatory procedures and some States for example Poland, Belgium, Italy, Germany, United Kingdom, France, Spain and Switzerland, have already made headway in their U-space implementation programmes, while others have waited for the regulations to be enacted before commencing work in this area and are still in the preliminary stages, resulting in fragmented and spasmodic U-space service areas in Europe [18]. One can only question how far the objective of having a common approach and establishing and implementing a uniform set of rules which are consistently applied throughout the European Union by all its member states, is in fact being met. In addition, questions regarding awareness, resources, commitment, collaboration from internal stakeholders as well as the length of time needed to develop complex U-space programmes that meet both operational requirements and a standards level, are raised. Tremendous work still needs to be done regarding the validation and certification of U-space service providers, and in stipulating and defining the level of acceptable means of compliance for national regulators certifying U-space areas [18]. The Notice of Proposed Amendment (NPA) which proposes comprehensive regulations is undeniably a positive way forward, however, the timeframe for proposals to be accepted, adopted and eventually to become effective, may be a serious impediment, resulting in perhaps, once again, the regulatory and legal framework lagging behind and not being as proactive as it perhaps could, and should be in order to keep pace with the speedy developments, technological advancements and rapidly increasing potential usage of drones and the growing drone industry.

2.3 FAA and USA

The Federal Aviation Administration (FAA) established in 1958 regulates all aspects of civil aviation activity in the USA and over the surrounding international waters and also administers the USA airspace.

The FAA was mandated by Congress with integrating unmanned aircraft systems into the national airspace system by September 2015 [19]. In 1990 the FAA granted authority for limited use of unmanned aerial systems in the national airspace for important public missions including firefighting, disaster relief, search and rescue, law enforcement, border patrol, scientific research, and testing and evaluation [22, 23]. In 2016 the FAA issued the Small UAS Rule for Small Drones which is the primary law for operating unmanned aircraft systems or drones (under 55 pounds) in the National Airspace System. Part 107 [21], commonly referred to as the Small UAS Rule for Small Drones provides that drones can be operated for work or business, recreation, education, or for public safety missions. However, an exemption (49 U.S.C. Section 44,809) permits recreational drones which are flown purely for leisure and personal enjoyment to be flown without complying with Part 107, if certain conditions are met. Non-recreational purposes include flights for some compensation, including goodwill and other kinds of non-monetary value which would constitute indirect compensation.

The FAA provided safety guidelines for flying recreational drones in Advisory Circular 91-57B which include: [27]

- keeping the drone within the visual line of sight or using a visual observer who is co-located and physically next to the operator and in direct communication with the operator;
- giving way to and not interfering with manned aircraft;
- flying in Class B, Class C, or Class D airspace or within the lateral boundaries of the surface area of Class E airspace designated for an airport, the operator obtains prior authorization from the Administrator or designee before operating and complies with all airspace restrictions and prohibitions;
- flying in Class G (uncontrolled) airspace at or below 400 feet and complying with all airspace restrictions and prohibitions;
- taking the recreational UAS safety test;
- having current registration which is clearly shown on the outside of the drone;
- flying the drone in a safe manner without interfering with law enforcement activities or emergency response; and
- not flying under the influence.

Operators who do not comply or who fly their drones dangerously may be faced with FAA enforcement action [28].

The law also provides that education and research uses of drones for educational purposes can be operated under the rules for recreational flyers (P.L. 115-254, Section 350, as amended by P.L. 116-283, Section 10,002). Most states have a group of trained volunteers who work closely with the FAA to make the public aware of the rules and to promote safety in their area.

The latest in the FAAs incremental approach aimed at accelerating the integration of UAS into the national airspace system is the passing of two Final Rules in April 2021, the remote identification (Remote ID) rule which establishes a standard for the remote identification of drones by requiring them to broadcast identifiers and the location of their control stations, and the Operations Over People Final Rule which permits drones to be flown at night, over people and moving vehicles in certain conditions.

The Remote ID Rule is aimed at reducing the risk of drones interfering with other aircraft or posing a risk to people and property on the ground by providing crucial information to law enforcement as well as the general public. The rule applies to all unmanned aircraft, whether they are operating for recreational or commercial purposes and makes it a requirement for unmanned aircraft systems operating in the airspace of the United States to comply with the operational requirements for remote identification. (Remote ID

Rule 2021) [24]. The three ways to comply with the remote identification requirements are to:

(i) Operate a standard remote identification unmanned aircraft that broadcasts identification, location, and performance information of the unmanned aircraft and control station;
(ii) Operate an unmanned aircraft with a remote identification broadcast module which broadcasts identification, location, and take-off information. This may be a separate device that is attached to an unmanned aircraft, or a feature built into the aircraft; and
(iii) Operate unmanned aircraft without any remote identification equipment, where the UAS is operated at specific FAA-recognized identification areas.

The two final rules are perhaps the most important regulations since the Part 107 regulation in 2016, making substantial progress towards achieving the safe integration of drones into nonsegregated airspace and which will also undoubtedly enhance drone operations.

The Remote ID rule that UAS operators are required to have a remote pilot certificate and provide identification, location and performance information enhances both safety and security, however, rogue operators, operators who fly for illegal activities or with malicious intent, and who may not comply with such requirements, or who fraudulently obtain certification and/or false identification, continue to pose a serious safety and security risk when operating in nonsegregated airspace.

The 2016 Part 107 in Title 14 of the Code of Federal Regulations for remote pilot certification and operating rules for civil small unmanned aircraft weighing less than 55 pounds, did not allow small drones to fly at night or over people without first having to obtain a waiver. Many concerns were raised regarding the safety of flights over people beyond visual line of sight and a rulemaking committee was created to provide recommendations by December 2016 for a federal regulatory framework for safe drone operations over people. This was prolonged largely as a result of law enforcement agencies who insisted that a remote identification standard had to be implemented first before such operations were permitted. In February 2019, the interim Notice of Proposed Rulemaking (NPRM) to permit operations of small USA over people and at night, was published. The final NPRM3 on remote identification was published, followed by additional proposed modifications, leading to the Operations Over People Final Rule which became effective in April 2021 [45].

The Operations Over People Final is the next step by the FAA towards the further integration of unmanned aircraft in the national airspace system. It provides drone operators the right to fly over people or vehicles based on the level of the risk that the small drone poses to people on the ground. In addition, it also permits operations at night under certain conditions provided that the remote pilots have completed the required training and passed the knowledge tests which stipulates that the operator must be 16 years old, be able to read, write, speak and understand English, and be in a physical and mental condition to safely fly an UAS.

The Final Rule establishes the following four new categories of small unmanned aircraft for routine operations over people: [25].

Category 1 eligible small unmanned aircraft must weigh less than 0.55, including everything on board or otherwise to the aircraft at the time of take-off and throughout the duration of each operation, and also contain no exposed rotating parts that would cause lacerations to human skin. No remote pilot in command may operate a small unmanned aircraft in sustained flight over open-air assemblies unless the operation is compliant with Remote ID, namely, the ability of the drone in flight to provide identification and location information. There is no requirement for FAA-accepted means of compliance or declaration of compliance. Category 2 eligible small unmanned aircraft must not cause injury to a human being that is equivalent to or greater than the severity of injury caused by a transfer of 11 foot-pounds of kinetic energy upon impact from a rigid object, contain no exposed rotating parts that would cause lacerations to human skin upon impact, and must not contain any safety defects. This category does require a FAA-accepted means of compliance and FAA-accepted declaration of compliance. This category also requires that the small unmanned aircraft be operated in sustained flight over open-air assemblies with Remote ID. Category 2 and Category 3 provide performance-based eligibility and operating requirements when conducting operations over people using unmanned aircraft that weigh more than 0.55 pounds but do not have an airworthiness certificate under part 21. Category 3 eligible small unmanned aircraft must not cause injury to a human being that is equivalent to or greater than the severity of injury caused by a transfer of 25 foot-pounds of kinetic energy upon impact from a rigid object, must not contain any exposed rotating parts that could lacerate human skin upon impact with a human being, and must not contain any safety defects. FAA-accepted means of compliance and FAA-accepted declaration of compliance is required in this category. The following additional operating restrictions also apply to this category. A remote pilot in command may not operate a small unmanned aircraft over open-air assemblies of human beings. Additionally, a remote pilot in command may only operate a small unmanned aircraft over people if:

- The operation is within or over a closed- or restricted-access site and all people on site have been notified that a small UAS may fly over them; or
- The small unmanned aircraft does not maintain sustained flight over any person unless that person is participating directly in the operation or is located under any covered structure or inside a stationary vehicle that can provide reasonable protection in the event the small unmanned aircraft will fall.

Category 4 allows small unmanned aircraft who have been issued with an airworthiness certificate under Part 21 of FAA regulations to operate over people, as long as the operating limitations specified in the approved Flight Manual or as otherwise specified by the Administrator do not prohibit operations over people and the maintenance, preventive maintenance, alterations, or inspections are performed in accordance with specific requirements in the final rule. In addition, the operation must be in compliance with the Remote ID requirements.

The operations over people rule goes on to explain that a sustained flight over an open-air assembly includes 'hovering above the heads of persons gathered in an open-air assembly, flying back and forth over an open-air assembly, or circling above the assembly in such a way that the small unmanned aircraft remains above some part of the assembly. 'Sustained flight' over an open-air assembly of people in a Category 1, 2, or 4 operation does not include a brief, one-time transiting over a portion of the assembled gathering, where the transit is merely incidental to a point-to-point operation unrelated to the assembly' [25].

The Final Rule permits operations over moving vehicles when they are transitory in nature, at a closed or restricted-access site, or by an unmanned aircraft that has received an FAA-type certificate. Unmanned aircraft must meet the requirements of Categories 1, 2, and 3 and must either remain within or over a closed or restricted access site. Additionally, all people inside a moving vehicle within the closed or restricted-access site must be on notice that a small unmanned aircraft may fly over them or may not maintain sustained flight over moving vehicles. In addition, a remote pilot may also conduct operations over moving vehicles with a small unmanned aircraft eligible for Category 4 operations as long as such operations are not prohibited by the applicable operating limitations in the approved Flight Manual or as otherwise specified by the Administrator.

The Final Rule allows routine operations of small UAS at night under two conditions, namely, the remote pilot in command must have completed the updated initial test or an updated recurrent online training, and the small unmanned aircraft must be equipped with flashing lights at a flash rate sufficient to avoid a collision and anti-collision lights which must be visible for at least three statute miles. A remote pilot, owner, or person manipulating the controls of an unmanned aircraft must have their remote pilot certificate in their physical possession at all times and must produce the certificate for inspection when requested to do so by the FAA, National Transportation Safety Board, Transportation Security Administration, or any federal, state, or local law enforcement officer.

2.3.1 Unmanned Aircraft Traffic Management—UTM

The Unmanned Aircraft System Traffic Management (UTM) system is to enable and manage multiple beyond visual line-of-sight drone operations in order to integrate drones safely and efficiently into air traffic. The UTM is an independent but complementary system to the existing FAA's air traffic management system aimed at safely integrating and managing UAS in non-segregated airspace. The UTM system will allow parties from government and the industry to work together to explore and devise concepts of operation, information architecture, data exchange requirements, software functions, infrastructure, performance requirements and a supporting framework to enable and manage multiple drone operations conducted beyond visual line-of-sight at low altitudes, namely, under 400 feet above ground level, in airspace where FAA air traffic services are not provided.

The FAA, NASA, other federal partner agencies and industry stakeholders established a research team to coordinate the UTM initiative to explore and devise concepts of operations, performance requirements, data exchange and information architecture, communications and navigation, sense and avoid, and a supporting framework for the safe operation of low-altitude uncontrolled drone operations. Four phases of research, with increasing complexity and technical capability levels (TLCs), were categorised, as follows:

TCL1 which was completed in 2015 focused on drones in agriculture, firefighting and infrastructure monitoring, and technologies to assist drones flying;

TCL2, completed in 2016, concentrated on monitoring drones flown in sparsely populated areas where an operators cannot see the drones they are flying and technologies to adjust flight zones or to clear airspaces in case of an emergency, search-and-rescue or for loss of communications with a small aircraft;

TCL3 conducted in 2018 was directed at technologies to ensure separation between drones, namely, detection and avoidance over moderately populated areas, and ensuring that drones stay in their designated flight zones;

TCL4 conducted from May to August 2019 was centred on demonstrating how the UTM system can integrate drones into urban areas and drone technical capabilities technologies onboard the drones and on the ground. These included incorporating more localized weather forecasting into flight planning, using cell phone networks to enhance drone traffic communications and relying on cameras, radar and other ways to ensure drones can manoeuvre around buildings and land when needed, while communicating with other drones and users of the UTM system [40]. The results of the UTM's research are given to the FAA for continuous testing and implementing the UTM system which is expected to be implemented incrementally over the next few years.

Crucially important to testing the UTM system is the ability to identify and contact the drone and its control station when necessary. Manned aircraft are required to have tail numbers on the vehicle, and drones that operate in dense airspace are required to have transponders and VHF radio equipment onboard in order for air traffic control to be able to track or contact them. Similarly, the Remote ID together with the registration and communication connectivity will provide a link between UAS pilots, UAS service suppliers, other aircraft, and, where applicable, air traffic control [26].

The Federal Aviation Administration (FAA) has planned to commence field testing new drone traffic management capabilities in 2022 to further integrate drones into national airspace. The filed test is also hoped to result in new policies and improving standards to permit drones to fly beyond visual line sight of the operator [20, 22, 23, 28].

Table 3 highlights some of the initiatives by the FAA as well as an indication of the research conducted and progress made towards fully integrating unmanned aircraft in the national airspace of the United States of America, before discussing some concerns and a discussion of the application of using the new systems of unmanned traffic management in urban air mobility.

Table 3 Some initiatives towards a fully integrated USA airspace [27]

FAA: 2016—Small UAS Rule for Small Drones	Primary law for operating unmanned aircraft systems under 55 pounds in the National Airspace System Part 107—drones can be operated for work or business, recreation, education, or for public safety missions Exemption—recreational drones can be flown without complying with Part 107, if certain conditions are met
Advisory Circular 91-57B	Provides safety guidelines for flying recreational drones, inter alia: Flying within visual line of sight or using a visual observer; requirements for flying in Class B, C, D, E. G airspace
Final Rules in April 2021 i. Remote Identification	Applies to all unmanned aircraft—recreational or commercial purposes Establishes a standard for the remote identification of drones Drones are required to broadcast identifiers and the location of their control stations Stipulates 3 ways to comply
ii. Operations Over People final rule	Permits drones to be flown at night, over people and moving vehicles in certain conditions Based on the level of the risk posed Permits operations at night under certain conditions provided the remote pilots have completed the required training and passed the knowledge test Establishes 4 new categories of small unmanned aircraft for routine operations over people: Category 1, 2, 3 & 4
UTM	FAA, NASA, other federal partner agencies and industry stakeholders - Research team to coordinate the UTM initiative Four phases of research

2.4 General Concerns Regarding UTM

Each jurisdiction will undoubtedly continue to develop and amend its legal and regulatory framework around UAS and the incorporation and management of UAS into national airspace. This may lead to additional challenges for UAS operators and companies who may have to work through a complex set of different, and perhaps contradictory rules, regulations and procedures. As most regulatory authorities and agencies are adopting an incremental approach to fully integrating UAS into their national airspace, some countries have made great progress in developing and testing their system to integrate and manage the unmanned traffic in their airspace.

A study conducted by Vee Weiland examined the interconnection between the air traffic controller and UAS to identify possible human factor implications when UAS enter national airspace. By studying the human performance of the controllers, the UAS operators, the commercial pilots and the equipment, the study determined that as air traffic controllers do not fully comprehend human factors in the integration of UAS, the controllers could make a number of human errors when UAS operate in national airspace and therefore, further research, education, and training is required to mitigate the possible errors. A further recommendation made is that the Safety Management System program for manned aircraft continue to be developed in the context of unmanned aircraft [49].

A major concern for all traffic management systems and all the services required to support the system is the cost involved in the research, development, creating, testing and implementation of a safe, secure, sustainable and effective system to manage unmanned air traffic and further, whether this research should be funded by the relevant stakeholders of unmanned aircraft and not by manned aviation.

2.5 Use of UTM Systems

A much-anticipated application of using the new systems of unmanned traffic management is urban air mobility. Urban air mobility is a safe, secure and more sustainable air transportation system using electric vertical take-off and landing aircraft that will operate at lower altitudes within urban and suburban areas for transporting cargo or passengers for inter alia, air taxis, sightseeing tours, emergency transport, and smaller drones for delivering parcels, goods or transporting medical equipment. In urban environments, urban air mobility will help to address the pollution and mobility difficulties that large cities face by enhancing the connectivity of people to cities and regions in a sustainable way. By using new technologies such as electric propulsion and enhanced battery capacity applied to vertical take-off and landing systems, urban air mobility will use highly automated aircraft and be performed by electric aircraft taking off and landing vertically, remotely piloted or with a pilot on board, and will be integrated into a transportation system encompassing different modes of transportation. Commercial operations are expected to begin in Europe within the next 3–5 years commencing firstly with goods being delivered by drones and then moving on to transporting passengers, firstly with a pilot on board and then remotely or autonomously [9].

Many stakeholders are focusing on developing U-space and urban air mobility through technological advancements, research, and innovation and pilot projects. European manufacturers, for example, are working on obtaining airworthiness for piloted vehicles for passenger transport and have already applied for certification [9]. Initiatives such as the AMU-LED project funded by the SESAR Joint Undertaking under the European Union's Horizon 2020 research and innovation programme are making tremendous progress. The aim of the project which started in January 2020 is to perform flight demonstrations within different scenarios and in different situations in urban environments to demonstrate the safe integration of manned and unmanned aircraft through the deployment of U-space, and ultimately, to achieve sustainable smart cities. During the demonstrations, large electrical vertical take-off and landing platforms for passenger and cargo transport, combined with smaller unmanned aircraft performing delivery of goods and medical supplies, surveillance or support for emergency services will be used. The demonstrations will focus on feasibility, viability and desirability and a total of seven demonstrations will take place in the United Kingdom, Netherlands and Spain form June to October 2022.

The first demonstration which took place in June in Cranfield, United Kingdom, focused on feasibility and illustrating the readiness of the AMU-LED solution, technologies, and systems to show that the AMU-LED solution is ready and safe to be tested in more complex environments. A second demonstration will take place in September. Three demonstrations will be held in the Netherlands, starting in August in Amsterdam where the desirability of its solution will be tested by focusing on public acceptance and social impact. By using focus groups, a number of flight demonstrations will be performed to test different aspects of U-space and to determine the public's perception with regards to safety, trust, noise and privacy issues and to then make suggestions and recommend mitigation measures for the concerns raised. The social impact of urban air mobility will then be tested in Enschede by focusing on socially relevant cases such as medical delivery, firefighting or police surveillance. The economic viability of urban air mobility will be demonstrated August, in Rotterdam, by examining inter alia, the viability of transporting ships' crews directly from the vessel to the hotel and flying different unmanned aircraft and vertical take-off and landing systems. The final demonstration will be held in September and October in Santiago de Compostela, Spain, which will focus on the correct implementation of all the aspects in urban environments to show how U-space can enable urban air mobility [41].

Inevitably to successfully implement a UTM system, the interaction between UTM systems and ATM systems will increase and this requires additional research, testing and training as well as a safe and secure infrastructure and services to allow unmanned aircraft to operate in existing or new airspace. The ability to operate in a non-segregated airspace requires a robust infrastructure and services, as well as clear and harmonised laws and regulations which prioritise safety and security which take into account the unique and distinctive aspects of unmanned aircraft and their systems.

To safely allow new aircraft, including air taxis and delivery drones to be integrated and operate in common non-segregated airspace, a robust, effective legal and regulatory framework as well as accurate planning, designing, developing and building the necessary critical infrastructure is required. However, as this critical part of the aviation industry is still developing, the technological advancements are continuing and the uses of UAS are rapidly increasing, legislators and policy makers seem to be more reactive as they often seem to lag behind this exponential growth of the UAS industry.

A number of important unprecedented operational, technical, regulatory challenges and concerns as well as legal complexities arise in integrating UAS into no-segregated airspace.

3 Challenges and Legal Complexities

3.1 Security

The sophisticated technology involved in unmanned aircraft systems increases vulnerabilities and potential security weaknesses which may be exploited by multiple actors including criminals, terrorists, competitors and many others having criminal intent. Unmanned aircraft can be used by multiple actors for a range of threats including using the drone as a weapon for a range of unlawful or malicious actions or attacks or flying the drone into restricted areas. Largely as a result of the fact that small drones are readily available, relatively inexpensive, easy to adapt and provide to some extent, anonymity, they can, and have been used in a wide range of illegal and criminal activity such as inter alia, smuggling and delivering drugs and other contraband, collecting intelligence and other sensitive data, obtaining private data and pin codes inserted into automated teller machines, stealing intellectual property, espionage, illegal filming, using them as weapons to fly into or over a target carrying traditional or improvised explosive devices [37].

The highly sophisticated technology used to operate unmanned aircraft systems renders them vulnerable to a number of flaws that can be exploited rendering commercial or recreational drones targets of cyber-attacks.

The majority of cybersecurity breaches are related to: [48]

- using a frequency transmitting device to block or interfere with radio communications, namely, global positioning system (GPS) jamming and spoofing
- video interception
- hijacking attacks via communication
- sensor spoofing
- exploiting the digital part's vulnerabilities

– disturbing unmanned aircraft operation or taking over control of the system for malicious objectives such as injuring people on the ground, violating privacy, or damaging infrastructure.

On-going research in this area needs to continue and be expanded in an attempt to provide effective defences and mitigate the risks of security breaches and security attacks on aviation, airspace users and the general public when integrating unmanned aircraft into non-segregated airspace.

3.2 Safety and Safety Risks

Conventional aircraft are legally required to be registered and to have an identification number. In the case of unmanned aircraft, there is no similar uniform universal requirement. Some jurisdictions however, have already established requirements for the remote identification of unmanned aircraft. In the USA, as outlined above, the FAA requires that all unmanned aircraft operating in US airspace, irrespective of the purpose of the operation, must comply with the Remote ID rule and provide certain identification, location, and performance information.

The vacuum created by the fact that unmanned aircraft which are small in size and fly at low levels cannot be detected, tracked and guided in the same manner that conventional manned aircraft can through the radio and electromagnetic waves emitted by primary and secondary RADAR (RAdio Detection And Ranging), is of crucial importance. In addition they are not equipped with a transponder which allows aircraft to communicate with air traffic control. Such technical intricacies may result in serious safety risks to all aircraft operating in non-segregated airspace and technologies need to be further explored and tested before unmanned aircraft are fully integrated into existing air traffic management or new unmanned traffic management airspace.

Much of the framework outlined above places the burden on the operators to demonstrate to the regulators that they do in fact meet the stipulated requirements and procedures and that they have established an adequate level of safety to operate their unmanned vehicle. To do so, they need to firstly identify and assess the safety risks to their operation and then demonstrate to the relevant civil aviation authority that they have put the necessary mitigating actions in place in order to establish a safe as is reasonably practicable operation.

As unmanned aircraft are not equipped with anti-collision systems such as the traffic collision avoidance system (TCAS) which is used in large manned aircraft, perhaps one of the most important safety concerns is the ability for such aircraft to avoid possible conflicts and collisions between other airspace users. An important safety risk that needs to be mitigated is that of causing harm to other manned or unmanned aircraft, damaging property or sensitive infrastructures, and injuring persons.

The first reported case of a drone hitting a passenger plane occurred in April 2016 when a British Airways A320 aircraft, on a scheduled passenger flight from Geneva, Switzerland, to London, UK, with 132 passengers and five crew members on board, was hit by an object believed to be a drone while it was on approach to land at London's Heathrow airport. The aircraft was flying at 1700 feet when it was struck at the nose. No injuries were sustained and the aircraft landed safely [43]. There are many similar incidents reported across the globe, including the Envoy Air flight which was scheduled to fly from Chicago O'Hare to Detroit in August 2021 which had to return to the O'Hare airport shortly after taking off after it collided with a drone at 2500 feet. There were no injuries or damage to the aircraft in this incident, however, in another incident which occurred in the same month, a Cessna plane suffered major damage after it was struck by a York Regional police drone [5].

3.3 Legal Complexities

The legal complexities inherent in integrating unmanned aircraft operations **into non-segregated airspace** are not limited to safety, security, operational and technical issues. The rapidly advancing cutting edge technology employed and the uniqueness of flying a drone present a number of important legal issues.

3.3.1 Accident Liability

A vital concern that has not yet been addressed is with regards to accidents or serious incidents such as a mid-air collision with a manned aircraft, or a collision with another unmanned aircraft, a structure or a building, or even an unmanned aircraft falling onto the ground causing property damage and injury to persons.

In fact, Annex 13, the relevant ICAO Standard and Recommended Practise (SARP) dealing with aircraft accident and incident investigation, defines an accident as '[a]n occurrence associated with the operation of an aircraft which takes place between the time any person boards the aircraft with the intention of flight until such time as all such persons have disembarked, in which... [inter alia] a person is fatally or seriously injured". No mention is made in the entire Annex of an unmanned or pilotless aircraft. A collision between a manned aircraft and an unmanned aircraft would constitute an accident, however there is a legal vacuum as any collision or damage caused to persons or property by a single unmanned aircraft or between two unmanned aircraft would technically not be covered under Annex 13. The risk, in terms of probability and likelihood, of human or machine failures, the intentional or unintentional non-compliance or non-adherence to existing operational, technical or safety requirements, wilful negligence or operating an unmanned aircraft with criminal intent, is high, and in the event that it caused damage and injury, the existing legal framework applicable to manned aircraft would not apply to unmanned aircraft. This serious vacuum needs to be addressed urgently.

Legal issues with regards to who has the legal responsibility, and hence, the legal accountability for any loss, damage or injury incurred in the operation of an unmanned aircraft remain unresolved. The technological sophistication and complications, the novel situations arising in integrating unmanned aircraft into normal air traffic management or an unmanned traffic management system, human factor issues, training, certification and airworthiness complexities, possible system failures and programming errors or errors in design and manufacturing, can lead to thorny legal issues of liability, accountability, compensation and insurance coverage.

3.3.2 Privacy

In accordance with Article 12 of the 1948 Universal Declaration of Human Rights, everyone has the right to privacy and not to be subjected to arbitrary interference with privacy, family, home or correspondence. The protection of a citizen's privacy is one of the most critical and complex issues that arises when permitting unmanned aircraft to operate in non-segregated national airspace. Privacy issues need to be adequately addressed in order to safeguard and protect the privacy of the public, the operators and their customers. In addition, there must be a balance between the unhindered growth in commerce and the aviation industry, and the rights of private property owners and individual citizens.

Drone surveillance is seen by many as a major threat to privacy as unmanned aircraft can have very advanced surveillance technologies such as, inter alia:

- Powerful lenses allowing substantial zooming resulting in individuals being observed or video recorded from afar, or having their photos taken without their knowledge or consent;
- Infrared or thermal imaging and ultraviolet imaging which enables night vision by capturing light which is not visible to the human eye. Thermal imaging detects heat being emitted and can thus identify humans or animals in the dark, ultraviolet imaging can detect some materials which are not visible in natural and infrared light is used to enhance detail not apparent in visible light;
- See-through imaging technology which can see through ceilings and walls and can track human targets who are inside buildings;
- Video analytics which applies artificial intelligence techniques, can recognize and respond to specific people, events, and objects and by using facial recognition can be used to continually track individuals or vehicles on the move or to identify any suspicious patterns or changes in routines, buildings or grounds [39].

Unmanned aircraft are being used positively in difficult and dangerous missions such as search and rescue, fighting wildfires, distributing aid and medical supplies to areas hit by disasters, and it is even been suggested by police in Houston, Texas, that, in addition to monitoring traffic, drones could also be used for writing tickets [6]. By the same token, they can also be used for controversial and illegal purposes such as illegal surveillance or tracking, dispersing crowds or protesters by dispersing tear gas, blocking an individual resisting arrest or using weapons and obstructing a car from fleeing. The American Civil Liberties Union, in their statement for a Senate Judiciary Committee Hearing, called for stricter laws, limits and regulations in order to better protect citizen's rights to privacy, and recommended that as a minimum, Congress enact adequate provisions related to inter alia, restricting the use of drones, the retention of images of individuals who can be identified, public notice of the procedures for using aerial surveillance technologies by unmanned aircraft and the auditing and effective tracking of the use of aerial surveillance technology [39].

The existing legal framework, specifically on an international level, does afford some protection regarding the collection, use, storing, and dissemination of a citizens personal details and photos legitimately obtained by the government or companies operating commercial drones. However, any laws regulating surveillance by individuals such as inter alia, the paparazzi, business competitors, those seeking to obtain a financial or other benefit, a neighbour, a tourist, etc., are largely enacted on a regional or national level or come about as a result of court decisions and judicial precedent. There is therefore, little or no protection provided against private individuals who, largely as a result of the ease of obtaining and using drones, obtain or collect personal data from unlawful surveillance, stalking and harassing. Legislators at all levels should therefore review the existing laws and make the necessary enactments and/or amendments to firstly ensure that unmanned aircraft technology falls in the ambit of exiting privacy and surveillance laws and secondly, create more stringent laws that provide a greater and wider scale of protection.

Due regard must also be given to federal and state privacy laws and the different approaches and varying scope of laws prohibiting surveillance and privacy issues. In America for example, the California Civil Code Section 1708.8(b) prohibits using a drone "in a manner that is offensive to a reasonable person" to capture or attempt to capture "any type of visual image, sound recording or other physical impression" of a person "engaging in a private, personal or familial activity." In circumstances where an individual has the reasonable expectation of privacy, a drone cannot capture any private, personal or familial activity which includes intimate details of an individual's personal life; any interaction with the individual's family or significant others; any activity occurring on residential property and any other aspects of private concerns. In the event that such surveillance is conducted for financial gain, the law affords the aggrieved party the right to claim for three times the amount of damages as well as punitive damages, and the offender may be ordered to also pay a civil fine [35].

In Florida however, state law prohibits the infringement of a reasonable expectation of privacy through surveillance. A reasonable expectation of privacy is defined in terms of individuals on privately owned real property and the property itself. Surveillance of individuals on privately owned real property refers to watching such individuals with sufficient visual clarity to obtain information regarding the identity, habits, conduct or movements of individuals. The surveillance of privately owned real property refers to observing such property's physical improvements with sufficient visual clarity to determine unique identifying features or its occupancy. Infringements may result in paying compensatory and punitive damages. It should be noted that any business or profession which is licensed by the state is exempted from the law on the condition that the surveillance is reasonably "within the scope of practice or activities permitted under such person's or entity's license" [35].

The rapid technological developments, the exponential growth of the drone industry and the increasing expanding range of potential uses of unmanned aircraft as well as the diverse interest of the many stakeholders and the limited legal safeguards of an individual's right to privacy clearly show that concerns regarding privacy issues are a long way from being adequately and effectively addressed by regulators and legislators. In addition, awareness needs to increase and the public needs to become cognizant of the laws regarding the operation of unmanned aerial vehicles and also of any penalties and punishment that will be administered in violation of privacy laws.

3.3.3 Nuisance and Trespass

Many legal complexities arise when a commercial or recreational drone is flown over or near one's private property and the question that needs to be considered is whether the drone flying in the airspace above private property would constitute trespass to land or private nuisance, both of which protect the legal rights of individual's exclusive use and enjoyment of land. Trespass to land refers to direct interference with the owner's exclusive right of possession and control of the land and nuisance refers to indirect interference.

In this context, the courts would need to determine whether a drone overflying private property would amount to trespass of the airspace of the land [44], and if so, whether landowners would be entitled to claim any civil or criminal remedies as a result. The basis of any legal action would be the legal maxim maxim cujus est solum ejus est usque ad coelum (for whoever owns the soil owns to the heavens), however, how the courts apply this in different jurisdictions may not always provide, satisfactory and equivalent remedies, if any at all, to aggrieved landowners. Courts may approach this issue by referring to the legal principles of nuisance when determining whether an overhead flight causes a substantial impairment of the use and enjoyment of one's property.

Irrespective of whether an aggrieved party can proceed with a claim based on trespass or nuisance, the real question is whether existing laws adequately protect landowners from infringing their rights by unwanted drones who fly over their property.

4 Conclusion

Tremendous progress has already been done towards achieving one of the biggest challenges of the aviation industry, namely the integration of unmanned aircraft vehicles in non-segregated airspace in a safe, secure and sustainable manner. Great inroads have been made in developing technical and regulatory standards, policy guidance, and operational procedures for the safe integration of unmanned aircraft. However, there is still work to be done by policy-makers, regulators, and all industry stakeholders.

Regulators need to carry on the on-going collaboration with all stakeholders to review and update existing regulations and enable new and novel unmanned aircraft operations and work towards establishing a standard, uniform and universal set of requirements as well as an effective means to assess compliance. The responsibilities and liabilities for all stakeholders need to be clarified. Regulations, fines and other penalties for violations need to be improved and enforced. Security concerns must be adequately addressed. Safety improvements need to be continuously advanced and the safety culture amongst drone operators must be improved. Awareness must be increased and sharing lessons learned must be encouraged. Risk management and safety management processes for integration need to be developed and improved. The highly technological advanced and rapid changes in the drone industry need to be explored and effectively managed. Research and pilot programmes need to be maintained and also enhanced. National and regional regulatory authorities and other stakeholders need to continue their work towards achieving a comprehensive unmanned aircraft traffic management system. There is also a need for increased awareness and additional and expanded training of current and prospective drone users. In addition, the public's trust and confidence in the system needs to be expanded, particularly in light of the increased ways that drones are being used and long term initiatives such as introducing air taxis as part of the urban air mobility initiative.

Perhaps one of the greatest challenges the aviation industry is facing is to safely integrate unmanned aircraft into non-segregated airspace. Some of the many novel situations and unprecedented legal complexities arising from this unparalleled task have been highlighted and discussed in this chapter. Undoubtedly, there will be rapid changes and developments in this area and one can only monitor how future progress and developments by the relevant authorities, the industry and all stakeholders involved will address them.

References

1. Air Transport Action Group (ATAG), Aviation: Benefits Beyond Borders, 2020 Report. https://aviationbenefits.org/economic-growth/adding-value-to-the-economy/
2. Allied Market Research, 'Commercial Drones Market: Global Opportunity Analysis and Industry Forecast, 2021–2030', 25 May 2022. https://www.alliedmarketresearch.com/commercial-drone-market
3. R.I.C. Bartsch, 'Unmanned And Uncontrolled: The commingling theory and the legality of unmanned aircraft system operations' (2016). https://ses.library.usyd.edu.au/bitstream/handle/2123/14844/bartsch_ri_thesis.pdf?sequence=2&isAllowed=y
4. Chicago Convention on International Civil Aviation (1944). https://www.refworld.org/pdfid/3ddca0dd4.pdf

5. B. Crumley, 'Envoy Air passenger plane hits drone after Chicago takeoff', 24 August 2021. https://dronedj.com/2021/08/24/envoy-air-passenger-plane-hits-drone-after-chicago-takeoff/
6. S. Dean, 'Police line up to use drones on patrol after Houston secret test', Houston Examiner, 11 January 2010. http://www.examiner.com/page-one-in-houston/police-line-up-to-use-drones-on-patrol-after-houston-secret-test
7. EASA 204696, Concept of Operations for Drones A risk based approach to regulation of unmanned aircraft. https://www.easa.europa.eu/sites/default/files/dfu/204696_EASA_concept_drone_brochure_web.pdf
8. EASA, 'Our Mission: Your safety' (2022a). https://www.easa.europa.eu/light/easa
9. EASA, 'Urban Air Mobility (UAM)' (2022b). https://www.easa.europa.eu/domains/urban-air-mobility-uam
10. EASA, 'EASA publishes world's first rules for operation of air taxis in cities', 30 June 2022c. https://www.easa.europa.eu/newsroom-and-events/press-releases/easa-publishes-worlds-first-rules-operation-air-taxis-cities
11. EASA, ED Decision 2022/002/R on geographical zones, cross-border operations and predefined risk assessment for beyond visual line of sight operations, Amendment 2 to Issue 1 of the Acceptable Means of Compliance and Guidance Material to Commission Implementing Regulation (EU) 2019/947 and to its Annex 'AMC and GM to Regulation (EU) 2019/947—Issue 1, Amendment 2' 'AMC and GM to the Annex to Regulation (EU) 2019/947—Issue 1, Amendment 2', EASA, 7 February 2022. https://www.easa.europa.eu/downloads/135909/en
12. EASA, Notice of Proposed Amendment (NPA) 2022–06, 'Introduction of a regulatory framework for the operation of drones', 30 June 2022. https://www.easa.europa.eu/downloads/136705/en
13. EU Commission Implementing Regulation (EU) 2021/664 of 22 April 2021 on a regulatory framework for the U space. https://eur lex.europa.eu/legal content/EN/TXT/PDF/?uri–CELEX:32021R0664&from=EN
14. EU Commission Implementing Regulation (EU) 2021a/664 of 22 April 2021a on a regulatory framework for the U-space (2021a). https://eur-lex.europa.eu/legal-content/EN/TXT/PDF/?uri=CELEX:32021R0664&from=EN
15. EU Commission Implementing Regulation (EU) 2021b/666 of 22 April 2021b amending Regulation (EU) No 923/2012 as regards requirements for manned aviation operating in U-space airspace. https://eur-lex.europa.eu/legal-content/EN/TXT/PDF/?uri=CELEX:32021bR0666&from=EN
16. EU Commission Implementing Regulation (EU) 2021c/665 of 22 April 2021c amending Implementing Regulation (EU) 2017/373 as regards requirements for providers of air traffic management/air navigation services and other air traffic management network functions in the U-space airspace designated in controlled airspace. https://eur-lex.europa.eu/legal-content/EN/TXT/PDF/?uri=CELEX:32021cR0665&from=EN
17. ECAC-CEAC, 'UAS European Civil Aviation Conference Bulletin on Unmanned Aircraft Systems', No 2, December 2021. https://www.ecac-ceac.org/images/news/uas-bulletin/UAS_Bulletin_2021_02.pdf
18. Eurocontrol, 'Europe is now in the fast lane to implementing UAS traffic management systems', 23 June 2021. https://www.eurocontrol.int/article/europe-now-fast-lane-implementing-uas-traffic-management-systems
19. FAA Modernization and Reform Act of 2012, P.L. 112–95, 126 Stat. 11 (2012). https://www.congress.gov/112/plaws/publ95/PLAW-112publ95.pdf
20. FAA, 'FAA Begins New Phase of Testing to Safely Integrate Drones into the National Airspace' 11 February 2022a. https://www.faa.gov/newsroom/faa-begins-new-phase-testing-safely-integrate-drones-national-airspace

21. FAA, 14CFR Part 107 (2016). https://www.ecfr.gov/current/title-14/chapter-I/subchapter-F/part-107
22. FAA, 'Operations Over People General Overview' (2021a). https://www.faa.gov/uas/commercial_operators/operations_over_people
23. FAA, 'Timeline of Drone Integration', 2 June 2022b. https://www.faa.gov/uas/resources/timeline
24. FAA, 'Remote Identification of Unmanned Aircraft', 14 CFR Parts 1, 11, 47, 48, 89, 91, and 107 Docket No.: FAA–2019–1100 RIN 2120–AL31 (2021). https://www.faa.gov/news/media/attachments/RemoteID_Final_Rule.pdf. https://public-inspection.federalregister.gov/2021-04882.pdf
25. FAA 'Operation of Small Unmanned Aircraft Systems Over People', 14 CFR Parts 11, 21, 43, and 107 [Docket No. FAA-2018–1087] RIN 2120-AK85 (2021). https://public-inspection.federalregister.gov/2021-04881.pdf
26. FAA, 'Integration of Civil Unmanned Aircraft Systems (UAS) in the National Airspace System (NAS) Roadmap, 3rd edn. (2020). 2019_UAS_Civil_Integration_Roadmap_third_edition.pdf. https://www.faa.gov/uas/resources/policy_library/media/2019_UAS_Civil_Integration_Roadmap_third_edition.pdf
27. FAA Advisory Circular, Exception for Limited Recreational Operations of Unmanned Aircraft, 2019. https://www.faa.gov/documentLibrary/media/Advisory_Circular/AC_91-57B.pdf
28. FAA, 49 USC 44809: Exception for limited recreational operations of unmanned aircraft, those laws in effect on 21 July 2022c. https://uscode.house.gov/view.xhtml?req=granuleid:USC-prelim-title49-section44809&num=0&edition=prelim
29. ICAO, Unmanned Aviation Bulletin, 2020/1. https://www.icao.int/safety/UA/Documents/ICAO%20UA%20Bulletin%202020%201a.pdf
30. ICAO, Global Air Traffic Management Operational Concept Doc 9854 AN/458 (2005). https://www.icao.int/Meetings/anconf12/Docu-ment%20Archive/9854_cons_en[1].pdf
31. ICAO Cir 328, Unmanned Aircraft System, Cir 328 AN/190 (2011). https://www.icao.int/meetings/uas/documents/circular%20328_en.pdf
32. ICAO Model UAS Regulations, 23 June 2020. https://www.icao.int/safety/UA/Documents/Model%20UAS%20Regulations%20-%20Parts%20101%20and%20102.pdf
33. ICAO Remotely Piloted Aircraft System (RPAs) Concept of Operations (CONOPS) For International IFR Operations' (2017). https://www.icao.int/safety/UA/Documents/ICAO%20RPAS%20CONOPS.pdf
34. ICAO UTM, 'Unmanned Aircraft Systems Traffic Management—A Common Framework with Core Principles for Global Harmonization, 3rd edn. September 2020. https://www.icao.int/safety/UA/Documents/UTM%20Framework%20Edition%203.pdf
35. M. Kohen, et al., 'National Companies Considering Commercial Drones Must Consider State Privacy Laws' (2015). https://casetext.com/analysis/national-companies-considering-commercial-drones-must-consider-state-privacy-laws?sort=relevance&resultsNav=false&q=
36. Markets and Markets, 'Unmanned Aerial Vehicle (UAV) Market (2021–2026),'Report Code: AS 2801, published June 2021. https://www.marketsandmarkets.com/Market-Reports/unmanned-aerial-vehicles-uav-market-662.html?gclid=CjwKCAjwt7SWBhAnEiwAx8ZLau_Nfo5soGBHDm8GAtq8eA42cZXo70CJtq14jdA2rD7m-_WUZr-p-RoC6y4QAvD_BwE
37. S. Mateou, A. Mateou, 'UASs as an Aviation Security Threat', Aviat. Space J. **1** (2021). http://www.aviationspacejournal.com/wp-content/uploads/2021/04/The-Aviation-Space-Journal-Year-XX-January-March-2021-1.pdf
38. S. Michaelides-Mateou, C. Erotokritou, 'Flying into the future with UAVs: the Jetstream 31 Flight'. Air & Space Law **39**, 111 (2013/2014)
39. L.W. Murphy, et al., 'The Future of Drones in America: Law Enforcement and Privacy Considerations', ACLU Statement for the Record for a Senate Judiciary Committee Hearing, American

Civil Liberties Union, 20 March 2013. https://www.aclu.org/other/future-drones-america-law-enforcement-and-privacy-considerations-aclu-statement-record-senate
40. NASA, 'What is Unmanned Aircraft Systems Traffic Management?' 27 May 2021. https://www.nasa.gov/ames/utm
41. Netherlands Aerospace Centre, 'Seven demonstrations across Europe of safe integration of unmanned aircraft' 30 June 2022. https://www.nlr.org/news/seven-demonstrations-across-europe-of-safe-integration-of-unmanned-aircraft/
42. NFP, 'Hospitals launch Poland's first long-haul drone delivery connection', 17 February 2022. https://notesfrompoland.com/2022/02/17/hospitals-launch-polands-first-long-haul-drone-delivery-connection/
43. P. Paganini, "A small drone hit a British Airways plane over the Heathrow Airport". Security Affairs. 18 April 2016. http://securityaffairs.co/wordpress/46442/security/drone-hit-british-airways-plane.html
44. Restatement (Second) Of Torts § 159(2) (Am. Law. Inst. 1965) which states that "Flight by aircraft in the air space above the land of another is a trespass if, but only if, (a) it enters into the immediate reaches of the air space next to the land, and (b) it interferes substantially with the other's use and enjoyment of his land."
45. J.E. Roberson, A.M. Mosner, A.M, M.V. O'Brien, B. Slocum, O.R.J Roberson, et al., 'FAA releases two final rules to advance drone integration'. J. Robot., Artif. Intell. Law **4**(3), 205–212 (2021)
46. SESAR, Europe's first-ever beyond visual line of sight drone flight takes off in Belgium', 11 July 2022. https://www.sesarju.eu/news/europes-first-ever-beyond-visual-line-sight-drone-flight-takes-belgium
47. SESAR, European Drones Outlook Study Unlocking the value for Europe (2016). https://www.sesarju.eu/sites/default/files/documents/reports/European_Drones_Outlook_Study_20.pdf
48. T.D. Tran, 'Cybersecurity risk assessment for Unmanned Aircraft Systems'. Automatic. Université Grenoble Alpes (2021). https://hal.archives-ouvertes.fr/tel-03200719v2/document
49. L.V. Weiland, 'Implications between UAV and ATM systems in commercial airspace incorporation' (2021). https://commons.erau.edu/publication/1627

Dr. Sofia Michaelides-Mateou is an Associate Professor of Law and Air Law with over twenty-five years teaching experience. Sofia is currently Associate Professor in the Department of Aviation Management at the College of Business Administration, Prince Sultan University in Riyadh, Saudi Arabia. She lectures at both undergraduate and post-graduate levels and is a thesis supervisor at the Master's and Ph.D. level.

She has a Ph.D. from the University of Middlesex, UK, (dealing with the criminalization of aviation professionals subsequent to an aviation accident and the effect on aviation safety), and is the author of the book Air Law: A Practical Perspective, (Sakkoulas, Greece, 2010) and co-author of the book Flying in the Face of Criminalisation: The Safety Implications for Prosecuting Aviation Professionals for Accidents (Ashgate, UK, 2010). Sofia has written a number of book chapters focusing on aviation and law and has presented many papers at international aviation conferences. Her published articles include those on unmanned aircraft vehicles (UAVs), aviation security, safety, unruly passengers, the liability of aviation professionals subsequent to a serious aviation accident or incident, protection of aviation safety data and Just Culture.

As an Aviation-legal Consultant she has participated in a number of aviation litigation cases. Sofia is also an Associate Member of International Society of Air Safety Investigators (ISASI), a member of the Flight Safety Foundation and of the Eurocontrol Just Culture Task Force.

GPSR Compliance

The European Union's (EU) General Product Safety Regulation (GPSR) is a set of rules that requires consumer products to be safe and our obligations to ensure this.

If you have any concerns about our products, you can contact us on ProductSafety@springernature.com

In case Publisher is established outside the EU, the EU authorized representative is:

Springer Nature Customer Service Center GmbH
Europaplatz 3
69115 Heidelberg, Germany

Batch number: 10706330

Printed by Printforce, the Netherlands